LEARNING
RESOURCES
CENTER

A

First Course

in

Factor Analysis

A First Course in Factor Analysis

ANDREW L. COMREY

University of California, Los Angeles

ACADEMIC PRESS New York and London

ACADEMIC PRESS, INC.
111 Fifth Avenue, New York, New York 10003

United Kingdom Edition published by
ACADEMIC PRESS, INC. (LONDON) LTD.
24/28 Oval Road, London NW1

LIBRARY OF CONGRESS CATALOG CARD NUMBER: 72-82633

PRINTED IN THE UNITED STATES OF AMERICA

To the Memory of

Halcyon Goff Nagel

Contents

CHAPTER FOUR

The Principal Factor and Minimum Residual Methods of Factor Extraction

CHAPTER FIVE

Orthogonal Hand Rotations

CHAPTER SIX

Oblique Hand Rotations

CHAPTER SEVEN

Simple Structure and Other Rotational Criteria

CHAPTER EIGHT

Planning the Standard Design Factor Analysis

CHAPTER NINE

Alternate Designs in Factor Analysis

CHAPTER TEN

Interpretation and Application of Factor Analytic Results

CHAPTER ELEVEN

Development of the Comrey Personality Scales: An Example of the Use of Factor Analysis

CHAPTER TWELVE

Computer Programs

Preface

It is perhaps presumptuous for a nonmathematician to undertake the writing of a textbook on a topic such as factor analysis. After teaching the course in factor analysis for many years, however, I feel that I have been able to develop a clear picture of what students need in the way of an introduction to the topic. Furthermore, although good books on factor analysis have been published in recent years, they tend to be either too difficult for the typical student or else not sufficiently oriented toward providing an understanding of the underlying rationale for the technique.

I have tried to develop a book that will serve the needs of talented but mathematically unsophisticated advanced undergraduates, graduate students, and research workers who wish to acquire a basic understanding of the principles upon which factor analysis rests so that they can correctly apply these methods in their own work and intelligently evaluate what others are doing with them. I have tried to pitch the book at a level that will be a challenge for such readers but not so difficult as to discourage them. On the assumption that the emphasis in such a beginning text should be on more fundamental topics, I have minimized the amount of attention given to certain complex controversial issues, and I have omitted many recent developments. I feel, therefore, that this book represents not a substitute for such excellent advanced works as those of Harman and Horst but rather an introduction to them. After mastering the contents of this book, the typical student will find it much easier to read these more advanced books as well as new material in such technical journals as *Psychometrika*.

Another reason for my writing this book was to present in one volume some of the things I have learned about factor analysis that are not treated by other authors. For example, I have developed the minimum residual method of factor extraction and also an analytic method of factor rotation. These methods have been applied in my own research as integral parts of an overall research strategy that has given me good empirical results. I want to share these and other techniques I have developed with fellow research workers in the hope that they too will find them useful. Finally, I have taken this opportunity to present some of my own philosophical views on the use of factor analytic methods in scientific research.

I have been influenced by many people in developing my own conceptions about factor analysis: by my teacher, J. P. Guilford; my friends and colleagues, H. H. Harman, R. B. Cattell, and H. F. Kaiser; and many others. I would like to acknowledge this debt. Students from two separate classes have used the manuscript as a text and have offered many helpful suggestions. Some of the people who have been most instrumental in bringing about changes in the manuscript itself are Peter Bentler, Lee Cooper, Ken Ford, Richard Harshman, Diana Solar, and Mike Vandeman. Last but not least, I am indebted to my wife, Barbara Sherman Comrey, for encouraging me to write the book and for her continued support.

CHAPTER ONE

Introduction

Showing what is related to what and how represents a very important part of the content of any scientific field. In the well-established sciences, the variables to be related are precisely defined and widely accepted by the scientific community as *the* important variables to study. The relationships of these variables with each other are frequently specified mathematically, and these mathematical relationships between variables in some cases are further organized into elaborate mathematical theories involving several interrelated variables. In the newer scientific fields the variables are less precisely defined, there is not so much agreement among scientists concerning what variables should be related to each other, and the nature of the relationships between variables is less clearly specified. Factor analysis represents a rapidly growing body of statistical methods that can be of great value in these less developed sciences. Factor analytic methods can help scientists to define their variables more precisely and decide which variables they should study and relate to each other in the attempt to develop their science to a higher level. Factor analytic methods can also help these scientists to gain a better understanding of the complex and poorly defined interrelationships among large numbers of imprecisely measured variables. The purpose of this book is to help the reader achieve some familiarity with these powerful methods of analysis.

This book has been written for advanced undergraduate students, graduate students, and research workers who need to develop a knowledge of factor analysis, either for the purpose of understanding the published research of others or to aid them in their own research. It is presumed that most readers will have a reasonable understanding of high school geometry, algebra, trigonometry, and a course in elementary statistics, even though the concepts in these areas may not have been reviewed recently. Some sections of the book make use of more advanced mathematical techniques than the typical high school graduate has been exposed to, but this should not prove to be an obstacle of any significant proportions for the average reader. As a nonmathematician himself, the author has tried hard to present the material in a way that can be understood by interested students with only a modest mathematical background. Although the book has been designed so that it will serve as a textbook for classes in factor analysis, it was also written with the expectation that many if not most of the readers will be using it without the guidance of a teacher.

Factor analysis represents a class of procedures for treating data that are being applied with increasing frequency in the behavioral and health sciences and even in the physical sciences. Despite this popularity, however, factor analysis is not well understood by a very high proportion of those scientists who use it. There are many reasons for this phenomenon. First of all, factor analysis is based on mathematical foundations that go beyond the level of quantitative sophistication and training possessed by the bulk of behavioral scientists who use it the most. Although there are two excellent books (Harman, 1967; Horst, 1965) that present the mathematical background of factor analysis, the average behavioral scientist finds that he is unable or unwilling to wade through these books, largely because of his limited mathematical skills.

Another problem with understanding factor analysis, however, is that it rests on empirical science as much as it does on mathematics. The person who is well trained in mathematics will have little difficulty with the mathematical aspects of the topic, but if he is not trained in empirical science applications of factor analysis, it may be very difficult for him to make effective use of these methods in practice. It is also true that the textbooks that are comprehensive with respect to the mathematical aspects of factor analysis do not deal with the nonmathematical parts of the subject with the same degree of adequacy.

Thus, neither mathematical skill nor scientific sophistication alone is sufficient to understand and use factor analysis effectively. Knowledge in both these areas is needed. An attempt is made in this book to present in both these realms the ideas that are necessary for a basic understanding of factor analysis and how to use it. There is no attempt to cover all the methods used in factor analysis. In particular, newer technical developments that have not yet gained general acceptance are not a primary focus of the book. Rather, the focus in this book is on the correct use and understanding of some practical, proven procedures. Step-by-step instructions are given that will permit the reader to apply these techniques correctly. Beyond this, however, there is an attempt to explain the underlying rationale for these methods. Without a fundamental understanding of why certain procedures are supposed to work, application of factor analysis becomes a cookbook operation in which the user is not in a position to evaluate whether or not what he is doing makes any sense. The first goal of this volume, therefore, is to promote understanding as a basis for proper application. Understanding is also the basis for intelligent evaluation of published applications of factor analysis by others. Most scientists need this capacity today whether or not they ever do a factor analysis.

This book provides more emphasis on the nonmathematical aspects of factor analysis than either the Harman (1967) or Horst (1965) texts previously mentioned. Although this book presents the essential aspects of the mathematical basis of factor analysis, it does so on a simpler level than either of these more mathematical texts. It is hoped that the mastery of the content of this book will provide a ready avenue for progression to the more advanced mathematical treatments by Harman and Horst. The nonmathematical aspects of the present treatment will also provide a useful background for placing more advanced mathematical material in proper perspective as well as making it easier to absorb. Other textbooks on factor analysis that may offer valuable assistance to the reader who wishes to explore the topic further include the following:

1. Cattell (1952), *Factor Analysis.* This is a book of intermediate difficulty in which simple structure is emphasized as a means of carrying out factor rotations. Ways of combining factor analysis with experimental design are described. Many important advances have occurred since the book was written so it is somewhat out-of-date.
2. Fruchter (1954), *Introduction to Factor Analysis.* An elementary book that is perhaps the easiest introduction to the use of factor analysis. It is now somewhat out-of-date.
3. Guertin & Bailey (1970), *Introduction to Modern Factor Analysis.* This book avoids mathematics and concentrates on how to use the computer to do factor analysis. It is easy to read but does not emphasize an understanding of the underlying theory.
4. Lawley & Maxwell (1963), *Factor Analysis as a Statistical Method.* This is a little book that presents, among others, Lawley's own maximum likelihood method. Some mathematical sophistication is required to read this book.
5. Thomson (1951), *The Factorial Analysis of Human Ability.* An older book presenting the point of view of a popular British historical figure in factor analysis. It is easy to read.

6. Thurstone (1947), *Multiple Factor Analysis.* This is the classic work in the field. Although it is now out-of-date, the book still contains much of value to the advanced student.

The author would recommend that after mastering the material in the present text, the reader go on to read Harman's book and then Horst's. Following this, he can explore the more recent technical advances in factor analysis by reading journal articles in *Psychometrika.* Other articles on factor analysis, usually somewhat less technical, are to be found in such journals as *Educational and Psychological Measurement* and *Multivariate Behavioral Research.* Many journals report research in which factor analysis is used to treat data. If this volume appears to be too technical for the reader at his present stage of knowledge, he would be advised to read the books of Fruchter (1954) and Guertin and Bailey (1970) before returning to this book.

[1.1] FACTOR ANALYTIC GOALS AND PROCEDURES

There are many reasons why an investigator may undertake a factor analysis. Some of these are the following: (a) He may have measurements on a collection of variables and would like to have some idea about what constructs might be used to explain the intercorrelations among these variables; (b) he may wish to test a theory about the number and nature of the factor constructs needed to account for the intercorrelations among the variables he is studying; (c) he may wish to determine the effect on the factor constructs brought about by changes in the variables measured and in the conditions under which the measurements are taken; (d) he may wish to verify previous findings, either his own or those of others, using a new sample from the same population or a sample from a different population; (e) he may wish to test the effect upon obtained results produced by a variation in the factor analytic procedures used.

Whatever the goals of an analysis, in most cases it will involve all of the following major steps: (a) selecting the variables; (b) computing the matrix of correlations among the variables; (c) extracting the unrotated factors; (d) rotating the factors; and (e) interpreting the rotated factor matrix. The variables studied in a factor analysis may represent merely a selection based on what happens to be available to the investigator in the way of existing data. On the other hand, the variables chosen may represent the results of a great deal of very careful planning.

The factor analysis proper typically begins with a matrix of correlation coefficients between data variables that are being studied. In psychology, for example, these variables might consist of test scores on several short tests of personality, such as 1. Lack of Reserve, 2. Lack of Seclusiveness, 3. No Loss for Words, 4. Lack of Shyness, 5. Lack of Stage Fright, 6. Gregariousness, and 7. Sociability. These tests might be administered to several hundred students. A correlation coefficient would be computed between each pair of tests over the scores for all students tested. These correlation coefficients would then be arranged in a matrix as shown in Table 1.1. Numerical values would appear in an actual table rather than the symbols shown in Table 1.1. Thus, the correlation between variables 1 and 2 appears in the first row, second column of the correlation matrix, shown in Table 1.1 as r_{12}. The subscripts indicate the variable numbers involved in the correlation. It should be noted that the entry r_{21} (in row 2, column 1) involves the correlation between variables 2 and 1. This is the same as r_{12} since the same two variables are involved. The correlation matrix, therefore, is said to be "symmetric" since $r_{ij} = r_{ji}$, that is, the entries in in the upper right-hand part of the matrix are identical to the corresponding entries in the lower left-hand part of the matrix.

The symbols shown in Table 1.1 are replaced by hypothetical numerical values to give the correlation matrix shown in Table 1.2. The matrix shown in

TABLE 1.1

A Symbolic Correlation Matrix

Variables	1	2	3	4	5	6	7
1. Lack of Reserve		r_{12}	r_{13}	r_{14}	r_{15}	r_{16}	r_{17}
2. Lack of Seclusiveness	r_{21}		r_{23}	r_{24}	r_{25}	r_{26}	r_{27}
3. No Loss for Words	r_{31}	r_{32}		r_{34}	r_{35}	r_{36}	r_{37}
4. Lack of Shyness	r_{41}	r_{42}	r_{43}		r_{45}	r_{46}	r_{47}
5. No Stage Fright	r_{51}	r_{52}	r_{53}	r_{54}		r_{56}	r_{57}
6. Gregariousness	r_{61}	r_{62}	r_{63}	r_{64}	r_{65}		r_{67}
7. Sociability	r_{71}	r_{72}	r_{73}	r_{74}	r_{75}	r_{76}	

TABLE 1.2

A Numerical Correlation Matrix

Variables	1	2	3	4	5	6	7
1. Lack of Reserve		.46	.66	.53	.36	.50	.59
2. Lack of Seclusiveness	.46		.55	.57	.27	.45	.50
3. No Loss for Words	.66	.55		.82	.46	.37	.45
4. Lack of Shyness	.53	.57	.82		.43	.45	.45
5. No Stage Fright	.36	.27	.46	.43		.38	.40
6. Gregariousness	.50	.45	.37	.45	.38		.55
7. Sociability	.59	.50	.45	.45	.40	.55	

Table 1.2 is a very small one, of course, being only a 7×7 matrix involving just seven variables. Matrices treated in actual research work frequently run to 60×60 or more, that is, involving 60 variables or more. The symmetric character of the matrix in Table 1.2 is seen by observing that the numbers above the main diagonal (the imaginary line through the blank cells in Table 1.2) are the same as those below the main diagonal. Also, the entries in row 1 are the same as those in column 1. The same thing holds for all other rows and columns that have the same number if the matrix is symmetric.

When the correlation matrix has substantial correlation coefficients in it, this indicates that the variables involved are related to each other, or overlap in what they measure, just as weight, for example, is related to height. On the average, tall people are heavier and short people are lighter, giving a correlation between height and weight in the neighborhood of .60. With a large number of variables and many substantial correlations among the variables, it becomes very difficult to keep in mind or even to contemplate all the intricacies of the various interrelationships. Factor analysis provides a way of thinking about these interrelationships by positing the existence of underlying "factors" or "factor constructs" that account for the values appearing in the matrix of intercorrelations among these variables. For example, a "factor" of "Bigness" could be used to account for the correlation between height and weight. People could be located somewhere along a continuum of Bigness between very large and very small. Both height and weight would be substantially correlated with the factor of Bigness. The correlation between height and weight would be accounted for by the fact that they both share a relationship to the hypothetical factor of Bigness.

Whether or not it is more useful to use a single concept like Bigness or to use two concepts like height and weight is a question that cannot be answered by factor analysis itself. One common objective of factor analysis, however, is to provide a relatively small number of factor constructs that will serve as satisfactory substitutes for a much larger number of variables. These factor constructs themselves are variables that may prove to be more useful than the original variables from which they were derived.

A factor construct that has proved to be very useful in psychology is Extraversion–Introversion. It is possible to account for a substantial part of the intercorrelations among the variables in Table 1.2 by this single factor construct. This is because all of these variables are correlated positively with each other. With most large real data matrices, however, the interrelationships between variables are much more complicated, with many near-zero entries, so that it usually requires more than one factor construct to account for the intercorrelations that are found in the correlation, or **R**, matrix.

After the correlation matrix **R** has been computed, the next step in the factor analysis is to determine how many factor constructs are needed to account for the pattern of values found in **R**. This is done through a process called "factor extraction," which constitutes the third major step in a factor analysis. This process involves a numerical procedure that uses the coefficients in the entire **R** matrix to produce a column of coefficients relating the variables included in the factor analysis to a hypothetical factor construct variable. The usual procedure followed is to "extract" factors from the correlation matrix **R** until there is no appreciable variance left, that is, until the "residual" correlations are all so close to zero that they are presumed to be of negligible importance.

There are many methods of extracting a factor but they all end up with a column of numbers, one for each variable, that represent the "loadings" of the variables on that factor. Some factor analysts, mostly British, prefer the term "saturation" to the term "loading." These loadings represent the extent to which the variables are related to the hypothetical factor. For most factor extraction methods, these loadings may be thought of as correlations between the variables and the factor. If a variable has an extracted factor loading of .7 on the factor, then its correlation is to the extent of .7 with that hypothetical factor construct. Another variable might have a substantial negative loading on the factor, indicating that it is negatively correlated with the factor construct.

After the first factor is extracted (factor extraction methods will be explained in Chapters 3 and 4), the effect of this factor is removed from the correlation matrix **R** to produce the matrix of first factor "residual" correlations. How this is done is explained in greater detail later but suppose that the first factor extracted from the matrix in Table 1.2 had loadings of .7 for 1. Lack of Reserve and .8 for 3. No Loss for Words. Multiplying $.7 \times .8$ gives .56 as the correlation between these two variables due to the first factor alone. Subtracting .56 from .66 leaves only .10 for the first factor residual correlation between these two variables. If all the other first factor residuals were also as small as this or smaller, it would probably be unnecessary to extract a second factor. If substantial values remain in the first factor residual correlations, however, it is necessary to extract a second factor. If substantial values remain in the second factor residual correlations, a third factor must be extracted, and so on, until the residuals are too small to continue.

Once the factors that are needed to account for the correlations in the **R** matrix have been extracted, the values are arranged in a table that is referred to as the "matrix of unrotated loadings." Such a matrix might look something like the example given in Table 1.3. The first factor in Table 1.3 is the largest factor in the sense that the sum of squares of the factor loadings is largest for the first column (2.37). The subsequent factors become progressively smaller, with the last one in this example being only about one-fifth as large as the first (.53). The last column of the matrix in Table 1.3, headed by h^2, contains the "communalities" for the variables. In this table, the communalities equal the sums of squares of the loadings for the variables over the four factors. That is,

TABLE 1.3

An Unrotated Factor Matrix

Variable	Factors				
	I	II	III	IV	h^2
1	.48	−.67	−.01	−.05	.68
2	.38	−.63	.12	.08	.56
3	.40	−.65	−.14	−.10	.61
4	.51	.27	.36	−.17	.50
5	.61	.26	.37	−.02	.57
6	.46	.22	.46	.09	.48
7	.41	.26	−.11	−.40	.41
8	.55	.28	−.26	−.25	.52
9	.41	.31	−.42	.32	.54
10	.47	.37	−.38	.38	.65
SSQ[a]	2.37	1.82	.91	.53	—

[a] The sum of squares.

the communality $h_1{}^2$ for variable 1, for example, is given by $.68 = (.48)^2 + (-.67)^2 + (-.01)^2 + (-.05)^2$, ignoring rounding errors. The communalities represent the extent of overlap between the variables and these four factors. If the communality for a variable were as high as 1.0, it would indicate that the variable overlaps totally with these four factors in what it measures. In this case, its scores could be predicted perfectly by a weighted combination of scores representing only these four factors. Another way of expressing this idea is to say that all the variance in this variable can be accounted for by scores for individuals representing their positions on the four factors. If one of these variables had a communality of zero, on the other hand, all four factor loadings for that variable would be zero and the variable would not share anything in common with any of the four factors. Communality values in between 1.0 and 0 indicate partial overlapping between the variables and the factors in what they measure.

In Table 1.3, the names of the variables have not been listed, contrary to what was done in Tables 1.1 and 1.2. Sometimes an author puts these names in the table of factor loadings and sometimes not. The sums of squares and the communalities also may or may not be shown. The decimal points on the correlations and factor loadings are included in all tables in this book, but these are often omitted for convenience in table preparation. Roman numerals are usually used to designate factors and arabic numerals to designate variables, although sometimes the factors also will be labeled with arabic numerals. The factors are almost always listed across the top of the table and the variables down the left side for tables of factor loadings.

The factor analysis does not stop, however, with the extraction of the factors and the preparation of a table of unrotated factor loadings. Although this table gives a factor solution based on mathematically acceptable factor constructs, the factor constructs represented in an unrotated factor matrix are rarely useful in scientific work. The reason for this is that most methods of factor extraction are designed to extract approximately as much variance as possible with each successive factor, resulting in a sharp drop off in factor size from the first to the last factor, as shown by the sum of squares of the columns of factor loadings. This phenomenon is very apparent in the unrotated factor matrix shown in Table 1.3. These unrotated factors obtained by extracting as much variance as possible from the matrix of correlations at each factor extraction step tend to be highly complex factor constructs that relate to or overlap with many of the variables rather than with just a few. Factor I, for example (in Table 1.3), has appreciable loadings for all the variables. Factor II has high negative loadings for three variables and all the other variables have positive loadings of some importance. Five of the ten variables are related to factor III to about the same level. Such complex, overlapping factors are usually difficult to interpret and use for scientific description because they contain within them many unrelated parts. They are not homogeneous in character.

Using an unrotated factor for scientific description is not unlike describing human beings with a physical variable developed by adding equally weighted scores based on Intelligence, Weight, Bank Balance, and Number of Siblings. Such a composite variable would be complex indeed. Knowledge of a person's score on such a variable would be virtually useless because it would be impossible to tell from knowledge of the total score where the individual falls with respect to the component parts.

Fortunately, however, it is possible to "rotate" the factor matrix into another form that is mathematically equivalent to the original unrotated matrix but which represents factor constructs that are often much more useful for scientific purposes than the unrotated factor constructs. The unrotated factor matrix in Table 1.3, for example, can be rotated into the form shown in Table 1.4. The values shown in Table 1.4 are decimal proportions of 1.0 with either a positive or a negative sign attached. This does not apply to the case where covariances are analyzed instead of correlation coefficients. This practice is relatively rare, however, so the present text is not concerned with the factor analysis of covariance matrices (see Harman, 1967, or Horst, 1965, for a treatment of this topic). Certain types of factor solutions, called "oblique" solutions, do permit the factor loadings to go over 1.0 but this is not often encountered.

The factor constructs represented by the rotated factor matrix in Table 1.4 are very different from those represented by the unrotated matrix in Table 1.3

even though the two matrices are mathematically equivalent in the sense that they both account equally well for the original correlation coefficients from which they were derived. Notice that factor I in Table 1.4 has high loadings for the first three variables and low loadings for all the rest. Factor II has high loadings for variables 4, 5, and 6 and much lower loadings for all the other variables. Factor III has high loadings for variables 7 and 8 with much lower loadings for all the other variables. Finally, factor IV has high loadings only for variables 9 and 10. Each factor, then, in Table 1.4 is related highly to only a few variables and the variables related to each factor are different.

TABLE 1.4

A Rotated Factor Matrix

Variable	Factors				
	I	II	III	IV	h^2
1	.82	.04	.08	.01	.68
2	.73	.10	−.10	−.02	.56
3	.76	−.10	.14	.03	.61
4	.03	.64	.28	−.05	.50
5	.09	.72	.20	.08	.57
6	.05	.69	.00	.02	.48
7	−.02	.22	.60	.06	.41
8	.04	.24	.60	.31	.52
9	−.06	.12	.16	.70	.54
10	−.08	.21	.14	.76	.65

By inspecting the content of variables 1, 2, and 3, it would be possible to gain some inkling about the nature of the underlying factor construct. Similarly, factors II, III, and IV could be tentatively identified and described on the basis of the variables that are related to them. These factors will all be well differentiated from each other since each is represented by a different set of variables. The factor constructs derived from the rotated matrix in Table 1.4, therefore, are much more likely to have some scientific utility than those derived from the unrotated matrix in Table 1.3.

The communalities in the last column are the same in both Tables 1.3 and 1.4. This is one indication of the mathematical equivalence of the two matrices for descibing the original correlation matrix. The procedures for rotating a matrix of unrotated factors into a mathematically equivalent rotated matrix that yields more interpretable factors are described and explained in later chapters.

Following the computation of the correlations, extraction of the unrotated factors, and the rotation of the unrotated factors, the factor analyst attempts to interpret what his factors are like, using the knowledge he has about the variables that went into the factor analysis and any other pertinent information at his disposal. He picks out the variables in each rotated factor that have high loadings, studies all of them carefully, and tries to come up with some kind of hypothesis concerning what they share in common. On the basis of this analysis, he will try to provide an appropriate name for each factor that has been identified.

[1.1.1] *FOLLOWING UP FACTOR ANALYTIC RESULTS*

In many cases carrying out a factor analysis and naming the factors may represent only an exercise if we do not go beyond these procedures. The rotated factor matrix and its implied factor constructs provide one interpretation of the data, but there is no guarantee that this interpretation is the "correct" one. Factor analysis could be considered a way of generating hypotheses about nature. The factor constructs that emerge from a factor analysis may be very useful as variables for understanding and describing the relationships within a given scientific domain, but the correctness of interpretations based on factor analytic results must be confirmed by evidence outside the factor analysis itself.

There are many different methods of carrying out a factor analysis. Several different factor analysts can take the same data and come up with as many different solutions. This fact has led to a great deal of misunderstanding about factor analysis and not a little distrust of the method. It must be recognized, however, that all of these different solutions for the same data by different analysts represent interpretations of the original correlation matrix that may be equally correct from the mathematical point of view. The mathematical analyses merely show that the same data may be interpreted in a variety of ways. Some of these interpretations undoubtedly are more useful than others from the standpoint of scientific objectives, but there is nothing in the factor analytic methods themselves that can demonstrate that one factor solution is more scientifically useful than another. The factor analyst, therefore, must use other means than the factor analysis he has carried out to demonstrate that his factor analytic results have scientific value. If he does not, or cannot, it is not the fault of the factor analytic tools he used to generate the hypotheses he failed to test.

[1.2] SOURCES OF DISAGREEMENT IN FACTOR ANALYTIC RESEARCH

There are many reasons why the solutions reached by different factor analysts do not always agree with each other. Some of the major reasons are outlined briefly in the following sections.

[1.2.1] *COMMUNALITIES*

The communality for a variable in a factor analysis has already been defined as the sum of squares of the factor loadings over all the factors, as shown in Tables 1.3 and 1.4. These are the communalities that emerge from the factor analysis. These values give an indication of the extent to which the variables overlap with the factors, or more technically, they give the proportion of variance in the variables that can be accounted for by scores in the factors. In many methods of factor extraction, however, estimates of the communality values must be inserted in the diagonal cells of the original correlation matrix before the factor analysis is carried out. Consequently, the factor loadings and the communalities that emerge from the factor analysis depend to some extent on the estimated communality values used at the beginning in the correlation matrix. Since these values are at best approximations to the "true" communalities, the resulting factor loadings are only approximately correct. Since there is no fully accepted mathematical solution to the "communality problem" at the present time, even using the same data, variations would be introduced into the results by different investigators by their choice of communality estimates.

[1.2.2] *METHODS OF FACTOR EXTRACTION*

There are many different ways to extract factors. Even with the same communality estimates, these different methods do not always give the same results. In most cases, however, the results from different methods of factor extraction can be transformed by the rotation process so that they are reasonably comparable from method to method. The fact that this is possible, however, does not mean that it will be done in practice. The initially extracted factors may appear to be very different from one method to another and the rotations may never bring them into congruence.

[*1.2.3*] *NUMBER OF FACTORS EXTRACTED*

No precise solution to the problem of how many factors to extract from the correlation matrix has been achieved in the ordinary empirical factor analysis. Many guidelines have been proposed and some have gained a measure of acceptance. Investigators differ in practice, however, in how they deal with this problem, which leads to variations in the results obtained. Differences between investigators in this area can lead to substantial variations in their final interpretive solutions.

[*1.2.4*] *HOW TO ROTATE THE FACTORS*

Most factor analysts agree that the unrotated factors do not generally represent useful scientific factor constructs and that it is usually necessary to rotate, when there are two or more extracted factors, if useful and meaningful factor constructs are to be identified. All factor analysts agree that all rotated solutions, when the same number of factors are rotated, are mathematically equivalent to each other and to the unrotated solution in that all of them account for the original correlation matrix equally well. This is so because in the final analysis, different factor solutions, rotated or otherwise, merely represent different basis systems (or sets of coordinate axes) for the same vector space. Beyond this, however, there is little agreement on what method of rotation gives the "best" solution in some scientific sense. Every analyst has his own preferred procedures, and he is usually convinced that they are better than alternate procedures used by other investigators. The fact that these disagreements occur, however, is not a reflection on the technique. It does point out that a matrix of intercorrelations between variables can be interpreted in many ways. Factor analytic procedures reveal the various possibilities that are mathematically valid. Subsequent experience, not factor analysis *per se*, will reveal which, if any, of these possibilities has scientific utility. Certain factor analytic procedures may produce useful scientific solutions much more consistently than other methods, but it is impossible to establish that one method is always preferable to other available methods.

[1.3] PLAN OF THE BOOK

The remaining chapters are now described briefly so that the reader will have an idea of what to expect and what he can afford to delete if he is not planning at the moment to read the entire text.

[1.3.1]
CHAPTER 2. THE FACTOR ANALYTIC MODEL

In this chapter, the basic equations and assumptions of factor analysis are presented in algebraic and matrix form. The bare minimum of matrix algebra needed to understand the theory is presented. Instead of first devoting a section or chapter to matrix theory, however, these ideas have been introduced as they are needed. This may not be very neat and orderly, but it does permit the reader to try immediately to grasp the ideas of factor analysis without first having to wade through a treatise on matrix theory which may have very little intrinsic interest for him. The author hopes that this manner of presentation will raise reader interest sufficiently to offset any loss of organizational precision.

A good part of Chapter 2 is devoted to showing that the correlation matrix can be decomposed into the matrix product of two other matrices, the unrotated factor matrix **A** and its transpose matrix **A**′, where the rows of **A**′ equal the columns of **A**, and vice versa. In schematic matrix form, this matrix equation looks like the following:

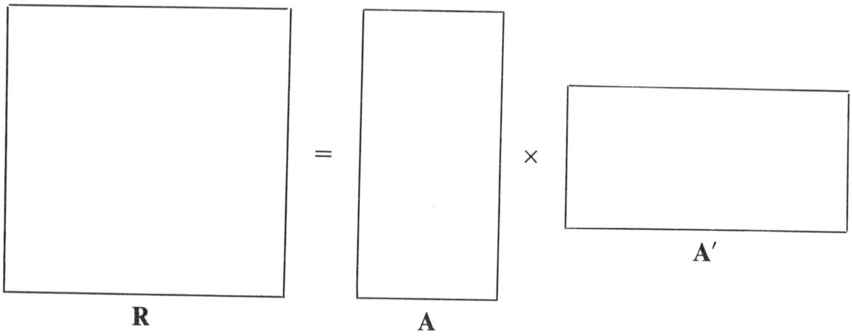

$$\mathbf{R} = \mathbf{A} \times \mathbf{A}'$$

The factor extraction process that yields matrix **A**, the matrix of unrotated factors, is explained further, and an example is presented. The process of factor rotation, where matrix **A** is transformed into a matrix of rotated factors, is also explained further using a miniature example. Both factor extraction and rotation are treated in much greater depth in later chapters, but the intent in this chapter is to paint the broad outlines of the picture before filling in the details. This should help to maintain reader interest and facilitate understanding.

[1.3.2]
CHAPTER 3. FACTOR EXTRACTION BY THE CENTROID METHOD

This chapter is devoted to explaining the basis for the centroid method of extracting factors and to presenting detailed how-to-do-it instructions.

Although this method would not ordinarily be used today if the investigator has access to an electronic computer, it is a good method to use if he has only a desk calculator to work with. It is also a good method to use for instructional purposes, particularly for individual class projects. If the student of factor analysis does his own factor analysis problem from beginning to end, he obtains a much better grasp of what it is all about. The centroid method is ideal for this purpose because the computing involved is not too extensive. First the chapter presents a small four-variable problem to illustrate the ideas and then a completely worked 12-variable example to outline practical computing procedures.

[1.3.3] *CHAPTER 4. THE PRINCIPAL FACTOR AND MINIMUM RESIDUAL METHODS OF FACTOR EXTRACTION*

Prior to the advent of the computer, the centroid method was the most popular way of extracting factors. Today, however, the principal factor method is probably the most widely used. Some underlying theory and the Jacobi method of computing a principal factor solution are presented in this chapter. In addition, theory and computation procedures for extracting factors by the minimum residual method are given. This method does not require the use of communality estimates, and it also provides an automatic rule for terminating factor extraction. Although it has not gained the popularity of the principal factor method, it does have certain advantages in its favor.

[1.3.4] *CHAPTER 5. ORTHOGONAL HAND ROTATIONS*

When the unrotated factors are rotated to obtain the rotated factor matrix, the investigator may elect to leave the factor axes at right angles to each other, in which case the factors are uncorrelated and the solution is called an "orthogonal" solution; or he may elect to allow the angles between the factor axes to depart from 90°. If the angles between the factor axes depart from 90°, the factors are no longer uncorrelated with each other, and the solution is referred to as an "oblique" solution. Chapter 5 presents the traditional method of rotating the factor axes orthogonally, leaving the factor axes at right angles to each other. The mechanics of making plots of factor axes, determining what rotations to perform, and then carrying them out are explained by means of an actual example. Although the mechanics of making the rotations apply to any principle used in deciding what rotations to carry out, Thurstone's principle of "simple structure" is used as a criterion for deciding what rotations to perform in this example. The logic underlying the principle of simple structure is explained.

Although the rotational methods described in Chapter 5 are not used as much today as they were before the advent of computers, these techniques are as good today as they ever were and probably should be used much more than they are. Too many investigators today wish to take short cuts by using computerized solutions that do not require them to make judgments about the rotations that should be performed. Unfortunately, such computerized solutions do not always represent a satisfactory substitute for the more traditional methods.

[*1.3.5*] *CHAPTER 6. OBLIQUE HAND ROTATIONS*

One of the reasons why many investigators prefer orthogonal rotations is that they are much easier to understand and simpler to compute than oblique solutions. This chapter takes up the geometric principles underlying the representation of data variables with respect to oblique factor axes and explains the procedures used in actually computing an oblique solution. An example is worked out to illustrate the principles presented. This chapter is one of the more difficult chapters in the book and can be omitted on the first reading of the book if desired. As long as the investigator sticks to orthogonal solutions, he does not need to use the material in this chapter for his own work. On the other hand, this material should be mastered eventually if the reader expects to achieve a good understanding of factor analytic methodology.

[*1.3.6*] *CHAPTER 7. SIMPLE STRUCTURE AND OTHER ROTATIONAL CRITERIA*

How to carry out rotations to achieve the most scientifically meaningful rotated factor solution has always been an area of heated controversy in factor analytic circles. For many years, simple structure was the dominant method. With the arrival of the electronic computer, however, many new methods have been proposed. In this chapter, the underlying rationale for rotation is considered in some detail to give the reader sufficient understanding of the principles involved so that he will be able to make reasonable decisions about what rotational methods should be employed in a given situation. Simple structure and several other approaches to the rotation process, including the so-called "analytic" methods, are discussed. The underlying theory for two of these analytic methods will be presented in greater detail, the Kaiser Varimax method and the author's own Tandem Criteria method. The Kaiser Varimax method has become perhaps the most popular rotation method in use today. The Tandem Criteria method uses the correlation coefficients between the variables as an additional aid in positioning the factor axes. The method also provides a technique for determining how many factors should be used in the final rotated solution.

[*1.3.7*]

CHAPTER 8. PLANNING THE STANDARD DESIGN FACTOR ANALYSIS

It is often said of factor analysis that you only get out of it what you put into it. This is an oversimplification, but it is certainly true that a poorly planned factor analysis is much less likely to yield useful results than one that is well planned. The purpose of this chapter is to explain the principles of good factor analytic design. Different methods of obtaining the correlation coefficients are compared, and finally a number of common errors in the design and execution of a factor analysis are discussed. Attention is given to the concept of the factor hierarchy and its implications for factor analytic design and interpretation.

[*1.3.8*]

CHAPTER 9. ALTERNATE DESIGNS IN FACTOR ANALYSIS

The most common type of factor analytic design is that involving analysis of a matrix of correlations between data variables. Each correlation between two data variables is computed over a substantial sample of data objects or individuals. This design is referred to as "*R*-technique," following Cattell's terminology. Occasionally, however, other designs, such as "*Q*-technique," are used. In *Q*-technique, the correlation matrix is obtained by correlating pairs of data objects or individuals over a substantial sample of data variables. This method can be used in locating "types" or clusters of similar data objects. There are also other designs that are even less used than *Q*-technique but which are useful for special purposes. This chapter treats a specialized topic, however, and may readily be omitted in the first reading.

[*1.3.9*]

CHAPTER 10. INTERPRETATION AND APPLICATION OF FACTOR ANALYTIC RESULTS

Even if it were possible to know that a particular factor analysis had been designed and executed perfectly, interpretation of the results would still leave room for disagreement between investigators. Typical procedures involved in factor interpretation are presented in this chapter and the meaning of factor constructs is discussed. Also, factor analysis is compared with multiple regression analysis to clarify their respective functions. In addition, this chapter presents several methods of computing factor scores and tells the reader how to write up reports of factor analytic research. Finally, there is a brief discussion of the purposes for which factor analysis can be used.

[*1.3.10*]

CHAPTER 11. DEVELOPMENT OF THE COMREY PERSONALITY SCALES: AN EXAMPLE OF THE USE OF FACTOR ANALYSIS

This chapter presents the author's approach to psychological test development and summarizes the factor analytic research based on that approach which has led to the development of a factor personality inventory, the Comrey Personality Scales. It gives the history and end results of the use of factor analysis in a programmatic research project leading to a tangible end product. This chapter should give the reader a good appreciation for the way factor analysis can be employed in test development, one of its most common applications.

[*1.3.11*]

CHAPTER 12. COMPUTER PROGRAMS

Over the years the author has developed a set of computer programs to carry out the numerical operations needed in his own applications of factor analytic methods. The descriptions of these computer programs and how to use them are given in this chapter. The programs themselves are available from the publisher on tape. These programs can be used to compute the correlation matrix, extract the factors, rotate them by traditional "hand" methods (either orthogonally or obliquely), rotate analytically by the Varimax or the Tandem Criteria method, compute total scores, prepare frequency distributions, obtain scaled scores, do multiple regression analysis, and prepare matrix printouts suitable for photocopying. Using a copy of the tape, the reader can prepare these programs for use at the computer center available to him. This chapter serves as a manual for use of the programs to carry out the various factor analytic procedures described in this book.

[*1.3.12*]

SUMMARY

The mathematical basis of the factor analytic procedures treated in this book is presented in Chapters 2–6, and for analytic rotation in Chapter 7. Instructions on the steps necessary for carrying out a factor analysis if the simplest procedures are to be used are given in Chapters 3–5. These instructions can be followed without mastering the mathematical basis of the methods. Non-mathematical discussions of important aspects of correct factor analytic practice are presented in Chapters 7, 8, 10, and 11. If the reader does not have a great deal of mathematical training and, therefore, many parts of the book do not appear clear on the first reading, he is urged to read all the way through without being discouraged. He should then go back over the text again. Against the background of the first reading, many more pieces of the puzzle will fall into place on the second reading. Many students will find it necessary to reread the entire text several times before gaining new insights or arriving at a greater depth of understanding. Persistence in rereading the text will be sufficient to compensate for inadequate mathematical background for most students. Others are urged to start with the simpler books recommended earlier.

CHAPTER TWO

The Factor Analytic Model

In this chapter, some of the fundamental ideas underlying factor analysis are developed in order to provide the reader with a theoretical introduction to the subject. Many of the ideas treated are highly abstract. As a consequence the student may have difficulty relating them to his own experience. It is also assumed that the student has a working knowledge of elementary trigonometry, algebra, and statistics. For these reasons, the reader may need to review some mathematical concepts and read the chapter several times before everything falls into place. The other chapters should be less difficult to follow once the material in this chapter is mastered. Some of the later chapters consider in greater detail the topics that are introduced here.

[2.1] THE SPECIFICATION EQUATION

In factor analysis, it is customary to start with a matrix of correlations among data variables and wind up with a matrix of factor loadings that can be interpreted in the orthogonal factor model as correlations between the data variables and certain hypothetical constructs, called "factors." At the basis of this whole procedure, however, lies a fundamental assumption which can be stated as follows: A standard score on a data variable can be expressed as a linear combination of common factor scores, specific factor scores, and error factor scores. That is,

$$z_{ik} = a_{i1} F_{1k} + a_{i2} F_{2k} + \cdots + a_{im} F_{mk} + a_{is} S_{ik} + a_{ie} E_{ik} \tag{2.1}$$

where

z_{ik} is a standard score for person k on data variable i,
a_{i1} is a factor loading for data variable i on common factor 1,
a_{i2} is a factor loading for data variable i on common factor 2,
a_{im} is a factor loading for data variable i on the last common factor,
a_{is} is a factor loading for data variable i on specific factor i,
a_{ie} is a factor loading for data variable i on error factor i,
F_{1k} is a standard score for person k on common factor 1,
F_{2k} is a standard score for person k on common factor 2,
F_{mk} is a standard score for person k on common factor m, the last common factor,
S_{ik} is a standard score for person k on specific factor i,
E_{ik} is a standard score for person k on error factor i.

The z, F, S, and E scores in Eq. (2.1) are all standard scores that have a mean (M) of zero and a standard deviation (σ) of 1.0. Each a value in Eq. (2.1) is a numerical constant, or weight, called a factor loading, which will usually fall between -1.0 and $+1.0$. The z score on the left-hand side of Eq. (2.1) is empirically obtained as a data-variable score, whereas the F, S, and E scores are hypothetical factor scores not obtained by data collection. The F scores are sometimes computed or estimated as part of a factor analysis, but the S and E scores are rarely computed in practice. The a values, the factor loadings, are found by the process of factor analysis itself. Standard scores rather than raw scores are used in Eq. (2.1) and in other equations to follow because the equations are simpler when standard scores are used.

Raw scores obtained in data collection may be converted to standard scores by the formula $(X - M)/\sigma$, where X is a raw score, M is the mean of the X scores, and, as before, σ is the standard deviation of the X scores. Conversion from raw scores to standard scores may also be accomplished through the use of normal curve tables which give the standard scores corresponding to percentile scores associated with the raw scores. Conversion of raw scores to standard scores by means of normal curve tables not only standardizes the scores, that is, makes $M = 0$ and $\sigma = 1.0$, but also normalizes the distribution of scores as well. Equation (2.1) is called a "linear" combination of scores because the variables in the equation, the standard score variables, are all to the first power. The weights, the a values in Eq. (2.1), are different for each variable and for each factor but remain constant from person to person.

Equation (2.1) can be made more concrete by relating it to a specific example. Let z_{ik} represent the standard score of person k on an intelligence test, to be treated as a single data variable in this example. Further, suppose that there are just five common factors concerned with human intelligence, Verbal Ability (V), Numerical Ability (N), Memory (M), Reasoning Ability (R), and Perceptual Ability (P). Assume that the standard scores in these ability factors are known for every person in a given population. Then, in Eq. (2.1), F_{1k} represents person k's standard score on the Verbal (V) factor, F_{2k} represents person k's standard score on the Numerical (N) factor, and F_{5k} ($m = 5$ in this case) represents person k's standard score on the Perceptual (P) factor. The term S_{ik} represents person k's standard score on a specific factor associated only with this intelligence test; E_{ik} represents person k's standard score on the error factor associated only with this intelligence test. The values $a_{i1}, a_{i2}, \ldots, a_{i5}, a_{is}$, and a_{ie} are the factor loadings for the intelligence test on the five common ability factors plus the specific and error factors. In the orthogonal factor model described here, these a values may be thought of as correlations between variable i and the factors V, N, M, R, P, S_i, and E_i. The a values are also weights which indicate the relative importance of the various factors in performance on the intelligence test. Equation (2.1) states, therefore, that the intelligence test score of person k equals the weighted sum of his standard scores in common factors V, N, M, R, and P plus a specific factor component plus an error factor component.

Selecting a particular individual to represent person k might result in the following standard score substitutions in Eq. (2.1):

$$1.96 = a_{i1}(1.5) + a_{i2}(1.0) + a_{i3}(2.5) + a_{i4}(-1.0) + a_{i5}(-.2) + a_{is}(-.3) + a_{ie}(1.0)$$

The standard scores substituted in Eq. (2.1) with rare exceptions will be values between -3.0 and $+3.0$. If hypothetical factor loadings are now substituted for the a values, the equation becomes

$$1.96 = .50(1.5) + .40(1.0) + .40(2.5) + .37(-1.0) + .30(-.2) + .30(-.3) + .33(1.0)$$

Person k's standard score on the intelligence test, 1.96, is a relatively high one, with about 97.5 percent of the population below him. When a mean of 100 and a standard deviation of 16 is used this would correspond to an IQ of about

132. His best ability is Memory, with Verbal and Numerical also well above average. He is below average in Reasoning and Perceptual abilities. His below-average specific factor score is countered by an above-average error score which boosted his intelligence score higher than it should be. It is clear that the data-variable score cannot be reproduced exactly by the common factors alone. The specific and error factors must be taken into account. The more reliable the test, however, the smaller will be the error factor loading.

Equation (2.1) may be represented in schematic matrix form for all values of i and k simultaneously, that is, for all data variables and all persons or other data-producing objects, as shown on page 24.

The schematic matrix equation given on page 24 may be represented by the following matrix equation:

$$\mathbf{Z} = \mathbf{A}_u \mathbf{F}_u \tag{2.2}$$

Equation (2.2) states that the matrix of data-variable scores $\mathbf{Z}$ may be obtained by multiplying the matrix of factor loadings $\mathbf{A}_u$ by the matrix of factor scores $\mathbf{F}_u$. Matrix multiplication is explained below.

Each row of the data score matrix $\mathbf{Z}$ gives the standard scores for all the data-yielding subjects, persons, or objects, for one variable. A column of the $\mathbf{Z}$ matrix gives all the standard scores on all the data variables for a given person or other data-yielding object. The matrix of factor loadings $\mathbf{A}_u$ is partitioned into three sections for didactic purposes. The left-hand section contains the factor loadings for the n data variables on the m common factors. The middle section of $\mathbf{A}_u$ contains the factor loadings for the n data variables on the specific factors. The right-hand section of $\mathbf{A}_u$ contains the factor loadings for the n data variables on the error factors. The middle and right-hand sections of $\mathbf{A}_u$ contain nonzero entries only in the diagonal cells due to the assumption of the factor analytic model that each data variable has a loading in its own specific and error factors but not in the specific and error factors for any other data variables. Any factor with nonzero loadings for two or more variables would, by definition, be a "common factor" and therefore would belong in the first part of the $\mathbf{A}_u$ matrix. Specific and error factors, on the other hand, each have loadings for one and only one variable, and hence cannot be common factors. Each row of the matrix of factor scores $\mathbf{F}_u$ contains the factor scores for all data-yielding persons or objects for a given factor. The common factor scores appear in the first m rows of $\mathbf{F}_u$, the specific factor scores in the middle n rows, and the error factor scores in the last n rows.

By multiplying the matrix $\mathbf{A}_u$ by the matrix $\mathbf{F}_u$, then, Eq. (2.1) is represented for all persons, or all values of k, and for all observed data variables, or all values of i. The common factor portion of $\mathbf{A}_u$, the left-hand section, will be called matrix $\mathbf{A}$ (without the subscript u), and the common factor portion of $\mathbf{F}_u$, the top section, will be called matrix $\mathbf{F}$.

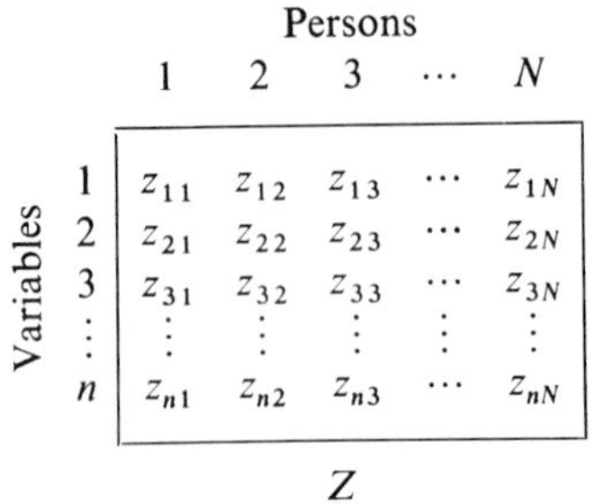

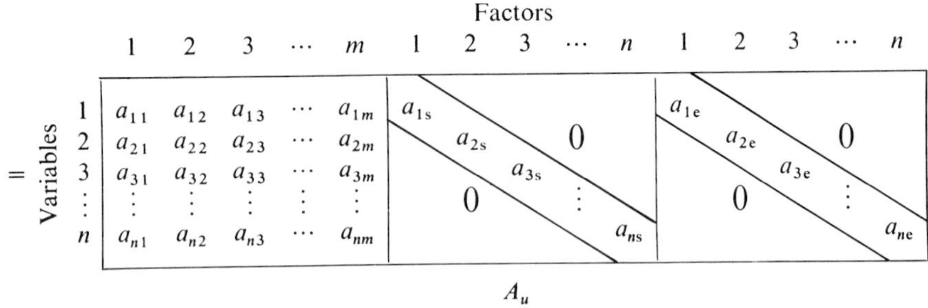

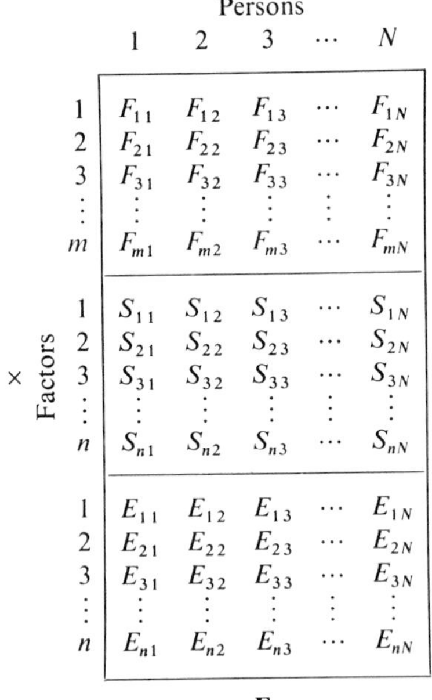

Schematic matrix notation for all values of i and k simultaneously.

[2.1.1]
MATRIX MULTIPLICATION

Matrix multiplication will be illustrated by an example. The first row of the matrix on the left is multiplied with the first column of the matrix on the right to give the entry in the first row and first column of the product matrix. In general, the ith row of the matrix on the left is multiplied with the jth column of the matrix on the right to give the ijth element of the product matrix. This row by column multiplication, called "inner multiplication," proceeds by multiplying the first element of the row on the left by the first element of the column on the right and adding this result to the products for the second elements, the third elements, and so on, as illustrated below:

$$\begin{bmatrix} 2 & 4 & 3 & 1 \\ 5 & 3 & 4 & 0 \end{bmatrix} \times \begin{bmatrix} 3 & 4 \\ 1 & 2 \\ 5 & 1 \\ 4 & 5 \end{bmatrix} = \begin{bmatrix} p_{11} & p_{12} \\ p_{21} & p_{22} \end{bmatrix} = \begin{bmatrix} 29 & 24 \\ 38 & 30 \end{bmatrix}$$

$$p_{11} = (2\times 3) + (4\times 1) + (3\times 5) + (1\times 4) = 6 + 4 + 15 + 4 = 29$$
$$p_{12} = (2\times 4) + (4\times 2) + (3\times 1) + (1\times 5) = 8 + 8 + 3 + 5 = 24$$
$$p_{21} = (5\times 3) + (3\times 1) + (4\times 5) + (0\times 4) = 15 + 3 + 20 + 0 = 38$$
$$p_{22} = (5\times 4) + (3\times 2) + (4\times 1) + (0\times 5) = 20 + 6 + 4 + 0 = 30$$

The number of columns in the matrix on the left must equal the number of rows in the matrix on the right; otherwise a row of the matrix on the left will not have the right number of elements to permit inner multiplication with a column of the matrix on the right. The product matrix will have the same number of rows as the matrix on the left and the same number of columns as the matrix on the right. Referring back to Eq. (2.2), the product $\mathbf{A}_u\mathbf{F}_u$ will have n rows and N columns, as shown in the product matrix **Z**. It should be noted that matrix multiplication is not commutative as a rule; that is, the product of matrix **A** times **B** is not in general equal to the product of matrix **B** times **A**. In fact, such reversals of the order of multiplication are often not even possible because the number of columns of the matrix on the left must equal the number of rows of the matrix on the right.

[2.2]
THE FUNDAMENTAL EQUATION OF FACTOR ANALYSIS

Returning to matrix equation (2.2), if we wish to reproduce the score on variable 5 for person number 9, the fifth row of matrix $\mathbf{A}_u$ is inner multiplied with the ninth column of matrix $\mathbf{F}_u$. This would give

(2.3)

$$z_{59} = a_{51} F_{19} + a_{52} F_{29} + \cdots + a_{5m} F_{m9} + 0 \cdot S_{19} + \cdots + 0 \cdot S_{49} + a_{5s} S_{59} + 0 \cdot S_{69} + \cdots + 0 \cdot S_{n9} + 0 \cdot E_{19} + \cdots + 0 \cdot E_{49} + a_{5e} E_{59} + 0 \cdot E_{69} + \cdots + 0 \cdot E_{n9}$$

Equation (2.3) is a form of Eq. (2.1), substituting 9 for k and 5 for i. Any value from 1 to N may be substituted for k in Eq. (2.1) and any value from 1 to n may be substituted for i. The upper case letter N will be used throughout the book to represent the number of persons or other data-yielding objects. The lower case letter n will always be used to represent the number of data variables for which scores are obtained from the data-yielding objects.

As ordinarily applied, factor analysis involves deriving a set of factor loadings from a matrix of correlation coefficients between the data variables. The following development will show that the correlation between a pair of data variables equals the sum of the products of their factor loadings, the a values from matrix A_u, on the common factors.

Substituting the subscript j for the subscript i in Eq. (2.1) yields an expression representing a standard score for person k in variable j:

(2.4) $$z_{jk} = a_{j1} F_{1k} + a_{j2} F_{2k} + \cdots + a_{jm} F_{mk} + a_{js} S_{jk} + a_{je} E_{jk}$$

If we now multiply Eq. (2.1) by Eq. (2.4), sum both sides over k from 1 to N, and divide both sides by N, the result after some simplification may be expressed as follows:

(2.5)

$$\frac{\sum_{k=1}^{N} z_{ik} z_{jk}}{N} = a_{i1} a_{j1} \left(\frac{\sum_{k=1}^{N} F_{1k}^2}{N} \right) + a_{i2} a_{j2} \left(\frac{\sum_{k=1}^{N} F_{2k}^2}{N} \right) + \cdots + a_{im} a_{jm} \left(\frac{\sum_{k=1}^{N} F_{mk}^2}{N} \right)$$
$$+ 2a_{i1} a_{j1} \left[\frac{\sum_{k=1}^{N} F_{1k} F_{2k}}{N} \right] + \cdots + 2a_{(m-1)i} a_{mj} \left[\frac{\sum_{k=1}^{N} F_{(m-1)k} F_{mk}}{N} \right]$$
$$+ a_{i1} a_{js} \left[\frac{\sum_{k=1}^{N} F_{1k} S_{jk}}{N} \right] + \cdots + a_{im} a_{js} \left[\frac{\sum_{k=1}^{N} F_{mk} S_{jk}}{N} \right]$$
$$+ a_{i1} a_{je} \left[\frac{\sum_{k=1}^{N} F_{1k} E_{jk}}{N} \right] + \cdots + a_{im} a_{je} \left[\frac{\sum_{k=1}^{N} F_{mk} E_{jk}}{N} \right]$$
$$+ a_{j1} a_{is} \left[\frac{\sum_{k=1}^{N} F_{1k} S_{ik}}{N} \right] + \cdots + a_{jm} a_{is} \left[\frac{\sum_{k=1}^{N} F_{mk} S_{ik}}{N} \right]$$

Equation (2.5) continues

$$+ a_{j1} a_{ie} \left[\frac{\sum_{k=1}^{N} F_{1k} E_{ik}}{N} \right] + \cdots + a_{jm} a_{ie} \left[\frac{\sum_{k=1}^{N} F_{mk} E_{ik}}{N} \right]$$

$$+ a_{is} a_{js} \left[\frac{\sum_{k=1}^{N} S_{ik} S_{jk}}{N} \right] + a_{is} a_{je} \left[\frac{\sum_{k=1}^{N} S_{ik} E_{jk}}{N} \right]$$

$$+ a_{ie} a_{js} \left[\frac{\sum_{k=1}^{N} E_{ik} S_{jk}}{N} \right] + a_{ie} a_{je} \left[\frac{\sum_{k=1}^{N} E_{ik} E_{jk}}{N} \right]$$

The formula for the correlation coefficient in standard score form is

$$r_{ij} = \frac{\sum_{k=1}^{N} z_{ik} z_{jk}}{N} \tag{2.6}$$

The formula for the variance, that is, the standard deviation squared, of a set of standard scores is

$$\sigma_i^2 = \frac{\sum_{k=1}^{N} z_k^2}{N} = 1.0 \tag{2.7}$$

The reader is reminded that the sign $\sum$ means to add up the terms with the subscripts which are placed immediately after this summation sign. Thus

$$\sum_{k=1}^{N} z_{ik} z_{jk} = z_{i1} z_{j1} + z_{i2} z_{j2} + \cdots + z_{iN} z_{jN} \tag{2.8}$$

That is, the two standard scores in data variables i and j for the first person are multiplied together; then, the two standard scores in data variables i and j for the second person are multiplied together, and so on, until the two scores for the last person are multiplied together; then, all these products are added up to get the summation total which is the single number symbolized by the expression in Eq. (2.8). Any numerical constant appearing in such a summation of terms may be moved from the right- to the left-hand side of the summation sign as in the following, where a_{i1} is an example of such a constant:

$$\begin{aligned} \sum_{k=1}^{N} a_{i1} F_{1k}^2 &= a_{i1} F_{11}^2 + a_{i1} F_{12}^2 + \cdots + a_{i1} F_{1N}^2 \\ &= a_{i1} [F_{11}^2 + F_{12}^2 + \cdots + F_{1N}^2] \\ &= a_{i1} \sum_{k=1}^{N} F_{1k}^2 \end{aligned} \tag{2.9}$$

Note that the summation is over the subscript k rather than the subscript i. For this reason, a_{i1} is a constant term as far as this summation is concerned since it contains no k subscript.

All the terms in brackets [ ] in Eq. (2.5) represent correlations between different sets of factor scores, and hence they are correlation terms like that in Eq. (2.6). In the orthogonal factor analytic model being considered, it is assumed that all sets of factor scores are uncorrelated with each other, and hence all terms in brackets, like those in Eq. (2.6), are equal to zero. The terms containing these zero correlations, therefore, vanish from Eq. (2.5). In a later chapter, a nonorthogonal, or oblique, factor analytic model is considered in which the factors are correlated with each other.

All the terms in parentheses ( ) in Eq. (2.5) represent variances (squared standard deviations) of factor scores, like those in Eq. (2.7), and hence are all equal to 1.0. Substituting zeros for the correlation coefficients in brackets and 1.0 for the variances in parentheses in Eq. (2.5) results in the simplification of Eq. (2.5) to

$$r_{ij} = a_{i1}a_{j1} + a_{i2}a_{j2} + \cdots + a_{im}a_{jm} \tag{2.10}$$

In other words, Eq. (2.10) states that the correlation between data variables i and j is the sum of the products of their factor loadings in the common factors only. All the specific and error factor cross-product terms drop out of Eq. (2.5) as long as the factors are orthogonal, that is, uncorrelated, with each other. The term on the left of Eq. (2.5) and (2.10) is not equal to zero because it is a correlation between data variables rather than between factor scores. Where i equals j, the correlation is that of a variable with itself due to common factor variance only, with the specific factor variance omitted.

For all values of i and j simultaneously, Eq. (2.10) may be represented by the accompanying schematic matrix notation.

$$\begin{array}{c|ccccc|}
 & 1 & 2 & 3 & \cdots & n \\ \hline
1 & r_{11} & r_{12} & r_{13} & \cdots & r_{1n} \\
2 & r_{21} & r_{22} & r_{23} & \cdots & r_{2n} \\
3 & r_{31} & r_{32} & r_{33} & \cdots & r_{3n} \\
\vdots & \vdots & \vdots & \vdots & \vdots & \vdots \\
n & r_{n1} & r_{n2} & r_{n3} & \cdots & r_{nn} \\ \hline
\end{array}$$

$$\mathbf{R}$$

$$= \begin{array}{c|ccccc|}
 & 1 & 2 & 3 & \cdots & m \\ \hline
1 & a_{11} & a_{12} & a_{13} & \cdots & a_{1m} \\
2 & a_{21} & a_{22} & a_{23} & \cdots & a_{2m} \\
3 & a_{31} & a_{32} & a_{33} & \cdots & a_{3m} \\
\vdots & \vdots & \vdots & \vdots & \vdots & \vdots \\
n & a_{n1} & a_{n2} & a_{n3} & \cdots & a_{nm} \\ \hline
\end{array}
\times
\begin{array}{c|ccccc|}
 & 1 & 2 & 3 & \cdots & n \\ \hline
1 & a_{11} & a_{21} & a_{31} & \cdots & a_{n1} \\
2 & a_{12} & a_{22} & a_{32} & \cdots & a_{n2} \\
3 & a_{13} & a_{23} & a_{33} & \cdots & a_{n3} \\
\vdots & \vdots & \vdots & \vdots & \vdots & \vdots \\
m & a_{1m} & a_{2m} & a_{3m} & \cdots & a_{nm} \\ \hline
\end{array}$$

$$\mathbf{A} \qquad\qquad\qquad \mathbf{A}'$$

Schematic matrix notation for all values of i and j simultaneously.

The matrix equation corresponding to this schematic representation is

(2.11) $$\mathbf{R} = \mathbf{A}\mathbf{A}'$$

Equation (2.11) states that the correlation matrix **R** with communalities in the diagonal cells may be represented as a product of the matrix of common factor loadings multiplied by the transpose of the matrix of common factor loadings. The transpose of a matrix is obtained by interchanging rows and columns; that is, row 1 of **A** is column 1 of **A**′. A numerical example follows:

If

$$\mathbf{P} = \begin{bmatrix} 1 & 2 \\ 4 & 5 \\ 7 & 3 \end{bmatrix}$$

then

$$\mathbf{P}' = \begin{bmatrix} 1 & 4 & 7 \\ 2 & 5 & 3 \end{bmatrix}$$

Just as the algebraic equation (2.10) could be derived by algebraic means from algebraic equation (2.1), so it is also possible to derive matrix equation (2.11), the matrix equivalent of algebraic equation (2.10), by matrix operations. First consider the following matrix equation:

(2.12) $$\mathbf{R}_u = \frac{1}{N}(\mathbf{Z}\mathbf{Z}')$$

Matrix equation (2.12) is represented schematically as shown on page 30. A row i of matrix **Z** inner multiplied with a column j of matrix **Z**′ gives a sum of cross products of standard scores. When this sum is divided by N, as required in Eq. (2.12), the resulting expression is a correlation coefficient between variables i and j as shown by the formula in Eq. (2.6). For the diagonal terms, the row by column multiplication gives variance terms, like those in Eq. (2.7), which are all equal to 1.0. When unities, rather than communalities, are placed in the diagonal cells, the correlation matrix will be designated by the symbol $\mathbf{R}_u$. With communalities in the diagonal cells, the correlation matrix will be designated by the symbol **R**.

Equation (2.2) states that $\mathbf{Z} = \mathbf{A}_u\mathbf{F}_u$; that is, the standard score matrix for the data variables is equal to the complete matrix of factor loadings multiplied by the complete matrix of factor scores. Substituting Eq. (2.2) into Eq. (2.12) gives

(2.13) $$\mathbf{R}_u = \frac{1}{N}(\mathbf{A}_u\mathbf{F}_u)(\mathbf{A}_u\mathbf{F}_u)'$$

$$
\begin{array}{c|ccccc|}
 & 1 & 2 & 3 & \cdots & n \\
\hline
1 & 1.0 & r_{12} & r_{13} & \cdots & r_{1n} \\
2 & r_{21} & 1.0 & r_{23} & \cdots & r_{2n} \\
3 & r_{31} & r_{32} & 1.0 & \cdots & r_{3n} \\
\vdots & \vdots & \vdots & \vdots & \vdots & \vdots \\
n & r_{n1} & r_{n2} & r_{n3} & \cdots & 1.0 \\
\hline
\end{array}
= \mathbf{R}_u = \frac{1}{N}(\mathbf{ZZ}') = \frac{1}{N} \times
\underset{\mathbf{Z}}{
\begin{array}{c|ccccc|}
 & 1 & 2 & 3 & \cdots & N \\
\hline
1 & z_{11} & z_{12} & z_{13} & \cdots & z_{1N} \\
2 & z_{21} & z_{22} & z_{23} & \cdots & z_{2N} \\
3 & z_{31} & z_{32} & z_{33} & \cdots & z_{3N} \\
\vdots & \vdots & \vdots & \vdots & \vdots & \vdots \\
n & z_{n1} & z_{n2} & z_{n3} & \cdots & z_{nN} \\
\hline
\end{array}}
\times
\underset{\mathbf{Z}'}{
\begin{array}{c|ccccc|}
 & 1 & 2 & 3 & \cdots & n \\
\hline
1 & z_{11} & z_{21} & z_{31} & \cdots & z_{n1} \\
2 & z_{12} & z_{22} & z_{32} & \cdots & z_{n2} \\
3 & z_{13} & z_{23} & z_{33} & \cdots & z_{n3} \\
\vdots & \vdots & \vdots & \vdots & \vdots & \vdots \\
N & z_{1N} & z_{2N} & z_{3N} & \cdots & z_{nN} \\
\hline
\end{array}}
$$

Matrix equation (2.12).

Using the fact that the transpose of a product of two matrices equals the product of their transposes in reverse order and relocating the position of the constant divisor N, Eq. (2.13) can be rewritten as follows:

$$\mathbf{R}_u = \mathbf{A}_u \left[\frac{\mathbf{F}_u \mathbf{F}_u'}{N} \right] \mathbf{A}_u' \tag{2.14}$$

The matrix product in parentheses on the right-hand side of Eq. (2.14), $\mathbf{F}_u \mathbf{F}_u'/N$, represents a matrix of correlations between factor scores, since each element of this product matrix is of the same form as Eq. (2.6). The matrix of correlations between these factor scores contains only elements of 1 and 0, depending on whether the correlation is that of a factor score with itself or with some other factor. The correlation of a factor with itself becomes a variance term, like that in Eq. (2.7), and hence equals 1.0. Such a matrix is termed an identity matrix **I**. It has ones in the diagonal cells and zeros in all other cells. Multiplying a matrix by an identity matrix leaves it unchanged; it is like multiplying by 1 in ordinary scalar algebra. Since the factor scores in this model are all uncorrelated, or orthogonal, the matrix $\mathbf{F}_u \mathbf{F}_u'/N$ looks like that shown below. Equation (2.14) may be rewritten, therefore, to give

$$\mathbf{R}_u = \mathbf{A}_u \mathbf{I} \mathbf{A}_u' \tag{2.15}$$

	1 2 3 ⋯ m	1 2 3 ⋯ n	1 2 3 ⋯ n
1	1		
2	1 0		
3	1	0	0
⋮	0 ⋮		
m	1		
1		1	
2		1 0	
3	0	1	0
⋮		0 ⋮	
n		1	
1			1
2			1 0
3	0	0	1
⋮			0 ⋮
n			1

Matrix $\mathbf{F}_u \mathbf{F}_u'/N$.

Dropping the identity matrix leaves

$$\mathbf{R}_u = \mathbf{A}_u \mathbf{A}_u' \tag{2.16}$$

where $\mathbf{R}_u$ is the correlation matrix among data variables with 1's in the diagonal cells and $\mathbf{A}_u$ is the complete matrix of factor loadings, including common, specific, and error factors. The matrix $\mathbf{R}_u$ is symmetric; that is, an element r_{ij} equals the element r_{ji}. Putting it another way, the elements above the diagonal are the same as those below the diagonal so that $\mathbf{R}_u = \mathbf{R}_u'$. Matrix $\mathbf{A}_u$ values also represent correlations between the data variables and the factors. These correlations are called the factor loadings in the orthogonal factor model which requires all factors to be at right angles to one another, that is, to be uncorrelated. The matrix $\mathbf{A}$ consists of only the common factor portion of the factor loadings in matrix $\mathbf{A}_u$.

From Eq. (2.10) and (2.11), it is apparent that only the common factors enter into determining the off-diagonal elements of $\mathbf{R}$, that is, those for which the two subscripts i and j are different for an element r_{ij} of $\mathbf{R}$. For the diagonal elements, however, or those r_{ij} values for which $i = j$, the specific and error factors do make a contribution in Eq. (2.16) where $\mathbf{R}_u$ is reproduced as the product $\mathbf{A}_u \mathbf{A}_u'$. To show this, square both sides of Eq. (2.1) to get

$$\begin{aligned} z_{ik}^2 = {} & a_{i1}^2 F_{1k}^2 + a_{i2}^2 F_{2k}^2 + \cdots + a_{im}^2 F_{mk}^2 + a_{is}^2 S_{ik}^2 + a_{ie}^2 E_{ik}^2 \\ & + 2a_{i1} F_{1k} a_{i2} F_{2k} + \cdots + 2a_{is} S_{ik} a_{ie} S_{ek} \end{aligned} \tag{2.17}$$

Summing both sides of Eq. (2.17) and dividing by N gives

$$\begin{aligned} \left(\frac{\sum_{k=1}^N z_{ik}^2}{N}\right) = {} & a_{i1}^2 \left(\frac{\sum_{k=1}^N F_{1k}^2}{N}\right) \\ & + a_{i2}^2 \left(\frac{\sum_{k=1}^N F_{2k}^2}{N}\right) + \cdots + a_{im}^2 \left(\frac{\sum_{k=1}^N F_{mk}^2}{N}\right) \\ & + a_{is}^2 \left(\frac{\sum_{k=1}^N S_{ik}^2}{N}\right) + a_{ie}^2 \left(\frac{\sum_{k=1}^N E_{ik}^2}{N}\right) \\ & + 2a_{i1} a_{i2} \left[\frac{\sum_{k=1}^N F_{1k} F_{2k}}{N}\right] + \cdots + 2a_{is} a_{ie} \left[\frac{\sum_{k=1}^N S_{ik} E_{ik}}{N}\right] \end{aligned} \tag{2.18}$$

Every term in parentheses in Eq. (2.18) is a variance of a standard score, like Eq. (2.7), in either a data variable or a factor, and hence equals 1.0. Every term in brackets in Eq. (2.18) is a correlation coefficient, like Eq. (2.6), but the correlation coefficient in each case is zero since it is a correlation between uncorrelated factors. Equation (2.18) simplifies, therefore, to

$$1 = a_{i1}^2 + a_{i2}^2 + \cdots + a_{im}^2 + a_{is}^2 + a_{ie}^2 \tag{2.19}$$

Equation (2.19) shows that all the diagonal elements of $\mathbf{R}_u = \mathbf{A}_u\mathbf{A}_u'$ are equal to 1, which is also the variance of any standard score variable. This variance may be divided up as follows:

$$1 = h_{ii}^2 + u_{ii}^2 \tag{2.20}$$

where

$$h_{ii}^2 = a_{i1}^2 + a_{i2}^2 + \cdots + a_{im}^2 \qquad \text{and} \qquad u_{ii}^2 = a_{is}^2 + a_{ie}^2$$

The h_{ii}^2 is called the communality of variable i. It is equal to the sum of squares of the loadings for variable i in the common factors. The u_{ii}^2 is called the uniqueness of variable i. It is equal to the sum of squares of the loadings in the specific and error factors. The communality represents the proportion of the total variance of a data variable which is due to common factors. The uniqueness represents the proportion of the total variance of a data variable which can be accounted for by error and specific factors. The reliability of a data variable is given by the communality plus the proportion of variance due to its specific factor, excluding only the error variance. Thus

$$r_{tt_i} = h_{ii}^2 + a_{is}^2 \tag{2.21}$$

where r_{tt_i} is the reliability for variable i and a_{is} is the specific factor loading for variable i.

Equations (2.16) and (2.11) are two forms of a very important theorem in factor analysis which has been called by Thurstone (1947) the fundamental equation of factor analysis; they both state that the correlation matrix among the data variables can be decomposed into the product of a factor matrix times its transpose. Equation (2.16), $\mathbf{R}_u = \mathbf{A}_u\mathbf{A}_u'$, reproduces the correlation matrix with 1's in the diagonals, using a factor matrix $\mathbf{A}_u$ which contains common, specific, and error factors. Equation (2.11), $\mathbf{R} = \mathbf{AA}'$, reproduces the reduced correlation matrix $\mathbf{R}$ with communalities in the diagonal cells instead of 1's. All other entries are the same in the two matrices $\mathbf{R}$ and $\mathbf{R}_u$. To account for the correlations among data variables, therefore, only the common factors are needed. To account for the total data-variable variance, specific, error, and common factors are needed in this model.

[2.3] FACTOR EXTRACTION

As ordinarily carried out, the process of factor extraction starts with a matrix of correlations between data variables with communalities in the diagonals and ends up with a matrix of factor loadings $\mathbf{A}$ such that when multiplied by its transpose $\mathbf{A}'$ the correlation matrix $\mathbf{R}$ will be reproduced, at least approximately. The specific and error factor loadings are not obtained ordinarily and the **R** matrix is not usually reproduced exactly. Factor extraction, then, represents a problem in matrix decomposition, that is, the decomposition of **R** into the product of another matrix and its transpose. Many different methods have been devised to find a matrix **A** such that $\mathbf{R} = \mathbf{AA}'$. Only a few of these methods are discussed in this book.

If 1's are put in the diagonal cells to get $\mathbf{R}_u$, it is ordinarily possible to carry out a factor analysis obtaining a matrix **A** such that $\mathbf{AA}'$ reproduces $\mathbf{R}_u$ exactly, but it usually requires an **A** matrix the same size as $\mathbf{R}_u$. That is, there are as many factors in **A** as there are data variables in $\mathbf{R}_u$. In such a case, all the factors are common factors in that the error and specific factor variance has been totally absorbed by the common factors.

Traditionally, however, one of the major goals of factor analysis has been to account for a data matrix in terms of fewer factors than data variables. According to the principle of scientific parsimony, it is desirable to account for the data in a given domain making use of as few constructs as necessary. In factor analysis, this principle is translated into the goal of reproducing the **R** matrix with an **A** matrix containing as few factors as possible.

To reproduce the **R** matrix exactly with real data ordinarily requires as many factors as there are data variables. It is usually possible, however, to reproduce approximately the **R** matrix with $\mathbf{AA}'$ where **A** has a number of common factors m such that m is considerably smaller than n, the number of data variables in **R**. Since it is known that sampling error exists in most data, the correlation coefficients are considered to be in error to some extent, making it useless in practice to reproduce them exactly. It is considered better to tolerate some discrepancies between the computed data correlations in the matrix **R** and the reproduced correlation matrix $\mathbf{AA}'$ in order to reduce the number of explanatory constructs, or factors, needed to account for the data, that is, the **R** matrix.

In certain cases, it would be ideal to be able to account for the **R** matrix with a single factor, as in the following fictitious example:

$$(2.22)\qquad \underset{\mathbf{R}}{\begin{bmatrix} .16 & .32 & .28 & .24 \\ .32 & .64 & .56 & .48 \\ .28 & .56 & .49 & .42 \\ .24 & .48 & .42 & .36 \end{bmatrix}} = \underset{\mathbf{A}}{\begin{bmatrix} .4 \\ .8 \\ .7 \\ .6 \end{bmatrix}} \times \underset{\mathbf{A}'}{[\,.4 \quad .8 \quad .7 \quad .6\,]}$$

If one factor is sufficient to account for the correlations, the **R** matrix shows some interesting properties. The columns are proportional to each other and consequently all the minor determinants of order 2 vanish. That is, picking out any 2×2 matrix embedded in the larger matrix and evaluating its determinant results in a value of zero. Thus, from the **R** matrix in Eq. (2.22) one of the submatrices of order 2 is

$$\begin{bmatrix} r_{12} & r_{14} \\ r_{32} & r_{34} \end{bmatrix} = \begin{bmatrix} .32 & .24 \\ .56 & .42 \end{bmatrix}$$

A 2×2 submatrix is formed by omitting all but two rows and all but two columns of the matrix. This leaves only four elements arranged in a 2×2 matrix. The minor determinant of the foregoing submatrix may be evaluated numerically as follows:

$$r_{12}r_{34} - r_{32}r_{14} = (.32)(.42) - (.56)(.24) = (.1344) - (.1344) = 0$$

If all the minor determinants of order 2 equal zero, the matrix is said to have rank 1; then, one factor will account for the **R** matrix. This was the criterion used by Spearman in developing his famous *G* factor of intelligence. He was attempting to account for intelligence in terms of a single common factor plus a separate specific factor for each kind of test. Thurstone employed the generalization of this principle as follows: If all the minor determinants of size 3×3 have determinants equal to zero but those of size 2×2 are not all zero, then two factors are needed to account for the **R** matrix exactly and the matrix has rank 2. In general, if all the minor determinants of size $m+1$ equal zero, but not all those of order m, then m factors are needed to account for the **R** matrix and the matrix has rank m. Random errors in the **R** matrix and the complexities of real-life data ordinarily combine to render even the total **R** matrix nonsingular, that is, having a determinant not equal to zero. Thus, n factors are needed to account for the typical **R** matrix exactly, giving it a rank of n. More efficient methods than evaluating determinants are available, however, for deciding how many factors are needed to account for the **R** matrix.

Methods of factor extraction, designed to produce the **A** matrix, usually seek to account for as much of the total extracted variance as possible on each successive extracted factor. That is, a factor is sought at each step for which the sum of squares of the factor loadings is as large as possible. The total variance extracted in a factor analysis is represented by the sum of the computed communalities, that is, $\sum_{i=1}^{m} h_{ii}^2$, where m is the number of common factors. All the data-variable variance is not ordinarily extracted. Each variable has a variance of 1, so the total data-variable variance that could theoretically be extracted is $n \times 1$, or n, where n is the number of data variables. If $\sum_{i=1}^{n} a_{ik}^2$ represents the sum of squares of loadings on factor k, the proportion of the total extracted variance due to factor k is obtained by dividing this total by the sum of the communalities. Since n is the total variance for all variables combined, that is, the sum of the diagonal elements of R_u, then dividing the sum of the communalities by n gives the proportion of the total variance that is accounted for by common factors.

After the first factor has been determined, its contribution to reproducing the $\mathbf{R}$ matrix is removed from $\mathbf{R}$ by the operation $\mathbf{R}_1 = \mathbf{R} - \mathbf{A}_1 \mathbf{A}_1'$, where $\mathbf{A}_1$ represents the first factor vector, $a_{11}, a_{21}, \ldots, a_{n1}$, and $\mathbf{A}_1'$ is the transpose of $\mathbf{A}_1$; $\mathbf{R}_1$ is called the residual matrix after extraction of factor 1, or the first residual matrix. It contains the residual correlations after the contribution to those correlations by factor 1 has been removed, that is, the term on the right-hand side of Eq. (2.10) associated with factor 1. If one factor is insufficient to reproduce the correlations in $\mathbf{R}$, then $\mathbf{R}_1$ will have some values which are substantially different from zero. If this is the case, another factor will be extracted from the first residual matrix by the equation $\mathbf{R}_2 = \mathbf{R}_1 - \mathbf{A}_2 \mathbf{A}_2'$. Thus, the second extracted factor is removed from the first factor residuals. In general, this process is continued, extracting the mth factor from the residuals left after taking out factor $m-1$, until the residuals are too small to yield another factor. Since at each step, as much variance is extracted as possible, the successive factors become smaller and smaller from first to last as shown by the sum of squares of the loadings in the successive columns of $\mathbf{A}$. Some methods of factor extraction, however, do not extract factor vectors in order of size of the sum of squares of their loadings. With such methods, the columns of the $\mathbf{A}$ matrix would have to be rearranged to attain this decreasing order of size of factors from first to last. This initial $\mathbf{A}$ matrix does not represent the final factor solution, however. These factors are "rotated" from their original positions by methods which are explained later. After the factors have been rotated, they are more nearly equal in size than the initially extracted factors.

[2.4] FACTOR ROTATION

Factor analysis in general and factor extraction methods in particular do not provide a unique solution to the matrix equation $\mathbf{R} = \mathbf{AA}'$. One of the reasons is that the $\mathbf{R}$ matrix is only approximately reproduced in practice and experimenters may differ on how closely they feel they must approximate $\mathbf{R}$. This will lead to their using different numbers of factors. Also, different methods of determining $\mathbf{A}$ may give slightly different results. An even more important reason for lack of unique solutions, however, is the fact that even for $\mathbf{A}$ matrices of the same number of factors, there are infinitely many different $\mathbf{A}$ matrices which will reproduce the $\mathbf{R}$ matrix equally well. Consider the following.

$$\underset{\mathbf{A}}{\begin{bmatrix} a_{11} & a_{12} \\ a_{21} & a_{22} \\ a_{31} & a_{32} \\ a_{41} & a_{42} \end{bmatrix}} \times \underset{\mathbf{\Lambda}}{\begin{bmatrix} \cos\varphi & \sin\varphi \\ -\sin\varphi & \cos\varphi \end{bmatrix}} = \underset{\mathbf{V}}{\begin{bmatrix} v_{11} & v_{12} \\ v_{21} & v_{22} \\ v_{31} & v_{32} \\ v_{41} & v_{42} \end{bmatrix}}$$

This schematic matrix operation may be expressed as a matrix equation:

(2.23) $$\mathbf{A\Lambda} = \mathbf{V}$$

If $\mathbf{R} = \mathbf{AA}'$, then it is also true that $\mathbf{R} = \mathbf{VV}'$ since if we transpose the product $\mathbf{A\Lambda}$, Eq. (2.23) may be rewritten as

(2.24) $$(\mathbf{A\Lambda})' = \mathbf{V}'$$

Since the transpose of a product is the product of transposes in reverse order, Eq. (2.24) becomes

(2.25) $$\mathbf{\Lambda}'\mathbf{A}' = \mathbf{V}'$$

By using Eqs. (2.23) and (2.25) the product $\mathbf{W}'$ becomes

(2.26) $$\mathbf{VV}' = \mathbf{A\Lambda\Lambda}'\mathbf{A}'$$

But $\mathbf{\Lambda\Lambda}'$, included in the middle of the matrix product in Eq. (2.26), gives an identity matrix, as follows:

(2.27) $$\underset{\mathbf{\Lambda}}{\begin{bmatrix} \cos\varphi & \sin\varphi \\ -\sin\varphi & \cos\varphi \end{bmatrix}} \times \underset{\mathbf{\Lambda}'}{\begin{bmatrix} \cos\varphi & -\sin\varphi \\ \sin\varphi & \cos\varphi \end{bmatrix}} = \underset{\mathbf{I}}{\begin{bmatrix} 1 & 0 \\ 0 & 1 \end{bmatrix}}$$

The reason for this is that the diagonal terms of the product matrix are equal to $\cos^2\varphi + \sin^2\varphi$, which equals 1 for all φ, and the off-diagonal elements are equal to $\sin\varphi\cos\varphi - \sin\varphi\cos\varphi$, which equals zero. As a result, Eq. (2.26) simplifies to $\mathbf{R} = \mathbf{AA}'$, since multiplying by an identity matrix does not alter the matrix, that is, $\mathbf{AIA}' = \mathbf{AA}'$.

As long as the matrix $\mathbf{\Lambda}$ is of such a form that $\mathbf{\Lambda\Lambda}' = \mathbf{I}$, then $\mathbf{A\Lambda}$ will reproduce the $\mathbf{R}$ matrix as well as $\mathbf{A}$ itself. Since the value of φ is not specified, this means that there are as many $\mathbf{\Lambda}$ matrices that will do this as there are values of φ. This particular $\mathbf{\Lambda}$ matrix was of size 2×2, or order 2, because only two factors were involved in the $\mathbf{A}$ matrix. If there had been three factors in the $\mathbf{A}$ matrix, the $\mathbf{\Lambda}$ matrix required would be of size 3×3. In general, if there are m factors in the **A** matrix, the **Λ** matrix will be of size $m \times m$. Any such **Λ** matrix must meet the following requirements: (1) The sums of squares of the rows must equal 1; (2) the sums of squares of the column must equal 1; (3) the inner product of one row by another row must equal zero for all pairs of nonidentical rows; (4) the inner product of one column by another column must equal zero for all pairs of nonidentical columns. If these conditions are met, then $\mathbf{\Lambda\Lambda}' = \mathbf{I} = \mathbf{\Lambda}'\mathbf{\Lambda}$. If these conditions are not met, then $\mathbf{\Lambda\Lambda}'$ is not equal to the identity matrix and **AΛ** will not substitute for **A** in reproducing the **R** matrix in the same way that **A** will.

The matrix **A** represents an interpretation of the data, the correlation matrix, in terms of a set of explanatory constructs, or factors. The values in a column of the **A** matrix of orthogonal factors are correlations of the data variables with the particular explanatory construct, or factor, represented by that column. By studying these correlations between the known data variables and the hypothetical explanatory constructs (or factors), it is possible to infer something about the nature of the explanatory constructs themselves. Factor interpretation is treated at greater length in Chapter 10.

The values in a given column of **AΛ** will be different from those of **A** itself. This means that different constructs are involved. Matrix **A** represents one set of constructs for accounting for the data. Matrix **AΛ** represents a different set of constructs which account for the data equally well in the mathematical sense that both reproduce the **R** matrix equally well.

The so-called rotational process in factor analysis involves finding a **Λ** matrix such that **AΛ** will represent an optimum set of constructs for scientific purposes. Since what is optimum for one investigator may not be optimum for another, this particular phase of the factor analytic process provides a fertile source of differences among investigators in the way they view the data. But, just as the artist, the engineer, the geologist, and the farmer may all describe a given piece of real estate accurately in very different ways, so can various transformations of **A** provide equally accurate but different descriptions of a body of data. An attempt is made later in this book to provide some guidelines for choosing among alternate possible interpretations, all of which are equally good mathematically for reproducing **R**.

It is sometimes proposed that the **A** matrix should be used as it is, without performing any transformations; that is, use **A** instead of **AΛ**. This approach will yield factor constructs which are highly complex since each extracted factor accounts ordinarily for approximately as much variance as possible. The only way to make a factor account for this much variance is to make the factor overlap as much as possible with as many of the data variables as possible. Since such a factor correlates substantially with many variables that are essentially uncorrelated with each other, it becomes a complex composite of unrelated things. After this factor is removed, the second factor will become a complex composite of as much as possible of what is left in the matrix of residuals. The position is taken in this book that factor constructs are more useful if they are relatively univocal in character, that is, if they cannot be broken down into several essentially uncorrelated components. To derive univocal factor constructs from a factor analysis, if it is possible at all in a given situation, almost always requires that the original extracted **A** matrix be rotated or transformed to a **V** matrix by the transformation $\mathbf{V} = \mathbf{A\Lambda}$ where

$$R = \mathbf{AA}' = \mathbf{VV}' \tag{2.28}$$

It is understood, of course, that in practice, Eq. (2.28) is only approximate in the sense that the **R** matrix is only approximately reproduced by $\mathbf{AA}'$ and $\mathbf{VV}'$.

[2.5] GEOMETRIC REPRESENTATION OF THE FACTOR MODEL

Although the factor model can be developed without reference to geometric concepts, as done by Horst (1965) in his book on factor analysis, most students find it helpful to employ an additional medium of representation in trying to understand the basis of this complex statistical technique.

In the geometric representation of the factor model, a data variable may be represented as a vector in a space of as many dimensions as there are common factors (see Fig. 2.1). In this case, the length of the vector is h, the square root of h^2, the communality. It is also possible to represent the data variables in spaces of higher order, adding also a dimension for each specific and error factor. If all of these factors are included, h^2 rises to 1.0 for each data variable and so does the vector length. In the most extreme case, it is possible to represent each data variable by a vector in N dimensions, requiring one coordinate for each person or other data-yielding object.

As an illustration, Fig. 2.1 represents four data variables as vectors in a two-dimensional space. A vector is a line extending from the origin to some point in space, in this example a space of two dimensions, that is, a plane. The coordinates of the end points (or termini) of the vectors are given. There are two coordinates for each point because the vectors are represented in two dimensions. The vectors are represented in the common factor space because the vectors are of length h rather than of length 1.0. If more than two common factors were needed to account for the data, the vectors would have to be represented in more than two dimensions and more than two coordinates per vector would be needed.

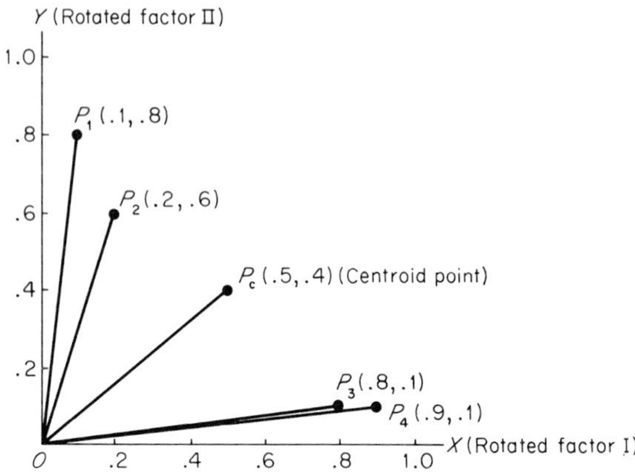

FIG. 2.1. Four data vectors.

By the Pythagorean theorem, the length of each vector in Fig. 2.1 is given by the square root of the sum of squares of its coordinates:

$$
\begin{aligned}
h_1 &= \sqrt{(.1)^2 + (.8)^2} = \sqrt{.01 + .64} = \sqrt{.65} = .806 \\
h_2 &= \sqrt{(.2)^2 + (.6)^2} = \sqrt{.04 + .36} = \sqrt{.40} = .632 \\
h_3 &= \sqrt{(.8)^2 + (.1)^2} = \sqrt{.64 + .01} = \sqrt{.65} = .806 \\
h_4 &= \sqrt{(.9)^2 + (.1)^2} = \sqrt{.81 + .01} = \sqrt{.82} = .906
\end{aligned} \tag{2.29}
$$

For more than two dimensions, for example, m dimensions, vector length is given by

$$h = \sqrt{{a_1}^2 + {a_2}^2 + {a_3}^2 + \cdots + {a_m}^2} \tag{2.30}$$

where the a_i values are the coordinates of the vector with respect to the m reference axes or dimensions. For more than three dimensions, of course, it is impossible to visualize the results.

The scalar product of two vectors may be defined as follows:

$$[\, a_{11} \quad a_{12} \quad a_{13} \quad \cdots \quad a_{1m} \,] \times \begin{bmatrix} a_{21} \\ a_{22} \\ a_{23} \\ \vdots \\ a_{2m} \end{bmatrix} = [c] \tag{2.31}$$

where $c = a_{11}a_{21} + a_{12}a_{22} + \cdots + a_{1m}a_{2m}$ is a constant. The values $a_{11}, a_{12}, a_{13}, \ldots, a_{1m}$ represent the coordinates of the first vector, and the values

$a_{21}, a_{22}, a_{23}, \ldots, a_{2m}$ are the coordinates for the second vector. Thus, the scalar products of all possible pairs of the four test vectors in Fig. 2.1 are given as follows:

(2.32)

$$\begin{array}{lll}
\text{Pair} & \text{Row} \times \text{Column} & \text{Scalar product} \\
1,2 & (.1\ .8) \times \begin{pmatrix} .2 \\ .6 \end{pmatrix} & = (.1 \times .2) + (.8 \times .6) = (.02 + .48) = .50 \\
1,3 & (.1\ .8) \times \begin{pmatrix} .8 \\ .1 \end{pmatrix} & = (.1 \times .8) + (.8 \times .1) = (.08 + .08) = .16 \\
1,4 & (.1\ .8) \times \begin{pmatrix} .9 \\ .1 \end{pmatrix} & = (.1 \times .9) + (.8 \times .1) = (.08 + .09) = .17 \\
2,3 & (.2\ .6) \times \begin{pmatrix} .8 \\ .1 \end{pmatrix} & = (.2 \times .8) + (.6 \times .1) = (.16 + .06) = .22 \\
2,4 & (.2\ .6) \times \begin{pmatrix} .9 \\ .1 \end{pmatrix} & = (.2 \times .9) + (.6 \times .1) = (.18 + .06) = .24 \\
3,4 & (.8\ .1) \times \begin{pmatrix} .9 \\ .1 \end{pmatrix} & = (.8 \times .9) + (.1 \times .1) = (.72 + .01) = .73
\end{array}$$

Let λ_{ij} represent the cosine of the angle between vector i and coordinate axis j. The value λ_{ij} is called the direction cosine of vector i with respect to coordinate (factor) axis j. If a_{ij} is the coordinate of data vector i with respect to factor axis j, and h_i is the length of vector i, then

$$\lambda_{11} = \frac{a_{11}}{h_1}, \ \lambda_{12} = \frac{a_{12}}{h_1}, \ \ldots, \ \lambda_{1m} = \frac{a_{1m}}{h_1}$$

or

$$a_{11} = h_1 \lambda_{11}, a_{12} = h_1 \lambda_{12}, \ldots, a_{1m} = h_1 \lambda_{1m}$$

and

$$a_{21} = h_2 \lambda_{21}, a_{22} = h_2 \lambda_{22}, \ldots, a_{2m} = h_2 \lambda_{2m}$$

Substituting these values in Eq. (2.31) gives the following representation of the scalar product of two vectors:

$$[\, h_i \lambda_{i1} \quad h_i \lambda_{i2} \quad \cdots \quad h_i \lambda_{im} \,] \times \begin{bmatrix} h_j \lambda_{j1} \\ h_j \lambda_{j2} \\ \vdots \\ h_j \lambda_{jm} \end{bmatrix} = [c]$$

or

$$h_i h_j \lambda_{i1} \lambda_{j1} + h_i h_j \lambda_{i2} \lambda_{j2} + \cdots + h_i h_j \lambda_{im} \lambda_{jm} = c$$

and

(2.33)

$$h_i h_j [\lambda_{i1} \lambda_{j1} + \lambda_{i2} \lambda_{j2} + \cdots + \lambda_{im} \lambda_{jm}] = c$$

A theorem from analytic geometry that will not be proved here states that the inner product of the direction cosines for two vectors equals the cosine of the angle between the vectors. Thus, Eq. (2.33) becomes

$$h_i h_j \cos \varphi_{ij} = c \tag{2.34}$$

The scalar product between vectors i and j is also equal to the correlation between them. The proof proceeds from Eq. (2.10):

$$r_{ij} = a_{i1} a_{j1} + a_{i2} a_{j2} + \cdots + a_{im} a_{jm}$$

Dividing both sides of this equation by h_i and h_j gives

$$\begin{aligned} \frac{r_{ij}}{h_i h_j} &= \frac{a_{i1}}{h_i} \cdot \frac{a_{j1}}{h_j} + \frac{a_{i2}}{h_i} \cdot \frac{a_{j2}}{h_j} + \cdots + \frac{a_{im}}{h_i} \cdot \frac{a_{jm}}{h_j} \\ &= \lambda_{i1} \lambda_{j1} + \lambda_{i2} \lambda_{j2} + \cdots + \lambda_{im} \lambda_{jm} = \cos \varphi_{ij} \end{aligned}$$

or

$$r_{ij} = h_i h_j \cos \varphi_{ij} \tag{2.35}$$

It should be pointed out that r_{ij} in Eq. (2.35) is the reproduced correlation based on the contributions of m common factors and hence is only approximately equal to the actual data correlation. The h_i, h_j, and $\cos \varphi_{ij}$ are the lengths of these vectors and the angle between them, respectively, in the common factor space.

Application of the law of cosines will show that the scalar products computed for all pairs of vectors in the series of equations in (2.32) are the same as the results of multiplying the lengths of two vectors by the cosine of the angle between them as on the right-hand side of Eq. (2.35). For two of the vectors, P_1 and P_2, for example, the law of cosines states that

$$h_1 h_2 \cos \varphi_{12} = \tfrac{1}{2}(h_1{}^2 + h_2{}^2 - c^2) \tag{2.36}$$

where c is the distance between the vector end points and is given by

$$c = \sqrt{(X_1 - X_2)^2 + (Y_1 - Y_2)^2} \tag{2.37}$$

where (X_1, Y_1) and (X_2, Y_2) are the coordinates for P_1 and P_2, respectively, in Fig. 2.1. For the first two vectors

$$c = \sqrt{(.1 - .2)^2 + (.8 - .6)^2} = \sqrt{.01 + .14} = \sqrt{.05}$$

Then, $h_1 h_2 \cos \varphi_{12} = \frac{1}{2}(.65 + .40 - .05) = .50$. This value is the same as that obtained for the scalar product of vectors 1 and 2 obtained in Eq. (2.32). Applying these same equations to the calculations for other vector pairs gives the same numbers as those obtained in (2.32) for the other scalar products.

The relationship between the correlation coefficient and the scalar product can be seen more clearly when the data variables are represented by vectors in a space of as many dimensions as persons, that is, in N dimensions. The lengths of such vectors are 1.0, so the coordinates for two vectors must be scaled by dividing the z scores by the square root of N. The sum of squares of such scores then equals 1.0, so the length of the vector scaled in this way is appropriately equal to 1.0. The scalar product of two such vectors is

$$\left[\frac{z_{i1}}{\sqrt{N}}\ \frac{z_{i2}}{\sqrt{N}}\ \frac{z_{i3}}{\sqrt{N}}\ \cdots\ \frac{z_{iN}}{\sqrt{N}}\right] \times \begin{bmatrix} \frac{z_{j1}}{\sqrt{N}} \\ \frac{z_{j2}}{\sqrt{N}} \\ \frac{z_{j3}}{\sqrt{N}} \\ \vdots \\ \frac{z_{jN}}{\sqrt{N}} \end{bmatrix} = \frac{\sum_{k=1}^{N} z_{ik} z_{jk}}{N} = r_{ij} \tag{2.38}$$

The right-hand side of Eq. (2.38) represents an expression like Eq. (2.6), and hence is a correlation coefficient. In this case, the vector lengths are 1.0, so $r_{ij} = h_i h_j \cos \varphi_{ij} = (1.0)(1.0) \cos \varphi_{ij}$, and the correlation equals the cosine of the angle between the two unit-length data vectors. In this case, the scalar product is between actual data vectors, so the correlation obtained equals the data correlation. With error-free data, or if m, the number of common factors, becomes as large as n, the scalar product in the common factor space will equal the real data correlation exactly.

Using the squared lengths of the vectors and scalar products for all non-identical pairs of vectors in Fig. 2.1, the correlation matrix with communalities, as shown in Table 2.1, is obtained for the four data variables.

TABLE 2.1

A Centroid Factor Analysis

R

	1	2	3	4
1	(.65)	.50	.16	.17
2	.50	(.40)	.22	.24
3	.16	.22	(.65)	.73
4	.17	.24	.73	(.82)

=

A

	I	II
1	.578	−.562
2	.531	−.344
3	.687	.422
4	.765	.484

×

A′

	1	2	3	4
I	.578	.531	.687	.765
II	−.562	−.344	.422	.484

In Table 2.1, **R** is the correlation matrix; **A** is a factor matrix, or a matrix of extracted factors derived from **R**; and **A**′ is the transpose of **A**. The values in parentheses along the main diagonal of **R** are the communalities h_{ii}^2 obtained by squaring the vector lengths h_i in Eq. (2.29). The off-diagonal elements of **R** are the correlations r_{ij} among the fictitious data variables, derived as scalar products of the vectors in Fig. 2.1.

Since the correlation matrix in Table 2.1 was derived from the coordinates of four vectors represented in a plane, which is a space of two dimensions, the correlation matrix is error free, and should yield exactly two factors. These two factors will give the coordinates for each variable in two dimensions.

Use of the centroid method of factor extraction, which is described in a later chapter, does indeed yield two factors which reproduce this correlation matrix perfectly, within the limits of rounding error. This reproduction of **R** is accomplished by the matrix operation $\mathbf{R} = \mathbf{AA}'$, as in Eq. (2.11). If more than two factors were necessary to reproduce **R**, the four vectors could not be drawn in a plane. A space of three or more dimensions would be needed. If fewer than two factors were needed, that is, only one, the vectors would all fall on the same line. This is the case for the matrix in Eq. (2.22) where the cosine of the angle between every pair of vectors is 1.0, meaning that the angle between any pair of vectors is zero. Applying Eq. (2.35) for one pair of these vectors gives $r_{12} = h_1 h_2 \cos \varphi_{12}$, or $.32 = \sqrt{.16}\sqrt{.64} \cos \varphi_{12}$ or $\cos \varphi_{12} = 1.0$, and φ_{12} equals 0°.

If the X and Y coordinates for the four vectors in Fig. 2.1 are averaged, the average X coordinate would be .5 and the average Y coordinate would be .4. These two coordinates locate the centroid point, P_c (.5,.4). Centroid factor analysis derives the factor loadings by obtaining the perpendicular projections of the test vectors onto a line extending from the origin through the centroid point P_c as in Fig. 2.1. The perpendicular projection of vector 1 on the centroid vector is shown as line OA in Fig. 2.2. Then, the $\cos \varphi_{1c}$ equals OA/OP_1 or a_1/h_1 where a_1 is the factor loading, or projection of vector 1 on the centroid vector. Using these equalities, the expression for the factor loading a_1 becomes

$$a_1 = h_1 \cos \varphi_{1c} \tag{2.39}$$

Cosine φ_{1c} here is the cosine of the angle between vector 1 and the centroid vector.

The scalar product between vector 1 and the centroid vector is given by[1]

$$(.1 \quad .8) \times \begin{pmatrix} .5 \\ .4 \end{pmatrix} = (.1 \times .5) + (.8 \times .4) = .37 = h_1 h_c \cos \varphi_{1c} \tag{2.40}$$

[1] The centroid factor vector has unit length, but the centroid vector, from the origin to the centroid point, with which it is coincident, does not.

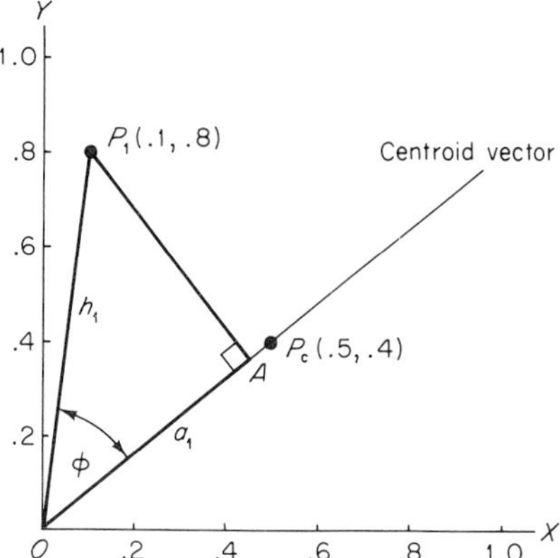

FIG. 2.2. Projection of a data vector on the centroid vector.

Using Eq. (2.39), Eq. (2.40) becomes

$$.37 = h_c(h_1 \cos \varphi_{1c}) = h_c a_1 \tag{2.41}$$

Solving for a_1 gives $a_1 = .37/h_c$. Since by the Pythagorean theorem h_c is given by $\sqrt{(.5)^2+(.4)^2} = \sqrt{.41} = .6403$, this gives $a_1 = .37/.6403 = .578$. Thus, dividing the scalar product of each vector with the centroid vector by the length of the centroid vector from the origin to the centroid point gives the projection of the data vector on the centroid vector. Performing this operation for the other three vectors in Fig. 2.1 gives $a_2 = .531$, $a_3 = .687$, and $a_4 = .765$.

These projections of the four data vectors in Fig. 2.1 on the centroid vector, derived from the coordinates of the vectors, are identical with the first factor loadings in matrix **A**, Table 2.1, derived by factor analysis of the correlation matrix **R**, also shown in Table 2.1.

The contribution of factor I to the **R** matrix is removed by the equation $\mathbf{R}_1 = \mathbf{R} - \mathbf{A}_1 \mathbf{A}_1'$ [see page 46, matrix equations (a) and (b)]. $\mathbf{R}_1$, the matrix of residuals after the extraction of factor I, is reproduced exactly by the product of the second factor times its transpose [see page 46, matrix equation (c)]. In matrix terms, $\mathbf{R}_1 - \mathbf{A}_2 \mathbf{A}_2' = 0$. Since $\mathbf{R}_1$ is reproduced exactly by the second factor multiplied by its transpose, no more than two factors are necessary to account for the original **R** matrix.

The sum of the original communality values is $.65 + .40 + .65 + .82 = 2.52$. The sum of squares of the first factor loadings, 1.6737, plus the sum of squares of the second factor loadings, .8463, also equals 2.52, showing that all of the common factor variance was extracted. The first factor was approximately twice as large as the second factor since 1.6737 is roughly twice as much as .8463.

(a)

$$\underset{\mathbf{R}}{\begin{bmatrix} (.65) & .50 & .16 & .17 \\ .50 & (.40) & .22 & .24 \\ .16 & .22 & (.65) & .73 \\ .17 & .24 & .73 & (.82) \end{bmatrix}} - \underset{\mathbf{A}_1}{\begin{bmatrix} .578 \\ .531 \\ .687 \\ .765 \end{bmatrix}} \times \underset{\mathbf{A}_1'}{\begin{bmatrix} .578 & .531 & .687 & .765 \end{bmatrix}}$$

giving

(b)

$$\underset{\mathbf{R}}{\begin{bmatrix} (.65) & .50 & .16 & .17 \\ .50 & (.40) & .22 & .24 \\ .16 & .22 & (.65) & .73 \\ .17 & .24 & .73 & (.82) \end{bmatrix}} - \underset{\mathbf{A}_1\mathbf{A}_1'}{\begin{bmatrix} .334 & .307 & .397 & .442 \\ .307 & .282 & .365 & .402 \\ .397 & .365 & .472 & .526 \\ .442 & .406 & .526 & .585 \end{bmatrix}}$$

$$= \underset{\mathbf{R}_1}{\begin{bmatrix} (.316) & .193 & -.237 & -.272 \\ .193 & (.118) & -.145 & -.166 \\ -.237 & -.145 & (.179) & .204 \\ -.272 & -.166 & .204 & (.234) \end{bmatrix}}$$

(c)

$$\underset{\mathbf{R}_1}{\begin{bmatrix} (.316) & .193 & -.237 & -.272 \\ .193 & (.118) & -.145 & -.166 \\ -.237 & -.145 & (.179) & .204 \\ -.272 & -.166 & .204 & (.234) \end{bmatrix}} = \underset{\mathbf{A}_2}{\begin{bmatrix} -.562 \\ -.344 \\ .422 \\ .484 \end{bmatrix}} \times \underset{\mathbf{A}_2'}{\begin{bmatrix} -.562 & -.344 & .422 & .484 \end{bmatrix}}$$

Calculating the matrix of first factor residuals $\mathbf{R}_1$.

The factor analysis was carried out through operations with the correlation matrix only, without using directly any information about the coordinates of the vectors with respect to the X and Y axes in Fig. 2.1. The correlation matrix shows the squared lengths and scalar products of the vectors but does not refer the vectors to any coordinate system. Factor analysis can be viewed as a technique for placing coordinate axes within the same framework created by orienting the data-variable vectors in space with the appropriate angles between them.

Although each data-variable vector has N coordinates, one for each subject, there are only n vectors; so n dimensions would be enough to accommodate these vectors, even if they were all orthogonal to one another. By appropriate location of the coordinate axes, it is possible to squeeze the dimensionality of the space down still further, from n to m, without distorting the geometric representation of the data too much. The factor analytic process determines just where the m factor vectors are to be placed with respect to the data-variable vectors and derives the coordinates which the data-variable vectors will have with respect to coordinate axes drawn collinearly with the factor vectors. The factor axes, all of unit length, provide a reference frame with respect to which the data-variable vectors can be located by giving coordinates or factor loadings. In Fig. 2.1, as an exercise, the coordinates with respect to the X and Y axes were used to derive a correlation matrix. Factor analysis is essentially the reverse of this process; that is, the given correlation matrix is used to find the coordinates of the vectors with respect to some system of coordinate axes.

The number of factor vectors needed is equal to the dimensionality of the space, that is, the number of dimensions in the space, into which the data-variable vectors are fitted. In the simple four-variable example given here, all the vectors lie in a plane; so two coordinate axes are adequate for a complete specification of the data-variable vectors.

The centroid method places the first reference axis through the centroid point at the center of the swarm of points that are the termini of the data-variable vectors (see Fig. 2.1). The second reference axis, or second centroid vector, is placed at right angles to the first centroid factor vector. In most methods of factor extraction, each succeeding factor extracted is placed at right angles to the preceding factors. The centroid loadings represent coordinates of the test vectors with respect to the orthogonal centroid axes. It is possible to transform these coordinates with respect to the centroid loadings into coordinates with respect to any other set of reference axes, such as those in Fig. 2.1, that is, the X and Y coordinate axes. This operation would represent the rotation of the centroid factors to new factor positions with greater value for descriptive, theoretical, or other scientific purposes.

The angle between the centroid axis and the Y axis in Fig. 2.1 may be obtained as follows: The Y-axis vector terminus has coordinates (0, 1). The scalar product with the centroid is given by

$$h_y h_c \cos \varphi_{yc} = (0 \quad 1) \times \begin{pmatrix} .5 \\ .4 \end{pmatrix} = (0 \times .5) + (1 \times .4) = .4$$

$$\cos \varphi = .4/h_y h_c = .4/(1.0 \times .6403) = .625$$

Hence, $\varphi = 51°19'$ and the sine of φ is .781. The angle of the centroid axis with the X axis in Fig. 2.1 is $90° - 51°19'$, or $38°41'$.

To obtain a configuration of points corresponding to those in Fig. 2.1 using the coordinates of the data points from the matrix of factor loadings **A** in Table 2.1, it is necessary to reverse the direction of factor II in matrix **A**. This is done by changing the signs of all the loadings in factor II, matrix **A**, Table 2.1. Any factor may be reversed in direction in this manner at any stage in the factor analytic process without affecting the property that **AA**′ reproduces the matrix **R**.

Reversing the direction of a factor by changing all the signs of the loadings on the factor, then, is perfectly acceptable. It merely inverts the meaning of the factor. For example, a factor named Lack of Intelligence would be converted to a factor properly named Intelligence merely by reversing the signs of all the loadings.

If the signs on the loadings for factor II in Table 2.1 are reversed in this manner, the coordinates for the four data vectors with respect to the centroid axes become P_1 (.578, .562), P_2 (.531, .344), P_3 (.687, −.422), and P_4 (.765, −.484). Plotting these data points with respect to centroid factor axes I and II yields Fig. 2.3. Only the end points of the data vectors are plotted here. This is a more usual practice than drawing in the full vector from the origin to each data-vector end point.

Rotation of factor I away from factor II by an angle of $38°41'$ will bring it into coincidence with the old X axis (see Fig. 2.3). At the same time, rotation of factor II an equivalent amount toward factor I will bring factor II into coincidence with the old Y axis. The matrix operations in Table 2.2 show how this rotation is carried out, transforming the coordinates with respect to centroid axes (after reversing axis II) into the coordinates with respect to the original X and Y coordinate axes, respectively.

In Table 2.2 **A** is the centroid matrix with factor II reversed; **Λ** is the transformation matrix with $\cos 38°41'$ in cells λ_{11} and λ_{22}, $-\sin 38°41'$ in cell λ_{21}, and $\sin 38°41'$ in cell λ_{12}; **V** is the matrix of coordinates of the data-variable vectors with respect to the original X and Y axes. The coordinates with respect to the original X and Y axes, matrix **V**, can also be transformed back to the

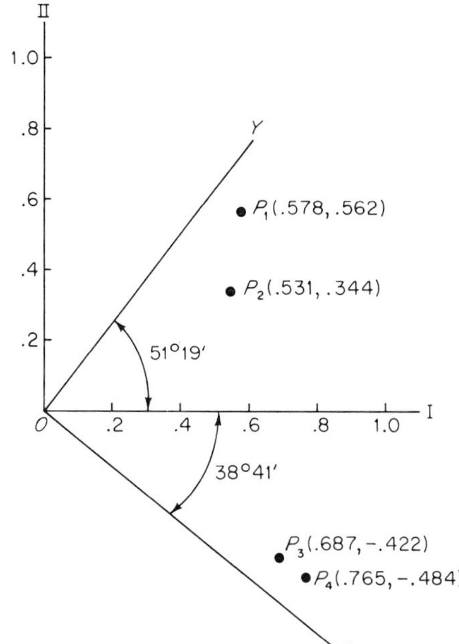

FIG. 2.3. Rotation from centroids to original coordinate axes.

centroid loadings. From Table 2.2 it is apparent that $\mathbf{V} = \mathbf{A\Lambda}$. Let the matrix $\mathbf{\Lambda}^{-1}$ be called the inverse of $\mathbf{\Lambda}$, a matrix that has the following property:

$$\mathbf{\Lambda}^{-1}\mathbf{\Lambda} = \mathbf{\Lambda\Lambda}^{-1} = \mathbf{I}$$

where $\mathbf{I}$ is the identity matrix as before. Then, if the matrix equation $\mathbf{V} = \mathbf{A\Lambda}$ is multiplied on both sides by $\mathbf{\Lambda}^{-1}$, the following matrix result is obtained:

$$\mathbf{V\Lambda}^{-1} = \mathbf{A\Lambda\Lambda}^{-1} = \mathbf{AI} = \mathbf{A} \tag{2.42}$$

TABLE 2.2

Rotation of the Centroid Axes

Unrotated factor matrix				Transformation matrix			Rotated factor matrix	
	I	II					I(X)	II(Y)
1	.578	.562				1	.1	.8
2	.531	.344	×	.781	.625	= 2	.2	.6
3	.687	−.422		−.625	.781	3	.8	.1
4	.765	−.484				4	.9	.1
	A			**Λ**			**V**	

Thus, if the matrix **V** is multiplied on the right by the inverse of the matrix which transforms the original centroid factors to **V**, that is, by the operation $\mathbf{V}\mathbf{\Lambda}^{-1}$, the **V** matrix will be transformed back to the centroid factors. In the case of an orthogonal matrix like $\mathbf{\Lambda}$, the inverse $\mathbf{\Lambda}^{-1}$ is very easily obtained because it is merely the transpose of $\mathbf{\Lambda}$. For nonorthogonal matrices, the inverse of a matrix is not so easily obtained and is not equal to the transpose.

Equation (2.42) for this example gives the following result:

$$\underset{\mathbf{V}}{\begin{bmatrix} .1 & .8 \\ .2 & .6 \\ .8 & .1 \\ .9 & .1 \end{bmatrix}} \times \underset{\mathbf{\Lambda}^{-1}}{\begin{bmatrix} .781 & -.625 \\ .625 & .781 \end{bmatrix}} = \underset{\mathbf{A}}{\begin{bmatrix} .578 & .562 \\ .531 & .344 \\ .687 & -.422 \\ .765 & -.484 \end{bmatrix}}$$

In this example, the centroid factors are complex composites of all the variables represented by the data vectors. A complex composite factor is one which has substantial loadings for variables that are slightly or not at all correlated with each other. By rotating matrix **A** to matrix **V**, the factor matrix is rotated to a form in which factor 1 of the **V** matrix is like variables 3 and 4 but not much like variables 1 and 2. The second factor of **V** is like variables 1 and 2 but not much like variables 3 and 4. The factor constructs represented by the **V** matrix are less complex than those represented by the **A** matrix, since in the **A** matrix all variables have substantial loadings on both factors. For purposes of scientific description and explanation, the factors in the **V** matrix are more useful because they are simpler, easier to understand, and better differentiated from each other since they are defined by different variables.

CHAPTER THREE

Factor Extraction by the Centroid Method

The centroid method of factor extraction (Thurstone, 1947; Fruchter, 1954) is probably the best known of all methods of factor extraction. Some form of the centroid method was used in most of the published factor analytic studies before the advent of computers. Since electronic computing equipment has become more generally available, the centroid method has been supplanted gradually by more precise methods, such as the principal factor method, which are too laborious for application with a desk calculator. The centroid method is well suited for teaching purposes, however, since it is feasible to do a problem of sufficient size with this method on a desk calculator in order to illustrate the factor analytic process. The centroid method also has the advantage of being easily conceptualized in terms of the geometric model of factor analysis. Because of its historical importance and its advantages for teaching purposes, the centroid method is presented here in some detail despite the fact that it is no longer widely used.

[3.1] EQUATIONS FOR THE CENTROID METHOD

In Fig. 2.1 the centroid point was located by averaging all the data-vector X coordinates to get the X coordinate of the centroid point and all the data-vector Y coordinates to get the Y coordinate of the centroid point. If there had been more than two dimensions involved, the third, fourth, and other coordinates would have been averaged to get the remaining coordinates of the centroid point.

Imagine a new set of coordinate axes at right angles to one another placed in such a way that one of the axes goes through the centroid point, as in Fig. 2.3. Suppose that the coordinates of the data vectors with respect to these new coordinate axes are known (actually they are the centroid loadings given in Table 2.2). With these new coordinates, it would be possible to recompute the coordinates of the centroid point with respect to the new coordinate axes. Since one of the axes, say the first one, goes right through the centroid point, however, the coordinates of the centroid point with respect to the new axes will be

$$\frac{1}{n}\sum_{i=1}^{n} a_{i1}, 0, 0, \ldots, 0 \tag{3.1}$$

That is, since the centroid point falls on the first coordinate axis, its coordinates with respect to all the other axes will be zero. In Fig. 2.3 there are only two factors, hence only two coordinates. Therefore, the coordinates of the centroid point can be found by averaging the loadings with respect to the new axes, the first of which goes through the centroid point. These loadings are given in Table 2.2. Thus, averaging the coordinates in the first column gives $\frac{1}{4}(.578+.531+.687+.765)=.64$ as the first coordinate of the centroid point. The second coordinate of the centroid point is the average of the second column of the **A** matrix in Table 2.2; that is, $\frac{1}{4}(.562+.344-.422-.484)=0$, which is in accordance with Eq. (3.1).

To use Eq. (3.1) for deriving an expression for the centroid loadings, Eq. (2.10) serves as a starting point:

$$r_{ij} = a_{i1}a_{j1} + a_{i2}a_{j2} + \cdots + a_{im}a_{jm} \tag{3.2}$$

Summing both sides of Eq. (3.2) over i gives

$$\sum_{i=1}^{n} r_{ij} = a_{j1}\sum_{i=1}^{n} a_{i1} + a_{j2}\sum_{i=1}^{n} a_{i2} + \cdots + a_{jm}\sum_{i=1}^{n} a_{im} \tag{3.3}$$

Summing both sides of Eq. (3.3) over j gives

$$\sum_{j=1}^{n}\sum_{i=1}^{n} r_{ij} = \sum_{j=1}^{n} a_{j1}\sum_{i=1}^{n} a_{i1} + \sum_{j=1}^{n} a_{j2}\sum_{i=1}^{n} a_{i2} + \cdots + \sum_{j=1}^{n} a_{jm}\sum_{i=1}^{n} a_{im} \tag{3.4}$$

But

$$\sum_{j=1}^{n} a_{jk} = \sum_{i=1}^{n} a_{ik} \tag{3.5}$$

since both terms in Eq. (3.5) are merely the sum of the entries in the kth column of the **A** matrix of factor loadings. Substituting (3.5) in (3.4), therefore, gives

$$\sum_{j=1}^{n}\sum_{i=1}^{n} r_{ij} = \left(\sum_{i=1}^{n} a_{i1}\right)^2 + \left(\sum_{i=1}^{n} a_{i2}\right)^2 + \cdots + \left(\sum_{i=1}^{n} a_{im}\right)^2 \tag{3.6}$$

By Eq. (3.1), however, all the sums on the right-hand side of Eq. (3.6) are zero except the first. Equation (3.6) reduces, therefore, to

$$\sum_{j=1}^{n}\sum_{i=1}^{n} r_{ij} = \left(\sum_{i=1}^{n} a_{i1}\right)^2 \tag{3.7}$$

Also by Eq. (3.1) the sums of loadings for the second and subsequent factors are zero, since the second and subsequent coordinates of the centroid point derived from these sums are zero, making Eq. (3.3) reduce to

$$\sum_{i=1}^{n} r_{ij} = a_{j1} \sum_{i=1}^{n} a_{i1} \tag{3.8}$$

Taking the square root of both sides of Eq. (3.7) and substituting in Eq. (3.8) gives

$$\sum_{i=1}^{n} r_{ij} = a_{j1} \sqrt{\sum_{j=1}^{n}\sum_{i=1}^{n} r_{ij}} \tag{3.9}$$

Solving for a_{j1} gives

$$a_{j1} = \frac{\sum_{i=1}^{n} r_{ij}}{\sqrt{\sum_{j=1}^{n}\sum_{i=1}^{n} r_{ij}}} \tag{3.10}$$

Equation (3.10) gives the formula for a centroid loading for variable j on factor 1 (or I, using roman numerals). To get the centroid factor loading for data variable 1, for example, Eq. (3.10) calls for the following steps:

1. Add up the entries in the first column of the correlation matrix, including the diagonal cell.
2. Divide this number by the square root (or multiply by the reciprocal of the square root) of the sum of all the entries in the entire correlation matrix, including the diagonal cells.

For the second data variable, compute the sum of the entries in the second column of the correlation matrix and divide by the same square root term as for the first data variable. Continue this for all columns of the **R** matrix. Thus, the computation steps involve computing the sums of the columns of the **R** matrix, adding these column sums to get the sum of all entries, taking the square root, and dividing this square root into each column sum.

[3.2]
A FOUR-VARIABLE, TWO-FACTOR EXAMPLE

Applying these steps to the four-variable correlation matrix in Table 2.1 gives the results in Table 3.1. The contribution of factor 1 must be removed

TABLE 3.1

Correlation Matrix

	1	2	3	4	
1	(.65)	.50	.16	.17	
2	.50	(.40)	.22	.24	
3	.16	.22	(.65)	.73	
4	.17	.24	.73	(.82)	$T = \sum\sum r = 6.56$
t	1.48	1.36	1.76	1.96	$\sqrt{T} = 2.56125$
a_i	.578	.531	.687	.765	$1/\sqrt{T} = .3904$

from the correlation matrix by the operation $\mathbf{R} - \mathbf{A}_1 \mathbf{A}_1'$, as shown on page 46. The results of this operation give the matrix $\mathbf{R}_1$, the matrix of first factor residuals shown in Table 3.2. Note that in Table 3.2 the columns in each case add up to zero, within the limits of rounding error. As a check, the rows should be added also, to make sure that the row totals equal the column totals as required for a symmetric matrix.

TABLE 3.2

First Factor Residuals

	1	2	3	4
1	(.316)	.193	−.237	−.272
2	.193	(.118)	−.145	−.166
3	−.237	−.145	(.179)	.204
4	−.272	−.166	.204	(.234)
Sum	.000	.000	.001	.000
$\sum_0$	−.316	−.118	−.178	−.234

It was established in Eq. (3.1) that the sum of the loadings on the second and subsequent centroid factors is zero, which provides the basis for a check on the computations of the first factor residuals used above. Since the sums of the columns are zero, however, it is clearly impossible to apply the steps used for computing the first centroid factor to the matrix of residuals given in Table 3.2. It is necessary first to carry out a process of reflecting the residuals to get rid of as many negative signs as possible in the matrix of first factor residuals.

The basis for this procedure is as follows: First, it is known that after they are computed, the second factor centroid loadings will sum to zero. This means that approximately half of the loadings will have negative signs. On the other hand, to extract a centroid factor, all the column sums of the residual correlation matrix should be positive and as large as possible. If this is the case, the loadings will be positive too, since they are the ratios of the positive column sums to the square root of the overall sum. To make the column sums positive and as large as possible, it is necessary to reflect, or change the direction, of those residual data-variable vectors that will ultimately have negative second factor loadings. This reflection is carried out by changing all the signs the variable has with every other variable in the matrix of residuals. As many variables are reflected as necessary to get the largest possible positive column sums in the matrix of reflected residuals. When all these sign changes have been effected, positive factor loadings can be extracted by the centroid method to obtain the second factor loadings. After extraction, however, with all positive signs, negative signs will be placed on the loadings for those variables that were reflected.

Any data variable that has been reflected, therefore, must have a negative sign attached to its second factor loading so that the second factor loadings with signs attached will add up algebraically to zero. Thus, any variable that will have the sign of its factor loading reversed on this factor from what it was on the previous factor will require reflection on this factor to make the sum of its column entries positive without including the communality value. Failure to carry out the reflection process far enough to get the maximum positive sums of the columns of reflected residuals will result in smaller column sums and hence in a smaller amount of variance being extracted on the factor than would be possible with complete reflection.

In picking a data variable to reflect, start with the variable that has the largest negative total without the communality added in. The communaltiy is not considered here because it may even change a total from negative to positive, thus stopping the reflection process too soon. In the example shown in Table 3.2, this would be variable 1, which has a negative total of $-.316$ without the h^2 added in. Reflecting variable 1, that is, changing the signs of the residuals for variable 1 with all the other variables, requires a reversal of all the signs in row 1 and column 1. This leaves the diagonal cell unchanged. The first factor residuals after reflecting variable 1 are shown in Table 3.3. In this table, only variable 2 now has a negative sum, exclusive of the communalities, so it is the next variable to be reflected. After reflecting variable 2, the values in Table 3.3 are changed to those in Table 3.4. The asterisk (*), appended to columns 1 and 2 and rows 1 and 2, indicates that the variable has been reflected and a negative sign must be attached to the second factor loadings for these variables.

TABLE 3.3

First Factor Residuals after Reflecting Variable 1

	* 1	2	3	4
* 1	(.316)	−.193	.237	.272
2	−.193	(.118)	−.145	−.166
3	.237	−.145	(.179)	.204
4	.272	−.166	.204	(.234)
Sums without communalities	.316	−.504	.296	.310

After reflecting variable 2, as shown in Table 3.4, no more negative signs remain in the matrix of reflected residuals. In practice, some negative signs will usually remain after reflection has been carried as far as possible. In some cases, it may even be necessary to reflect back again a variable that was previously reflected. This just cancels the effect of the previous reflection of that variable, leaving it as a variable with a positive second factor loading rather than a negative one.

The computed second factor loadings in Table 3.4, obtained by the same steps applied in Table 3.1 for the first factor loadings, are all positive. Negative signs must be attached to the first two loadings, however, that is, for variables 1 and 2, since these two variables were reflected. These second factor loadings reproduce the matrix of second factor residuals exactly, as shown in the last chapter. In equation form, $\mathbf{R}_1 - \mathbf{A}_2\mathbf{A}_2' = 0$. In that example, the matrix of residuals before reflection was reproduced by the product $\mathbf{A}_2\mathbf{A}_2'$ with signs attached to the second factor loadings for variables 1 and 2. In actual practice, the column of positive factor loadings $\mathbf{A}_2$ is multiplied by its transpose $\mathbf{A}_2'$ to give an $n \times n$ matrix that is subtracted from the matrix of first factor residuals after they have been reflected. Subtraction takes place by subtracting corresponding matrix elements.

TABLE 3.4

Reflected First Factor Residuals and Second Factor Calculations

	* 1	* 2	3	4	
* 1	(.316)	.193	.237	.272	
* 2	.193	(.118)	.145	.166	
3	.237	.145	(.179)	.204	
4	.272	.166	.204	(.234)	$T = 3.281$
t	1.018	.622	.764	.877	$\sqrt{T} = 1.81135$
a_i	.562	.343	.422	.484	$1/\sqrt{T} = .55207$

Thus, in summary, the residuals are reflected to make them positive. A factor with positive signs is extracted. This positively signed factor is removed from the residuals after they have been reflected. After this, the negative signs are attached to the loadings for the variables that were reflected. The new residual matrix has sums of columns that equal zero and hence it must be reflected to start the cycle anew.

Once a variable has been reflected, all of its factor loadings will have minus signs attached to them for that and all later factors unless it is reflected again. If it is reflected again, it goes back to positive from that point on, remaining that way until it is reflected once more. It should be noted that changing all the signs on a factor will not affect the reproduction of the **R** matrix by the equation $\mathbf{R} = \mathbf{AA}'$. Any variable or any factor may be reflected if appropriate changes are made in the signs of the correlations and factor loadings. If any variable has a negative column sum without the communality added in, it is necessary to carry out reflections of the data variables in the original correlation matrix before even the first factor is computed. It is also desirable in many cases to reflect a factor after it has been extracted or rotated so that the positive end of the factor vector will correspond with the positive side of a bipolar concept, for example, Adjustment versus Maladjustment, represented by the factor.

The process of reflecting a variable at any point amounts to changing the signs of all correlations of that variable with all other variables. This causes the reversal of all signs on the factor loadings for this variable from that point on. The correlations for which the signs are changed may be from the original correlation matrix or from any of the subsequent residual matrices. After extraction, the signs on the factor loadings must be adjusted to be consistent with these reflections. Once a matrix of original or residual correlations with positive column totals has been obtained, extraction of a centroid factor proceeds according to the same steps enumerated above for the extraction of the first centroid factor.

[3.3] A 12-VARIABLE, FOUR-FACTOR EXAMPLE

Understanding of the factor analytic extraction process by the centroid method is facilitated by considering a somewhat larger problem based on real data. Table 11.1 (page 260) gives the matrix of intercorrelations for the 40 variables which define the eight factors on the Comrey Personality Scales, a personality inventory developed by factor analytic methods (Comrey, 1970a,b). This 44×44 table of correlations shown in Table 11.1 provides a source of sample problems for class exercises. By selecting four of the eight factors and three variables with high factor loadings that define each factor, a 12-variable sample problem may be generated which can serve as an exercise for the student. If desired, each student may take a different sample of factors and/or variables so that no two students are working on the same exercise. It is important, however, to have at least four factors and preferably three variables defining each factor. In this way, the exercise will provide an appropriate learning experience for the student, giving him an opportunity to develop the skills needed for both factor extraction and rotation.

In choosing the variables to represent a given factor, reference to Table 11.4 (page 264) will be helpful. This table gives the final eight-factor orthogonal rotated solution for the 44-variable matrix shown in Table 11.1. To be chosen as a variable to represent a factor in a sample problem, the variable should have a factor loading over .50 on that factor in Table 11.4. Names and descriptions of the variables are given in the text of Chapter 11. From the standpoint of the student, there is no substitute for actually carrying out a sample factor analysis on his own. Failure to do so almost inevitably will result in a poorer grasp of the subject. A sample problem of 12 variables and four factors should give sufficient scope to the problem for an initial attempt. Fewer factors will not permit adequate practice in rotation. If the size of the student problem must be reduced, such as for a one-quarter course instead of a full semester course, it is better to use only two marker variables for some or even all of the factors rather than to reduce the number of factors.

A complete factor analytic solution is given in Chapter 11 for this 44-variable, eight-factor problem which will serve as a model solution against which solutions to sample problems may be compared. Only the rotated solutions can be compared, however, since the extracted factor loadings depend too heavily on the particular combination of variables included in the analysis. After rotation, sample problems should produce a pattern recognizably similar to the model solution to the extent that a solution for a subset of the variables can be expected to resemble the solution for the full set. The size of the factor loadings for a sample problem, however is expected to agree only approximately with the loadings for the solution based on the entire matrix, as shown in Table 11.4.

The 12-variable sample problem chosen to be used as an example in this book is given in Table 3.5. Using the numbers in the second column of Table 3.5, Table 11.1 may be entered to find the correlations among these 12 selected variables. This correlation matrix is shown in Table 3.6. The correlation coefficients are entered both above and below the main diagonal in Table 3.6 to facilitate calculation of the first centroid factor. The estimated communalities in parentheses along the main diagonal are added to the correlation matrix in the process of extracting the first centroid factor.

TABLE 3.5

12-Variable Sample Problem

Variable number	Number in Table 11.1	Variable name	Factor
1	11	Law Enforcement	C
2	12	Acceptance of Social Order	C
3	13	Intolerance of Nonconformity	C
4	21	Lack of Inferiority Feelings	S
5	22	Lack of Depression	S
6	23	Lack of Agitation	S
7	26	Lack of Reserve	E
8	27	Lack of Seclusiveness	E
9	28	No Loss for Words	E
10	36	Sympathy	P
11	37	Helpfulness	P
12	39	Generosity	P

[3.3.1]
STEPS IN COMPUTING THE FIRST CENTROID FACTOR

1. Add up algebraically (using signs) each column of the correlation matrix (Table 3.6) omitting the diagonal cell or h^2 value from the total. Record these totals in row $\sum_0$ at the bottom of the correlation matrix.

2. Add the rows across, computing each row total of **R**, again omitting the diagonal cells. Check the row totals against the column totals to make sure they are identical.

3. If any of these column–row totals is negative, the reflection process must be applied, as explained under the steps for computing the second centroid factor, before proceeding. If all the column totals are positive, proceed to step 4.

4. Place estimated communality values in the diagonal cells. Most commonly, the absolute values (h^2 values must be positive) of the highest correlations in the columns are used as the estimates.

5. Add the estimated communalities to the column totals and place these sums in row t.

6. As a check, subtract the communalities from the totals in row t to see if the column totals without communalities are recovered.

7. Add the column totals with communalities, that is, the elements of row t, to obtain T, the sum of all entries in the table.

8. Compute the square root of T to at least five decimal places.

9. Compute the reciprocal of the square root of T to at least five places.

TABLE 3.6

12-Variable Correlation Matrix and First Factor Computations

	1	2	3	4	5	6	7	8	9	10	11	12	$\sum_o$
1	(.64)	.61	.64	.07	.14	.07	.04	.08	.06	−.02	−.02	−.17	1.50
2	.61	(.61)	.50	.02	.13	.08	−.05	.02	−.02	−.07	−.08	−.15	.99
3	.64	.50	(.64)	.02	.04	−.03	−.01	.08	.01	−.01	−.02	−.04	1.18
4	.07	.02	.02	(.51)	.51	.46	.28	.29	.47	.10	.14	.05	2.41
5	.14	.13	.04	.51	(.51)	.51	.29	.31	.41	.18	.24	.09	2.85
6	.07	.08	−.03	.46	.51	(.51)	.13	.19	.31	.05	.18	.05	2.00
7	.04	−.05	−.01	.28	.29	.13	(.66)	.46	.66	.16	.16	.12	2.24
8	.08	.02	.08	.29	.31	.19	.46	(.55)	.55	.31	.39	.26	2.94
9	.06	−.02	.01	.47	.41	.31	.66	.55	(.66)	.20	.28	.17	3.10
10	−.02	−.07	−.01	.10	.18	.05	.16	.31	.20	(.60)	.60	.46	1.96
11	−.02	−.08	−.02	.14	.24	.18	.16	.39	.28	.60	(.61)	.61	2.48
12	−.17	−.15	−.04	.05	.09	.05	.12	.26	.17	.46	.61	(.61)	1.45
$\sum_o$	1.50	.99	1.18	2.41	2.85	2.00	2.24	2.94	3.10	1.96	2.48	1.45	25.10
t	2.14	1.60	1.82	2.92	3.36	2.51	2.90	3.49	3.76	2.56	3.09	2.06	$T = 32.21$
a_I	.377	.282	.321	.514	.592	.442	.511	.615	.662	.451	.544	.363	$\sqrt{T} = 5.67538$
													$1/\sqrt{T} = .17620$

10. Multiply the column totals with communalities, that is, the entries in row t, by $1/\sqrt{T}$ and enter the products in row a_i. These are the centroid loadings before any signs have been attached.

11. If any variable has been reflected prior to these computations, these centroid loadings must have the appropriate signs attached before they are entered as the first column in matrix **A**, the table of factor loadings (see Table 3.13).

12. Sum the factor loadings in row a_i and check to see that it equals $\sqrt{T}$ within the limits of rounding error, as called for by Eq. (3.7).

The calculations for the 12-variable sample problem called for in steps 1–12 are shown in Table 3.6. Since all the columns, exclusive of h^2 values, were positive, no reflection of the original **R** matrix was necessary.

[3.3.2] *COMPUTING THE FIRST FACTOR RESIDUALS*

The matrix of first factor residuals $\mathbf{R}_1$ is computed by the matrix equation

$$\mathbf{R}_1 = \mathbf{R} - \mathbf{A}_1 \mathbf{A}_1' \tag{3.11}$$

where **R** is the original matrix of correlations, reflected if necessary; $\mathbf{A}_1$ is the first factor, a column matrix, with all positive signs attached; and $\mathbf{A}_1'$ is the transpose of $\mathbf{A}_1$. For a given element of $\mathbf{R}_1$, r_{1ij}, the computation is as follows:

$$r_{1_{ij}} = r_{ij} - a_{i1} a_{j1} \tag{3.12}$$

where r_{ij} is an element of the original correlation matrix and a_{i1} and a_{j1} are the positive first factor loadings for variables i and j.

Some analysts prefer to compute an $n \times n$ matrix of the positive elements $a_{i1} a_{j1}$ and then subtract this matrix, element by element, from the original correlation matrix, reflected if that should be required. To do this, lay out a square work sheet with the n positive first factor loadings down the side and also across the top of the page. Fill in the $n \times n$ blank cells in the matrix by multiplying the loading at the side by the loading at the top. Each element of this $n \times n$ matrix, including the diagonals, is subtracted from the corresponding element of the original correlation matrix to obtain the first factor residuals. For the 12-variable sample problem, these first factor residuals are shown in Table 3.7.

[3.3.3] *COMPUTING THE SECOND CENTROID FACTOR*

Steps 1 and 2 of the first centroid factor extraction procedure are repeated; that is, the rows and columns of the matrix of residuals are added up, omitting

TABLE 3.7 First Factor Residuals and Second Factor Computations

	*1	*2	*3	4	5	6	7	8	9	10	11	12	$\sum_o$
*1	.519 (.498)	.504	.519	−.124	−.083	−.097	−.153	−.152	−.190	−.190	−.225	−.307	−.498
*2	.504	.504 (.530)	.410	−.125	−.037	−.045	−.194	−.153	−.207	−.197	−.233	−.252	−.529
*3	.519	.410	.519 (.537)	−.145	−.150	−.172	−.174	−.117	−.202	−.155	−.195	−.156	−.537
4	−.124	−.125	−.145	.232 (.246)	.205	.232	.017	−.026	.129	−.132	−.140	−.137	−.246
5	−.083	−.037	−.150	.205	.248 (.160)	.248	−.013	−.054	.018	−.087	−.082	−.125	−.160
6	−.097	−.045	−.172	.232	.248	.248 (.315)	−.096	−.082	.017	−.149	−.061	−.111	−.316
7	−.153	−.194	−.174	.017	−.013	−.096	.321 (.399)	.146	.321	−.070	−.118	−.065	−.399
8	−.152	−.153	−.117	−.026	−.054	−.082	.146	.153 (.172)	.143	.033	.055	.037	−.170
9	−.190	−.207	−.202	.129	.018	.017	.321	.143	.321 (.222)	−.099	−.081	−.070	−.221
10	−.190	−.197	−.155	−.132	−.087	−.149	−.070	.033	−.099	.354 (.397)	.354	.296	−.396
11	−.225	−.233	−.195	−.140	−.082	−.061	−.118	.055	−.081	.354	.412 (.314)	.412	−.314
12	−.307	−.252	−.156	−.137	−.125	−.111	−.065	.037	−.070	.296	.412	.412 (.478)	−.478
$\sum_o$	−.498	−.529	−.537	−.246	−.160	−.316	−.399	−.170	−.221	−.396	−.314	−.478	−4.264
3	−1.536	−1.349	.537	.044	.140	.028	−.051	.064	.183	−.086	.076	−.166	−2.116
1	1.536	−2.357	1.575	.292	.306	.222	.255	.368	.563	.294	.526	.448	4.028
2	2.544	2.357	2.395	.542	.380	.312	.643	.674	.977	.688	.992	.952	13.456
t	3.063	2.861	2.914	.774	.628	.560	.964	.827	1.298	1.042	1.404	1.364	
a_i	.728	.680	.692	.184	.149	.133	.229	.196	.309	.248	.334	.324	

$T = 17.699$

$\sqrt{T} = 4.20740$

$1/\sqrt{T} = .23768$

the diagonal elements, and the row sums are checked against the column sums. These column sums are entered in row $\sum_0$ and in turn are added up to give an overall sum of all elements in the table except the communality values (see Table 3.7). These column sums in row $\sum_0$ should be equal to the residual communality values but opposite in sign so that if the diagonal cells were added in, the column totals would be zero within the limits of rounding error. These totals should be negative in most cases, although occasionally, especially with later factors, a residual communality will go negative, requiring a positive column sum to balance it. The balancing of the column sums with the residual communality values must be satisfied before proceeding. If they do not balance, an error has been made and it must be corrected.

[3.3.4] REFLECTION OF VARIABLES

Sometimes on the first factor, and always on the second and subsequent factors, the sums of some of the columns, not counting the diagonals, are negative. This requires the reflection of variables with negative column sums (without communalities added in) so that all the column sums will be positive. The steps involved in this reflection process are as follows:

1. Locate the variable that has the largest negative total without the communality added in as the first variable to be reflected. Place an asterisk (*) over this column and to the left of this row to indicate that it is a variable that is being reflected.
2. Label the row immediately below the $\sum_0$ row with the number of the variable to be reflected. In Table 3.7, the first variable to be reflected is variable number 3, so the first row below the $\sum_0$ row is labeled 3.
3. Bring down the column sum from the $\sum_0$ row for this variable, changing its sign from negative to positive and place it in the next row, underlining the number. Since reflecting a variable changes all the signs in the column, the sum with signs changed is just the same as before the sign changes except that the sign on the total goes from minus to plus. In Table 3.7, the positive total of .537 is recorded for variable 3 in row 3 and underlined to indicate that the total is for a reflected variable.
4. The column totals for the other variables must now be adjusted to allow for the fact that variable 3 is being reflected. The column total for variable 3 is already taken care of, but these same residuals also appear in row 3, and hence adjustments are necessary to the other column totals when these values are reflected. Before reflection, these values were added into the total with one sign. After reflection, they are added in with the opposite sign. To adjust the total correctly, therefore, the entry must be doubled, its sign reversed, and the

result added to the previous column total. Thus, before reflection of variable 3, the entry in row 3 of column 4, Table 3.7, is $-.145$. When this entry is doubled, its sign changed, and the result (.290) added to the column 4 entry in row Σ_0, $-.246$, the new column total, 0.044, is obtained and entered under column 4 in row 3. The other entries in this row, except that for the variable being reflected, are obtained in a similar way.

5. As a check, the new column totals are added up by adding across row 3 at the bottom of Table 3.7. This value should equal the old sum of column totals plus four times the positive column sum for the variable being reflected, that is, $4 \times .537$, since both the row and column values must first be taken out of the old total and then added into the new total with different signs. Thus $-2.116 = -4.264 + 4(.537)$.

6. If any column totals are still negative after the first reflection, and they usually are, pick out the variable with the largest remaining negative total, put an asterisk (*) on its column and row, and proceed to reflect this variable in the same way the first one was reflected. Its total is brought down with a plus sign and underlined as the new total for that column. Proceed as before, doubling row values, changing their signs, and adding to the previous column totals. In Table 3.7 for example, the next variable to be reflected is variable 1. The entry in column 2, row 1, is .504. Doubling and changing signs gives -1.008 which is added to the previous column total, -1.349, to give a new column total of -2.357, which is placed in row 1 (below the main table), column 2.

Coming to column 3, however, brings us to a variable which has been previously reflected. Since this variable has been previously reflected, its effective sign is therefore opposite to that which shows in the table of residuals. Another reflection has the effect of bringing the sign back to what it is in the table as shown. It is unnecessary, therefore, to change the sign after doubling before adding to the old column total to get the new column total. Therefore, the entry in column 3, row 1, is doubled and added to the previous column total *without changing its sign*. Its new total is also underlined, indicating that it is a variable previously reflected, requiring the addition of the doubled row entry without sign change. Once a variable is reflected, its column totals are underlined in each successive row during the reflection process.

7. Add up the entries of row 1. This total, 4.028, should equal the previous row total, -2.116, plus four times the positive sum of the first column, that is, 4×1.536.

8. This process of reflecting variables is continued until no negative column total remains. Occasionally it may be necessary to re-reflect a variable that has already been reflected earlier. The asterisk (*) must be removed from the column and row but in carrying out the re-reflection, the row entries for previously unreflected variables are doubled and added without changing signs since the variable has been reflected previously. For the previously

reflected variable entries, the signs are changed after doubling and before adding. Table 3.7 shows the completed reflection process for the first factor residuals for the 12-variable sample problem.

[3.3.5] *COMPUTING THE SECOND FACTOR LOADINGS*

It would be possible to use the residual communalities (in parentheses) in Table 3.7 as the communality estimates for the second factor extraction. This procedure is used where good communality estimates were available at the beginning of the analysis or where a solution with these particular communality estimates is desired. In this sample problem, and in common practice, new estimated communalities are introduced for each factor extracted.

In reestimating the communalities, then, locate the largest residual in each column, make it positive in sign, and place it above the residual communality value, as shown in Table 3.7. Thus, in Table 3.7, column 1, .519 is the new communality estimate and .498 is the residual communality value that will not be used further.

In this example, the reestimated communalities in Table 3.7 are added to the positive column totals without communalities in row 2 (just above row t) to give the column totals with communalities in row t. The second factor loadings are then computed in the same way as for the first factor and the results entered in row a_i, Table 3.7. These a_i values from Table 3.7 are also entered in the **A** matrix, Table 3.13. Since variables 1, 2, and 3 were reflected in making the columns have positive totals, the signs of the factor loadings for variables 1, 2, and 3 are made negative on factor II in Table 3.13.

[3.3.6] *COMPUTING THE SECOND FACTOR RESIDUALS*

When the second factor residuals are computed by the formula $\mathbf{R}_2 = \mathbf{R}_1 - \mathbf{A}_2\mathbf{A}_2'$, $\mathbf{R}_1$ is a matrix of reflected first factor residuals and $\mathbf{A}_2$ is a single column with all positive signs. To get $\mathbf{R}_1$, variables 1, 2, and 3 in Table 3.7 must be reflected. This gives Table 3.8, the $\mathbf{R}_1$ matrix required with reflected residuals. In carrying out the reflections for a given variable, both row and column signs for that variable must be changed unless the residual is with another variable that is also reflected, in which case the sign is not changed, Thus, in reflecting variable 1, entries r_{16} and r_{61}, residual elements from Table 3.7 which equal $-.097$, are changed from minus to plus. Entries r_{13} and r_{31}, both equal to .519, are left unchanged as to sign, however, since variable 3 is also reflected. If both variables 1 and 3 are reflected, this means two sign changes which cancel each other; so the residual retains the same sign. The

TABLE 3.8

First Factor Residuals after Reflection with Reestimated Communalities

	1	2	3	4	5	6	7	8	9	10	11	12
1	(.519)	.504	.519	.124	.083	.097	.153	.152	.190	.190	.225	.307
2	.504	(.504)	.410	.125	.037	.045	.194	.153	.207	.197	.233	.252
3	.519	.410	(.519)	.145	.150	.172	.174	.117	.202	.155	.195	.156
4	.124	.125	.145	(.232)	.205	.232	.017	−.026	.129	−.132	−.140	−.137
5	.083	.037	.150	.205	(.248)	.248	−.013	−.054	.018	−.087	−.082	−.125
6	.097	.045	.172	.232	.248	(.248)	−.096	−.082	.017	−.149	−.061	−.111
7	.153	.194	.174	.017	−.013	−.096	(.321)	.146	.321	−.070	−.118	−.065
8	.152	.153	.117	−.026	−.054	−.082	.146	(.153)	.143	.033	.055	.037
9	.190	.207	.202	.129	.018	.017	.321	.143	(.321)	−.099	−.081	−.070
10	.190	.197	.155	−.132	−.087	−.149	−.070	.033	−.099	(.354)	.354	.296
11	.225	.233	.195	−.140	−.082	−.061	−.118	.055	−.081	.354	(.412)	.412
12	.307	.252	.156	−.137	−.125	−.111	−.065	.037	−.070	.296	.412	(.412)

results of the reflection process can be described in another way: If r'_{ij} represents an unreflected residual and r''_{ij} represents the reflected residual, then $r''_{ij} = r'_{ij}(-1)^{n_i}(-1)^{n_j}$, where n_i and n_j are the numbers of times variables i and j, respectively, have been reflected during this particular reflection process preparatory to extracting a factor.

After the matrix of reflected residuals is formed, Table 3.8, the matrix of products $\mathbf{A}_2\mathbf{A}_2'$ is obtained by multiplying the absolute values of the second factor loadings times each other. This matrix of products, representing the contribution of factor II to the correlations, is subtracted term by term from the $\mathbf{R}_1$ matrix of reflected residuals, Table 3.8, to get the second factor residuals which are shown in Table 3.9.

[3.3.7] *COMPUTING THE THIRD FACTOR LOADINGS*

The third factor loadings are obtained by essentially the same steps used for computing the second factor loadings except that a different set of variables must be reflected. For the third factor, variables 12, 11, 10, 2, 1, and 8 are reflected in that order before the column totals without communalities become all positive. These computations are shown in Table 3.9. The asterisks, for reflected variables 1, 2, and 3, are *not* carried over from Table 3.7 to Table 3.9. Each new reflection process for computing another factor starts with no asterisks on the table.

When the third factor loadings are entered in matrix **A** (Table 3.13), however, attaching the proper signs is more complicated than it was for the second factor. The sign of each variable that was reflected during the third factor computations must be reversed, but this is not just a matter of putting negative signs on the loadings for those variables in factor III. Rather, it is a matter of changing the sign from what it was on the immediately preceding factor II. If a variable has not been reflected in computing any previous factor, the signs of all previous factor loadings will be positive; so a reflection on the current factor will merely cause its sign to go negative. This is the case in this example for variables 8, 10, 11, and 12.

Variables 1 and 2, however, were previously reflected on factor II and therefore had negative signs there. When they are reflected on factor III in computing factor III, their signs are changed back to positive because they are re-reflected. Variable 3 has a negative loading on factor III not because it was reflected in computing factor III, but because it was reflected in computing factor II, making it negative, and it was not reflected back again in computing factor III. A variable maintains its sign, whatever it is, until a reflection changes it. When changed, it remains in that state until another reflection occurs to change it back again.

TABLE 3.9 Second Factor Residuals and Third Factor Computations

	* 1	* 2	3	4	5	6	7	* 8	9	* 10	* 11	* 12	$\sum_o$
* 1	.071 (−.011)	.009	.015	−.010	−.026	.000	−.014	.009	−.035	.010	−.018	.071	.011
* 2	.009	.065 (.041)	−.061	.000	−.065	−.046	.038	.020	−.003	.029	.006	.032	−.041
3	.015	−.061	.080 (.040)	.017	.046	.080	.015	−.019	−.011	−.017	−.037	−.068	−.040
4	−.010	.000	.017	.208 (.198)	.178	.208	−.025	−.063	.072	−.178	−.202	−.197	−.200
5	−.026	−.065	.046	.178	.228 (.226)	.228	−.047	−.083	−.028	−.124	−.132	−.173	−.226
6	.000	−.046	.080	.208	.228	.228 (.230)	−.126	−.108	−.024	−.182	−.105	−.154	−.229
7	−.014	.038	.015	−.025	−.047	−.126	.251 (.269)	.101	.251	−.127	−.195	−.140	−.269
* 8	.009	.020	−.019	−.063	−.083	−.108	.101	.108 (.115)	.082	−.016	−.010	−.027	−.114
9	−.035	−.003	−.011	.072	−.028	−.024	.251	.082	.251 (.226)	−.175	−.184	−.171	−.226
* 10	.010	.029	−.017	−.178	−.124	−.182	−.127	−.016	−.175	.272 (.293)	.272	.216	−.292
* 11	−.018	.006	−.037	−.202	−.132	−.105	−.195	−.010	−.184	.272	.304 (.301)	.304	−.301
* 12	.071	.032	−.068	−.197	−.173	−.154	−.140	−.027	−.171	.216	.304	.304 (.307)	−.307
$\sum_o$	.011	−.041	−.040	−.200	−.226	−.229	−.269	−.114	−.226	−.292	−.301	−.307	−2.234
12	−.131	−.105	.096	.194	.120	.079	.011	−.060	.116	−.724	−.909	.307	−1.006
11	−.095	−.117	.170	.598	.384	.289	.401	−.040	.484	−1.268	.909	.915	2.630
10	−.115	−.175	.204	.954	.632	.653	.655	−.008	.834	1.268	1.453	1.347	7.702
2	−.133	.175	.326	.954	.762	.745	.579	−.048	.840	1.326	1.465	1.411	8.402
1	.133	.193	.296	.974	.814	.745	.607	−.066	.910	1.346	1.429	1.553	8.934
8	.151	.233	.334	1.100	.980	.961	.405	.066	.746	1.314	1.409	1.499	9.198
t	.222	.298	.414	1.308	1.208	1.189	.656	.174	.997	1.586	1.713	1.803	
a_i	.065	.087	.122	.384	.356	.350	.193	.051	.293	.466	.504	.530	

$T = 11.568$

$\sqrt{T} = 3.40008$

$1/\sqrt{T} = .29411$

[3.3.8]
COMPUTING THE FOURTH FACTOR LOADINGS

After the third factor is extracted, it is necessary to carry out the reflections of the residuals before removing the contribution of the third factor by the formula $\mathbf{R}_3 = \mathbf{R}_2 - \mathbf{A}_3\mathbf{A}_3'$. The reflected $\mathbf{R}_2$ matrix is not shown, but it is obtained by the same method that was used to obtain the reflected $\mathbf{R}_1$ matrix in Table 3.8. In this case, however, the residuals for variables 1, 2, 8, 10, 11, and 12 had to be reflected. The third factor residuals (matrix $\mathbf{R}_3$), derived from the reflected second factor residuals, are shown in Table 3.10, along with the computations for the fourth factor loadings. For the fourth factor, variables 1, 7, 9, 10, 11, and 12 must be reflected to make all the column totals positive. In entering the fourth factor loadings in matrix **A** (Table 3.13), the signs for these newly reflected variables must be made opposite to what they were for factor III. The remaining fourth factor loadings will have the same signs that were present for these variables on the third factor, since they were not reflected in obtaining the fourth factor loadings.

The fourth factor residuals, computed by the formula $\mathbf{R}_4 = \mathbf{R}_3 - \mathbf{A}_4\mathbf{A}_4'$, are shown in Table 3.11 (page 71). Again, $\mathbf{R}_3$ is a matrix of reflected residuals and $\mathbf{A}_4$ has all positive signs; $\mathbf{R}_4$, of course, is a matrix of unreflected residuals. If a fifth were computed and the matrix equation $\mathbf{R}_5 = \mathbf{R}_4 - \mathbf{A}_5\mathbf{A}_5'$ applied to get fifth factor residuals, it is important to recognize that although $\mathbf{R}_4$ appears in both equations, in the first case it is unreflected and in the second case it is reflected, although the same symbol is used in both places.

No more than four factors were expected to appear in this analysis and indeed no fourth factor residual correlation exceeds .057 in absolute value. Factor extraction was terminated after four factors, therefore, since the matrix of fourth factor residuals contained only small residual correlations.

[3.3.9]
REPRODUCING THE CORRELATION MATRIX

If the computations have been carried out correctly, the following equation should hold:

$$\mathbf{R} = \mathbf{AA}' + \mathbf{R}_4 \tag{3.13}$$

That is, if the factor matrix, consisting of four factors, is multiplied by its transpose and added to the matrix of fourth factor residuals, the original correlation matrix should be reproduced within the limits of rounding error. The matrix $\mathbf{AA}'$ is shown in Table 3.12. Before $\mathbf{R}_4$ can be added to $\mathbf{AA}'$, however, the individual residuals must be returned to their original unreflected state. Otherwise, some residual elements would have to be added to the inner product of the factor loadings in $\mathbf{AA}'$, to reproduce the original correlation,

TABLE 3.10 Third Factor Residuals and Fourth Factor Computations

	* 1	2	3	4	5	6	* 7	8	* 9	* 10	* 11	* 12	$\sum_o$
*1	.051 (.066)	.003	−.023	−.015	.002	−.023	.002	.006	.016	−.020	−.051	.036	−.067
2	.003	.055 (.057)	.051	−.033	.034	.015	−.055	.015	−.023	−.012	−.038	−.015	−.058
3	−.023	.051	.051 (.065)	−.029	.003	.037	−.008	.013	−.047	−.040	−.025	.004	−.064
4	−.015	−.033	−.029	.099 (.060)	.041	.073	−.099	.043	−.040	−.002	.008	−.007	−.060
5	.002	.034	.003	.041	.132 (.102)	.104	−.115	.065	−.132	−.042	−.047	−.015	−.102
6	−.023	.015	.037	.073	.104	.194 (.106)	−.194	.090	−.127	.019	−.071	−.032	−.109
*7	.002	−.055	−.008	−.099	−.115	−.194	.194 (.214)	−.111	.194	.037	.098	.038	−.213
8	.006	.015	.013	.043	.065	.090	−.111	.111 (.106)	−.097	−.040	−.036	−.054	−.106
*9	.016	−.023	−.047	−.040	−.132	−.127	.194	−.097	.194 (.165)	.039	.036	.015	−.166
*10	−.020	−.012	−.040	−.002	−.042	.019	.037	−.040	.039	.042 (.054)	.037	−.031	−.055
*11	−.051	−.038	−.025	.008	−.047	−.071	.098	−.036	.036	.037	.098 (.050)	.037	−.052
*12	.036	−.015	.004	−.007	−.015	−.032	.038	−.054	.015	−.031	.037	.054 (.023)	−.024
$\sum_o$	−.067	−.058	−.064	−.060	−.102	−.109	−.213	−.106	−.166	−.055	−.052	−.024	−1.076
7	−.071	.052	−.048	.138	.128	.279	.213	.116	−.554	−.129	−.248	−.100	−.224
9	−.103	.098	.046	.218	.392	.533	.601	.310	.554	−.207	−.320	−.130	1.992
11	−.001	.174	.096	.202	.486	.675	.797	.382	.626	−.281	.320	−.204	3.272
10	.039	.198	.176	.206	.570	.637	.871	.462	.704	.281	.394	−.142	4.396
12	−.033	.228	.168	.220	.600	.701	.947	.570	.734	.219	.468	.142	4.964
1	.033	.222	.214	.250	.596	.747	.951	.558	.766	.179	.366	.214	5.096
t	.084	.277	.265	.349	.728	.941	1.145	.669	.960	.221	.464	.268	$T = 6.371$
a_i	.033	.109	.104	.138	.289	.373	.454	.265	.381	.087	.184	.106	$\sqrt{T} = 2.52252$
													$1/\sqrt{T} = .39643$

TABLE 3.11

Fourth Factor Residuals

	1	2	3	4	5	6	7	8	9	10	11	12
1	(.049)	−.007	.020	.010	−.012	.010	−.013	−.014	.003	−.023	−.057	.033
2	−.007	(.043)	.039	−.048	.002	−.025	.006	−.014	−.019	.002	.018	.003
3	.020	.039	(.040)	−.044	−.027	−.002	−.039	−.015	.007	.031	.005	−.015
4	.010	−.048	−.044	(.080)	.001	.022	.036	.006	−.012	−.010	−.033	−.008
5	−.012	.002	−.027	.001	(.049)	−.004	−.016	−.011	.022	.017	−.006	−.016
6	.010	−.025	−.002	.022	−.004	(.055)	.025	−.009	−.015	−.052	.002	−.008
7	−.013	.006	−.039	.036	−.016	.025	(−.011)	−.010	.021	−.002	.014	−.011
8	−.014	−.014	−.015	.006	−.011	−.009	−.010	(.040)	−.004	.017	−.013	.026
9	.003	−.019	.007	−.012	.022	−.015	.021	−.004	(.049)	.006	−.034	−.025
10	−.023	.002	.031	−.010	.017	−.052	−.002	.017	.006	(.034)	.021	−.040
11	−.057	.018	.005	−.033	−.006	.002	.014	−.013	−.034	.021	(.064)	.018
12	.033	.003	−.015	−.008	−.016	−.008	−.011	.026	−.025	−.040	.018	(.043)

TABLE 3.12

Reproduced Correlation Matrix (**AA**′)

	1	2	3	4	5	6	7	8	9	10	11	12
1	(.677)	.603	.620	.080	.128	.080	.053	.094	.057	−.043	−.077	−.137
2	.603	(.561)	.539	.068	.128	.105	−.044	.006	−.039	−.072	−.098	−.153
3	.620	.539	(.608)	−.024	.013	−.032	.029	.095	.003	.021	−.015	−.055
4	.080	.068	−.024	(.466)	.509	.438	.316	.296	.458	.110	.173	.058
5	.128	.128	.013	.509	(.583)	.514	.274	.299	.432	.163	.246	.106
6	.080	.105	−.032	.438	.514	(.474)	.155	.181	.295	.102	.178	.058
7	.053	−.044	.029	.316	.274	.155	(.557)	.470	.639	.158	.174	.109
8	.094	.006	.095	.296	.299	.181	.470	(.490)	.554	.327	.377	.286
9	.057	−.039	.003	.458	.432	.295	.639	.554	(.765)	.206	.246	.145
10	−.043	−.072	.021	.110	.163	.102	.158	.327	.206	(.490)	.579	.500
11	−.077	−.098	−.015	.173	.246	.178	.174	.377	.246	.579	(.695)	.592
12	−.137	−.153	−.055	.058	.106	.058	.109	.286	.145	.500	.592	(.529)

and others would have to be subtracted. To adjust the signs on the residuals, it is necessary to know how many times each matrix element has been reflected in sign.

Table 3.13 gives the number of reflections for each variable over all the factor extraction operations, that is, the number of times the factor loading sign of each variable has been reversed. For any given residual r_{ij} add up the number of sign changes for the two variables i and j. If this number is odd, the r_{ij} sign is out of phase and must be reversed from what it is in the fourth factor residual matrix, Table 3.11. If the sum of the two numbers of sign changes is an even number, however, the r_{ij} sign is correct as shown in Table 3.11. After these sign changes are made, Eq. (3.13) holds as expected, although not for the diagonal cells if communalities have been reestimated during the factor extraction process.

TABLE 3.13

Factor Matrix (A)[a]

	I	II	III	IV	h^2	Number of reflections
1	.377	−.728	.065	−.033	.677	3
2	.282	−.680	.087	.109	.561	2
3	.321	−.692	−.122	−.104	.608	1
4	.514	.184	.384	.138	.466	0
5	.592	.149	.356	.289	.583	0
6	.442	.133	.350	.373	.474	0
7	.511	.229	.193	−.454	.557	1
8	.615	.196	−.051	−.265	.490	1
9	.662	.309	.293	−.381	.765	1
10	.451	.248	−.466	.087	.490	2
11	.544	.334	−.504	.184	.695	2
12	.363	.324	−.530	.106	.529	2
SSQ	2.845	2.010	1.300	.739	[6.895]	

[a] The SSQ values for columns I, II, III, and IV are the sums of squares of the column elements. The value in brackets, 6.895, is the sum of the communalities. The proportion of the extracted variance accounted for by any factor is its SSQ total divided by the sum of the communalities.

[3.4] ITERATING THE COMMUNALITIES

In a large matrix, say 50 variables, errors in estimating the communalities usually do not affect the solution heavily since such discrepancies are typically small compared with the column sums after reflection. With smaller matrices, however, these errors can be of greater significance. For this reason, it is sometimes necessary to calculate the communalities derived from the first solution by summing the squares of the extracted factor loadings, and to use these values as estimates of the communalities for a second solution. The communalities derived from the second solution are used as estimates for a third solution, and so on, until the communalities which emerge from a given solution are close to those that were input as estimates. Usually 10 to 15 iterations will be sufficient for all practical purposes although many more may be required to achieve convergence.

In such an iterated series, after the first solution, the residual communalities are carried throughout the solution without reestimating communalities at each factor extraction step as was done in this sample problem. Thus, the residual communality from the first factor is used as the estimate of the communality for the second factor. The residual for the second factor is used as the estimate for the third factor, and so on. The same number of factors is extracted in each solution. Table 3.14 gives a solution after six iterations of the communalities in the sample problem used here as an illustration of centroid factor extraction. The differences in factor loadings between the first solution and the sixth iterated solution are not large, even for a small problem like this one. With matrices of 20 or more variables, iteration of the solution until stabilized

TABLE 3.14

Factor Matrix with Iterated Communalities[a]

	I	II	III	IV	h^2
1	.398	−.771	.037	−.012	.753
2	.265	−.651	.062	.075	.503
3	.309	−.673	−.100	−.063	.563
4	.504	.181	.368	.113	.434
5	.610	.135	.343	.304	.600
6	.443	.148	.374	.394	.514
7	.500	.242	.204	−.493	.593
8	.602	.194	−.042	−.244	.462
9	.686	.294	.276	−.382	.779
10	.428	.246	−.452	.068	.452
11	.567	.320	−.538	.162	.740
12	.350	.334	−.534	.078	.526
SSQ	2.855	2.000	1.307	0.756	[6.918]

[a] The SSQ values for columns I, II, III, and IV are the sums of squares of the column elements. The value in brackets, 6.918, is the sum of the communalities. The proportion of the extracted variance accounted for by any factor is its SSQ total divided by the sum of the communalities.

communalities are obtained may not be worth the extra computation time involved, particularly if the computations are being carried out on a desk calculator. In actual practice, most empirical problems with real data do not require a high degree of precision in the estimated communality values to obtain useful results. The "communality problem" is considered further in the next chapter where a method of factor extraction which does not require the use of communality estimates to obtain the factor loadings is presented.

CHAPTER FOUR

The Principal Factor and Minimum Residual Methods of Factor Extraction

The centroid method of factor extraction was developed by Thurstone as a practical method for use with the desk calculator. It is an approximation to the principal factor method which, though more demanding of computation time, does extract the maximum variance with each succeeding factor. The centroid method only approximates this ideal. With electronic computers, the additional computation time required for doing a principal factor solution, instead of a centroid solution, does not represent an unreasonable expenditure of resources. Consequently, the principal factor method in recent years has become much more widely used. With the principal factor method, it is still necessary to estimate the communalities. Because there is no clear solution to this problem, a certain indeterminacy is introduced by using estimated communalities since the solution depends to some extent on the values chosen. It is possible, however, to obtain a solution without using any communality estimates by using the minimum residual method of factor extraction (Comrey, 1962a; Comrey & Ahumada, 1964, 1965). This method is described later in this chapter.

[4.1]
THE PRINCIPAL FACTOR METHOD

A theorem from matrix algebra that will not be proved here states that a symmetric matrix can be "diagonalized" as follows:

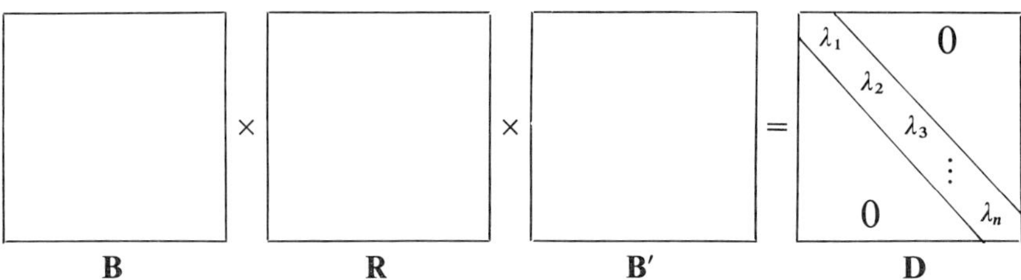

where **R** is the symmetric matrix to be diagonalized, **B** is an orthogonal matrix, **B**′ is the transpose of **B**, and **D** is a diagonal matrix. The reader will recall from Chapter 2 that an orthogonal matrix is one in which the sum of squares for each row and each column equals 1.0 and the inner products of any two non-identical rows or columns equal zero. The schematic matrix operation just shown may be represented by the matrix equation

$$\mathbf{BRB'} = \mathbf{D} \tag{4.1}$$

Since **B** is orthogonal with sums of squares of rows and columns equal to 1.0 and inner products of nonidentical rows or columns equal to zero, $\mathbf{BB'} = \mathbf{B'B} = \mathbf{I}$, the identity matrix. Thus, if **B** is orthogonal, **B**′ is the inverse of **B** and vice versa. This fact will be used to operate on Eq. (4.1) as follows:

$$\mathbf{B'BRB'} = \mathbf{B'D} \tag{4.2}$$

B′B equals the identity matrix on the left in Eq. (4.2); so it drops out. Then, multiplying both sides of Eq. (4.2) on the right side by **B** gives

$$\mathbf{RB'B} = \mathbf{B'DB} \tag{4.3}$$

But **B′B** is also an identity matrix; so Eq. (4.3) becomes

$$\mathbf{R} = \mathbf{B'DB} \tag{4.4}$$

Now, let $\sqrt{\mathbf{D}}$ be a diagonal matrix that has elements in the diagonals that are the square roots of the corresponding elements in the diagonals of the diagonal matrix **D**. Then, since $\sqrt{\mathbf{D}}\sqrt{\mathbf{D}} = \mathbf{D}$, Eq. (4.4) may be written as

$$\mathbf{R} = \mathbf{B'}\sqrt{\mathbf{D}}\sqrt{\mathbf{D}}\mathbf{B} = (\mathbf{B'}\sqrt{\mathbf{D}})(\sqrt{\mathbf{D}}\mathbf{B}) = \mathbf{AA'} \tag{4.5}$$

That is, the matrix $(\mathbf{B'}\sqrt{\mathbf{D}})$ is a factor matrix **A** and the matrix $(\sqrt{\mathbf{D}}\mathbf{B})$ is its transpose **A**′.

A series of steps that could produce a principal factor solution, then, may be set out as follows:

1. Form the $n \times n$ correlation matrix $\mathbf{R}$ with appropriate values selected for the diagonal cells as communality estimates.
2. Find an $n \times n$ orthogonal matrix $\mathbf{B}$ such that when $\mathbf{R}$ is premultiplied by $\mathbf{B}$ and postmultiplied by $\mathbf{B}'$, the result is an $n \times n$ diagonal matrix $\mathbf{D}$ with diagonal elements $\lambda_1, \lambda_2, \lambda_3, \ldots, \lambda_n$.
3. Multiply each element of column 1 of $\mathbf{B}'$ by $\sqrt{\lambda_1}$. Multiply each element of column 2 of $\mathbf{B}'$ by $\sqrt{\lambda_2}$, and so on, for the other columns until each element of the last column of $\mathbf{B}'$ is multiplied by $\sqrt{\lambda_n}$. These operations in step 3 are equivalent to multiplying the matrix $\mathbf{B}'$ on the right by a diagonal matrix with diagonal elements $\sqrt{\lambda_1}, \sqrt{\lambda_2}, \sqrt{\lambda_3}, \ldots, \sqrt{\lambda_n}$. Multiplying a matrix on the left by a diagonal matrix has the effect of multiplying each row by a different constant.
4. The matrix obtained in step 3 is a factor matrix $\mathbf{A}$ such that $\mathbf{R} = \mathbf{AA}'$. The sum of squares of column 1 of $\mathbf{A}$ equals λ_1, the sum of squares of column 2 of $\mathbf{A}$ equals λ_2, and so on, until the sum of squares of column n of $\mathbf{A}$ equals λ_n.

The proportion of variance extracted by factor i is $\lambda_i / \sum_{i=1}^{n} \lambda_i$. The sum of the λ_i values represent the total variance extracted since by a theorem in matrix algebra the sum of the λ_i values will equal the sum of the diagonal values of the $\mathbf{R}$ matrix. For each variable, the diagonal values of $\mathbf{R}$ give the proportions of variance h_i^2 that are input to the analysis, so their sum represents the total extractable variance in the matrix. If all the variance in the matrix were extracted, then $\sum_{i=1}^{n} h_i^2 = \sum_{i=1}^{n} \lambda_i$.

[4.1.1] *THE JACOBI METHOD*

There are various methods of obtaining the $\mathbf{B}$ matrix such that $\mathbf{BRB}' = \mathbf{D}$. Obtaining $\mathbf{B}$ is the major part of the work since once $\mathbf{B}$ is obtained, it is a simple matter to compute $\mathbf{B}'\sqrt{\mathbf{D}}$. One method of obtaining matrix $\mathbf{B}$ is that developed in the last century by the mathematician Jacobi. The procedure begins by finding a matrix $\mathbf{B}_1$ such that (see Fig. 4.1)

$$\mathbf{B}_1 \mathbf{R} \mathbf{B}_1{}' = \mathbf{D}_1 \tag{4.6}$$

where $\mathbf{B}_1$ is an orthogonal matrix that has nonzero elements in the cells b_{ii}, b_{ij}, b_{ji}, and b_{jj}, and also in the other diagonal cells. Except for b_{ii} and b_{jj}, however, the diagonal cells have the value 1.0 placed in them. The matrix $\mathbf{D}_1$ will have a zero in cells d_{ij} and d_{ji}. Thus, this matrix operation "annihilates" one off-diagonal value in the original $\mathbf{R}$ matrix, moving it closer to the ultimate diagonal form shown in Eq. (4.1). The r_{ij} chosen for "annihilation" is the largest off-diagonal element of the matrix.

$$\begin{vmatrix} 1 & 0 & 0 & 0 & 0 \\ 0 & b_{22} & 0 & b_{24} & 0 \\ 0 & 0 & 1 & 0 & 0 \\ 0 & b_{42} & 0 & b_{44} & 0 \\ 0 & 0 & 0 & 0 & 1 \end{vmatrix} \times \begin{vmatrix} h_1 & r_{12} & r_{13} & r_{14} & r_{15} \\ r_{21} & h_2 & r_{23} & r_{24} & r_{25} \\ r_{31} & r_{32} & h_3 & r_{34} & r_{35} \\ r_{41} & r_{42} & r_{43} & h_4 & r_{45} \\ r_{51} & r_{52} & r_{53} & r_{54} & h_5 \end{vmatrix}$$

$$\mathbf{B}_1 \qquad\qquad \mathbf{R}$$

$$\times \begin{vmatrix} 1 & 0 & 0 & 0 & 0 \\ 0 & b_{22} & 0 & b_{42} & 0 \\ 0 & 0 & 1 & 0 & 0 \\ 0 & b_{24} & 0 & b_{44} & 0 \\ 0 & 0 & 0 & 0 & 1 \end{vmatrix} = \begin{vmatrix} d_{11} & d_{12} & d_{13} & d_{14} & d_{15} \\ d_{21} & d_{22} & d_{23} & 0 & d_{25} \\ d_{31} & d_{32} & d_{33} & d_{34} & d_{35} \\ d_{41} & 0 & d_{43} & d_{44} & d_{45} \\ d_{51} & d_{52} & d_{53} & d_{54} & d_{55} \end{vmatrix}$$

$$\mathbf{B}_1' \qquad\qquad \mathbf{D}_1$$

Fig. 4.1. One Jacobi transformation. In this schematic representation, the annihilation of one pair of elements of the 5×5 correlation matrix is indicated, namely, r_{24} and r_{42}. The elements of $\mathbf{B}_1$ are found as follows: b_{22} and b_{44} equal $\cos\varphi_{24}$, $b_{24} = -\sin\varphi_{24}$, and $b_{42} = \sin\varphi_{24}$, where φ_{24} is found using Eq. (4.7) with $i = 2$ and $j = 4$.

The elements of matrix B_1 are b_{ii} and b_{jj} equal to $\cos\varphi_{ij}$; $b_{ji} = \sin\varphi_{ij}$; and $b_{ij} = -\sin\varphi_{ij}$ where

$$\tan 2\varphi = \frac{-2r_{ij}}{h_i^2 - h_j^2} \tag{4.7}$$

The effect of these matrix operations for a 2×2 correlation matrix may be shown in the following equations, which also demonstrate how Eq. (4.7) may be developed:

$$\underset{\mathbf{B}}{\begin{bmatrix} \cos\varphi & -\sin\varphi \\ \sin\varphi & \cos\varphi \end{bmatrix}} \times \underset{\mathbf{R}}{\begin{bmatrix} h_1^2 & r_{12} \\ r_{21} & h_2^2 \end{bmatrix}} \times \underset{\mathbf{B}'}{\begin{bmatrix} \cos\varphi & \sin\varphi \\ -\sin\varphi & \cos\varphi \end{bmatrix}} = \underset{\mathbf{D}}{\begin{bmatrix} \lambda_1 & 0 \\ 0 & \lambda_2 \end{bmatrix}}$$

$$\underset{\mathbf{BR}}{\begin{bmatrix} (\cos\varphi \cdot h_1^2 - \sin\varphi \cdot r_{21}) & (\cos\varphi \cdot r_{12} - \sin\varphi \cdot h_2^2) \\ (\sin\varphi \cdot h_1^2 + \cos\varphi \cdot r_{21}) & (\sin\varphi \cdot r_{12} + \cos\varphi \cdot h_2^2) \end{bmatrix}}$$

$$\times \underset{\mathbf{B}'}{\begin{bmatrix} \cos\varphi & \sin\varphi \\ -\sin\varphi & \cos\varphi \end{bmatrix}} = \underset{\mathbf{D}}{\begin{bmatrix} \lambda_1 & 0 \\ 0 & \lambda_2 \end{bmatrix}}$$

Computing the elements d_{21} and d_{12} of the product $\mathbf{BRB}'$ gives

$$d_{21} = (\sin\varphi\cdot h_1^2+\cos\varphi\cdot r_{21})\cos\varphi - (\sin\varphi\cdot r_{12}+\cos\varphi\cdot h_2^2)\sin\varphi = 0$$

$$d_{12} = (\cos\varphi\cdot h_1^2-\sin\varphi\cdot r_{21})\sin\varphi + (\cos\varphi\cdot r_{12}-\sin\varphi\cdot h_2^2)\cos\varphi = 0$$

$$d_{21} = \cos\varphi\sin\varphi\cdot h_1^2 + \cos^2\varphi\cdot r_{21} - \sin^2\varphi\cdot r_{12} - \sin\varphi\cos\varphi\cdot h_2^2 = 0$$

$$d_{12} = \sin\varphi\cos\varphi\cdot h_1^2 - \sin^2\varphi\cdot r_{21} + \cos^2\varphi\cdot r_{12} - \sin\varphi\cos\varphi\cdot h_2^2 = 0$$

Adding these last two equations together gives

$$2\sin\varphi\cos\varphi\cdot h_1^2 + 2(\cos^2\varphi-\sin^2\varphi)r_{12} - 2\sin\varphi\cos\varphi\cdot h_2^2 = 0$$

$$\sin 2\varphi\cdot h_1^2 + 2\cos 2\varphi\cdot r_{12} - \sin 2\varphi\cdot h_2^2 = \sin 2\varphi(h_1^2-h_2^2) + 2r_{12}\cos 2\varphi = 0$$

$$\tan 2\varphi = \frac{\sin 2\varphi}{\cos 2\varphi} = \frac{-2r_{12}}{h_1^2-h_2^2} \qquad (4.7')$$

After the matrix $\mathbf{B}_1\mathbf{RB}_1' = \mathbf{D}_1$ is computed with a zero in elements d_{ij} and d_{ji}, the largest remaining matrix element off the diagonal in D_1 is annihilated as follows:

$$\mathbf{B}_2\mathbf{D}_1\mathbf{B}_2' = \mathbf{B}_2(\mathbf{B}_1\mathbf{RB}_1')\mathbf{B}_2' = \mathbf{D}_2 \qquad (4.8)$$

Suppose the largest off-diagonal element in $\mathbf{D}_1$ were d_{13}. Then the result of the operation in Eq. (4.8) would be to place a zero in cells d_{13} and d_{31} of matrix $\mathbf{D}_2$. The only off-diagonal elements of $\mathbf{B}_2$ that would be nonzero in this case would be b_{13} and b_{31}. The elements of $\mathbf{B}_2$ of principal interest would be as follows: b_{11} and b_{33} equal to $\cos\varphi_{13}$, with other diagonal elements equal to 1.0; $b_{31} = \sin\varphi_{13}$; and $b_{13} = -\sin\varphi_{13}$ where

$$\tan 2\varphi_{13} = \frac{-2r_{13}}{h_1^2-h_3^2} \qquad (4.9)$$

It must be emphasized, however, that the elements r_{13}, h_1^2, and h_3^2 in Eq. (4.9) are taken from the matrix $\mathbf{D}_1$ rather than from the original $\mathbf{R}$ matrix.

This entire process is repeated over and over again, always with the largest remaining off-diagonal value selected, regardless of sign, for annihilation. Each element that is annihilated requires the addition of another $\mathbf{B}_i$ matrix. Unfortunately, a matrix element of $\mathbf{D}_i$ does not stay annihilated. It will return to a nonzero value as other matrix elements are annihilated. Each time a matrix element is annihilated, however, the overall sum of squares of the off-diagonal elements is reduced. Thus, by iterating over and over again, the off-diagonal element can be made as close to zero as the accuracy of calculation permits. At the end, the result looks like the following:

$$\mathbf{B}_k\cdots\mathbf{B}_3\mathbf{B}_2\mathbf{B}_1\mathbf{RB}_1'\mathbf{B}_2'\mathbf{B}_3'\cdots\mathbf{B}_k' = \mathbf{D} \qquad (4.10)$$

where k ordinarily is a number considerably larger than n. That is, many iterations are required to complete the diagonalization of $\mathbf{R}$. After the annihilation of each off-diagonal element, a check is made to see if the largest remaining off-diagonal element in $\mathbf{D}_i$ is greater than some selected value of ε, close to zero. If so, another iteration is run, and another check is made. Eventually, the largest off-diagonal element of the matrix $\mathbf{D}_i$ will be less than ε. At this point, iteration is terminated, and the $\mathbf{D}_i$ is the $\mathbf{D}$ of Eq. (4.1) achieved by diagonalizing $\mathbf{R}$. Then, the $\mathbf{B}$ matrix of Eq. (4.1) is given by

$$\mathbf{B} = \mathbf{B}_k \cdots \mathbf{B}_3 \mathbf{B}_2 \mathbf{B}_1 \tag{4.11}$$

and $\mathbf{B}'$ is the transpose of $\mathbf{B}$. Each of these intermediate $\mathbf{B}_i$ matrices individually is an orthogonal matrix since $\cos^2 \varphi + \sin^2 \varphi = 1$ and the sums of squares of all rows and columns equal this expression. The inner products of nonidentical rows and columns of each $\mathbf{B}_i$ equal zero. A theorem of matrix algebra that will be assumed here rather than proved states that the product of orthogonal matrices is an orthogonal matrix, so $\mathbf{B}$ in (4.11) is also an orthogonal matrix, satisfying the requirements of Eq. (4.1).

The $\mathbf{B}_i$ matrix required to annihilate any particular matrix element r_{ij} can be formed as follows:

1. Put 1 in all diagonal cells of $\mathbf{B}_i$ except for elements b_{ii} and b_{jj}.
2. Find $\tan 2\varphi_{ij} = -2r_{ij}/(h_i^2 - h_j^2)$, as in Eq. (4.9), where r_{ij}, h_i, and h_j are taken from the $\mathbf{R}$ matrix for the first transformation and from the $\mathbf{D}$ matrix in its present state of transition toward diagonal form for all subsequent transformations. From $\tan 2\varphi_{ij}$ determine the angle 2φ and then the angle φ itself.
3. Determine the following elements of $\mathbf{B}_i$: $b_{ii} = b_{jj} = \cos \varphi_{ij}$; $b_{ji} = \sin \varphi_{ij}$; and $b_{ij} = -\sin \varphi_{ij}$. Other off-diagonal elements of $\mathbf{B}_i$ equal zero.

If the tangent of 2φ is negative, then 2φ is a negative angle. Find 2φ, ignoring the sign on the tangent; find φ, and then put a negative sign on the sine of φ. This will have the effect of making b_{ji} negative and b_{ij} positive. If the tangent of 2φ is positive, however, 2φ and φ will be positive angles and the sine of φ will be positive. This leaves b_{ji} positive and b_{ij} negative. In a computer program application, it is more convenient to compute $\cos\varphi$ and $\sin\varphi$ algebraically from $\tan 2\varphi$ by trigonometric identities rather than by using tables.

A Jacobi principal factor solution to the four-variable problem treated in previous chapters by other methods is shown in Table 4.1. The $\mathbf{B}_i$ and $\mathbf{D}_i$ matrices for each off-diagonal element annihilation are shown in order. The largest off-diagonal element of the original correlation matrix (see Table 2.1) was $r_{34} = r_{43} = .73$. The first $\mathbf{B}_i$ matrix, then, annihilates r_{34}; this is apparent in the first $\mathbf{D}_i$ matrix which has cells d_{34} and $d_{43} = 0$. The largest element of $\mathbf{D}_1$

TABLE 4.1

Principal Factor Solution to the Four-Variable Problem

	$\mathbf{B}_i$					$\mathbf{D}_i$			
	1	2	3	4		1	2	3	4
1	1	0	0	0		.6500	.5000	.0065	.2334
2	0	1	0	0		.5000	.4000	.0047	.3255
3	0	0	.7469	−.6650		.0065	.0047	.0001	.0000
4	0	0	.6650	.7469		.2334	.3255	.0000	1.4699
1	.7882	.6154	0	0		1.0404	0	.0080	.3843
2	−.6154	.7882	0	0		0	.0096	−.0002	.1130
3	0	0	1	0		.0080	−.0002	.0001	0
4	0	0	0	1		.3843	.1130	0	1.4699
1	.8625	0	0	−.5060		.8149	−.0572	.0069	0
2	0	1	0	0		−.0572	.0096	−.0002	.0974
3	0	0	1	0		.0069	−.0002	.0001	.0040
4	.5060	0	0	.8625		0	.0974	.0040	1.6954
1	1	0	0	0		.8149	−.0571	.0069	−.0033
2	0	.9983	0	−.0575		−.0571	.0040	−.0005	0
3	0	0	1	0		.0069	−.0005	.0001	.0040
4	0	.0575	0	.9983		−.0033	0	.0040	1.7010
1	.9976	−.0699	0	0		.8189	0	.0069	−.0033
2	.0699	.9976	0	0		0	0	0	−.0002
3	0	0	1	0		.0069	0	.0001	.0040
4	0	0	0	1		−.0033	−.0002	.0040	1.7010
1	1	0	.0084	0		.8190	0	0	−.0032
2	0	1	0	0		0	0	0	−.0002
3	−.0084	0	1	0		0	0	0	.0040
4	0	0	0	1		−.0032	−.0002	.0040	1.7010
1	1	0	0	0		.8190	0	0	−.0032
2	0	1	0	0		0	0	0	−.0002
3	0	0	1	−.0023		0	0	0	.0002
4	0	0	.0023	1		−.0032	−.0002	.0002	1.7010
1	1	0	0	.0036		.8190	0	0	0
2	0	1	0	0		0	0	0	−.0002
3	0	0	1	0		0	0	0	.0002
4	−.0036	0	0	1		0	−.0002	.0002	1.7010
1	.7240	.4771	−.3250	−.3777	1	.655	0	0	.470
2	−.5883	.8042	−.0564	−.0634	2	.432	0	0	.462
3	−.0069	−.0048	.7484	−.6632	3	−.294	0	0	.751
4	.3601	.3545	.5755	.6430	4	−.342	0	0	.839
	$(\mathbf{B}_8 \cdots \mathbf{B}_3 \mathbf{B}_2 \mathbf{B}_1)$					$\mathbf{A}$ (Factor matrix)			

is .50, in elements d_{12} and d_{21}; so $\mathbf{B}_2$ annihilates this element, leaving zero in cells d_{12} and d_{21} of $\mathbf{D}_2$. The off-diagonal elements of $\mathbf{D}_i$ are gradually reduced and after eight iterations, the largest off-diagonal element is .0002 in absolute value. The accumulated **B** matrix, computed by Eq. (4.11), and the factor matrix are also shown in Table 4.1, after the last $\mathbf{B}_i$ and the final **D** matrix. The final **D** matrix has the eigenvalues in its diagonals, λ_1, λ_2, λ_3, and λ_4. Ignoring the rounding errors, only two nonzero eigenvalues were obtained, 1.7010 in d_{44} and .8190 in d_{11}. Since there are only two nonzero eigenvalues, there are only two factors, as expected. This is evident in the factor matrix which has only two nonzero columns. It will be noted that the factors do not emerge in order of size in this method. The factor matrix obtained may be rearranged, if desired, to put the largest factor in the first column, the next largest in the second column, and so on.

The centroid solution to this problem (see Tables 3.1 and 3.4) gives factor loadings that appear to be very different from those for the final factor matrix in Table 4.1. Judging from this information alone, it might appear that the principal factor and centroid solutions are not comparable. A rotation of the fourth principal factor away from the first principal factor by an angle of approximately 27°, however, recovers the initial configuration in Fig. 2.1, just as a rotation of the centroid axes recovered the same configuration (see Table 2.2). Thus, although the centroid and principal factor solutions look different, they are actually nearly equivalent in the sense that they both produce essentially the same rotated configuration if they start with the same communality estimates. The transformation of the principal factor solution to the original configuration is shown as follows:

$$\underset{\mathbf{A}}{\begin{bmatrix} .470 & .655 \\ .462 & .432 \\ .751 & -.294 \\ .839 & -.342 \end{bmatrix}} \times \underset{\mathbf{\Lambda}}{\begin{bmatrix} \cos 27° & \sin 27° \\ -\sin 27° & \cos 27° \end{bmatrix}} \doteq \underset{\mathbf{V}}{\begin{bmatrix} .1 & .8 \\ .2 & .6 \\ .8 & .1 \\ .9 & .1 \end{bmatrix}}$$

Application of the principal factor method, Jacobi-type solution, to the 12-variable sample problem, treated by the centroid method in the previous chapter (see Table 3.6), gives the results shown in Table 4.2. The four factors that were associated with positive eigenvalues are shown in the first four columns. The sums of squares of the factor loadings, the derived communalities, are shown in the column headed by h^2. The values in the column headed by R^2 are the initial communality estimates inserted in the diagonal cells of the **R** matrix. The principal factor solution obtained, therefore, was to some extent determined by these estimated communalities.

TABLE 4.2

Principal Factor Solution to 12-Variable Problem for Four Factors with Positive Eigenvalues[a]

	I	II	III	IV	R^2	h^2
1	.092	.768	.236	−.035	.547	.653
2	.012	.671	.185	.054	.413	.489
3	.043	.645	.317	−.081	.448	.524
4	.545	.139	−.325	.208	.387	.464
5	.589	.187	−.213	.288	.415	.509
6	.446	.133	−.263	.383	.342	.435
7	.575	.016	−.211	−.413	.467	.544
8	.642	.015	.047	−.213	.397	.457
9	.742	.052	−.249	−.270	.588	.688
10	.467	−.229	.427	.050	.390	.455
11	.590	−.263	.473	.133	.552	.660
12	.414	−.353	.433	.058	.423	.488
SSQ[b]	2.884	1.777	1.113	0.595	(5.369)	(6.369)

[a] The SMCs R^2 and communalities were computed by the BMD 03M program. This solution used these R^2 values as communality estimates for a Jacobi method solution. The BMD solution varied from this no more than .002 at any point.

[b] The sum of squares.

In the four-variable fictional data example, the exact communalities were known and utilized in the solutions to obtain precise results, namely, exactly two factors with essentially zero residuals, except for rounding errors. The 12-variable sample problem is based on real data, however; so the communalities are not known.

The R^2 values used here as estimates of those communalities are the squared multiple correlations (SMCs) which represent the square of the multiple correlation (Guilford, 1965, Chap. 16) of each variable with all the other variables. These SMCs have been shown under certain conditions to represent lower bound estimates of the unknown communalities (see Harman, 1967, for a more extended discussion of this topic). To get these SMCs, find $\mathbf{R}^{-1}$, the inverse of **R** with 1.0 inserted in each diagonal cell. Take the reciprocals of the diagonal elements of $\mathbf{R}^{-1}$ and subtract each such reciprocal from 1.0. The resulting values are the SMCs entered in Table 4.2 under column R^2. These R^2 values were computed using program 03M of the BMD series of computer programs (Dixon, 1970a) which also computes a principal factor solution. There is also another factor analysis program in the "X" series of programs by the same author (Dixon, 1970b). The factor solution in Table 4.2 was computed by means of a computer program written by the author, using the SMCs as communality estimates. The BMD factor analysis programs, however, offer a variety of options for efficient factor analytic computations by the principal factor method. These programs are available for use at many computing centers throughout the world.

In comparing the principal factor solution in Table 4.2 with the centroid solution in Table 3.13, certain similarities can be seen between the two, although the signs of factors II and III are reversed in the two solutions. These two factors can be arbitrarily reversed by changing all the signs in the two columns for one of the two solutions. There would still remain discrepancies of some magnitude between the two solutions, however. Part of this is due to the fact that the SMCs represented communality estimates that were lower generally than the high column values used in the centroid solution. This difference is reflected in the greater sum of the h^2 values after extraction for the centroid method (for example, 6.895) than for the principal factor method (6.369). Despite the greater total for the centroid method, however, the amount of variance extracted on the largest principal factor was actually greater than that for the first centroid, 2.884 compared with 2.845. This is due to the fact that the principal factor method extracts the maximum amount of variance at a given step given the particular communalities used, whereas the centroid method only approximates this result. Discrepancies between two solutions due to differences in communality estimates will persist through the rotation process since rotation does not affect the communalities. Differences between solutions by different factor extraction methods using the same communalities tend to disappear, however, in the ultimate rotated solution. This phenomenon has already been illustrated by the centroid and principal factor solutions to the four-variable sample problem. Although the unrotated factors were different for the two solutions, the two rotated solutions were essentially identical.

In the principal factor solution to the 12-variable problem using SMCs as h^2 estimates, only four of the eigenvalues obtained were positive, for example, 2.884, 1.777, 1.113, and .595, as shown in the last row of Table 4.2. The remaining eight eigenvalues were all negative: $-.226$, $-.194$, $-.189$, $-.136$, $-.112$, $-.066$, $-.051$, and $-.016$. Since the sum of squares of the factor loadings for a factor is supposed to equal its eigenvalue, because the factor loadings equal the normalized eigenvector multiplied by the square root of the eigenvalue as called for by Eq. (4.5), a negative eigenvalue poses something of a dilemma. A sum of squares of real numbers cannot be negative. There can be no legitimate factors, therefore, associated with negative eigenvalues. With SMCs as communality estimates, then, only four factors can be extracted from the 12-variable sample matrix. The general problem of determining how many factors to extract from a matrix is considered in greater detail later in this chapter.

[4.2]

THE MINIMUM RESIDUAL METHOD

Both the centroid and the principal factor methods require communality estimates which in turn influence the nature of the results. Since there is no way known at present to determine the "correct" communalities, factor analysis results are subject to some distortion by the use of inaccurate communality estimates. To avoid this problem, the author developed a method of factor extraction, called the "minimum residual" method, that utilizes only the off-diagonal elements of the correlation matrix and hence requires no communality estimates (Comrey, 1962a).

Let $\mathbf{A}$ represent a column of the factor matrix with elements $a_1, a_2, \ldots, a_n$. Let $F(\mathbf{R}, \mathbf{A})$ be a function of the correlation matrix $\mathbf{R}$ and the column of factor loadings $\mathbf{A}$ such that

(4.12) $$F(\mathbf{R}, \mathbf{A}) = \sum_{i=1}^{n} \sum_{j=1}^{n} (r_{ij} - a_i a_j)^2, \qquad j \neq i$$

$F(\mathbf{R}, \mathbf{A})$ is the sum of squares of the off-diagonal residuals after the factor $\mathbf{A}$ has been removed from the $\mathbf{R}$ matrix. The basis of the minimum residual method, as the name suggests, is to determine the factor loadings $a_1, a_2, \ldots, a_n$ of the factor $\mathbf{A}$ such that $F(\mathbf{R}, \mathbf{A})$ will be as small as possible. That is, a factor is sought which, after it is removed, will leave the sum of squares of the residuals as small as possible, ignoring the diagonal elements in the sum. This is indicated by the $j \neq i$ notation in Eq. (4.12). This procedure minimizes the sum of squares of the residuals at each factor extraction step, not after m factors.

Solving for a formula to obtain the elements a_i of $\mathbf{A}$ requires some calculus, so the reader unfamiliar with this mathematical tool may wish to skip the following derivation and go directly to the end result, Eq. (4.18) below. Solution of this problem requires partial differentiation of the function $F(\mathbf{R}, \mathbf{A})$ with respect to each unknown a_i, setting these n equations equal to zero, and solving them simultaneously for the n unknown a_i values. Thus

(4.13)
$$F(\mathbf{R}, \mathbf{A}) = \sum_{i=1}^{n} \sum_{j=1}^{n} (r_{ij}^2 + a_i^2 a_j^2 - 2r_{ij} a_i a_j), \qquad j \neq i$$

(4.14)
$$F(\mathbf{R}, \mathbf{A}) = \sum_{i=1}^{n} \sum_{j=1}^{n} r_{ij}^2 + \sum_{i=1}^{n} \sum_{j=1}^{n} a_i^2 a_j^2 - 2 \sum_{i=1}^{n} \sum_{j=1}^{n} r_{ij} a_i a_j, \qquad j \neq i$$

Differentiating partially with respect to a_1 and setting the result equal to zero gives

(4.15)
$$\frac{\partial F}{\partial a_1} = 0 + 2a_1 a_2^2 + 2a_1 a_3^2 + \cdots + 2a_1 a_n^2 - 2r_{12} a_2 - 2r_{13} a_3 - \cdots - 2r_{1n} a_n = 0$$

$$\frac{\partial F}{\partial a_1} = a_1 \sum_{j=2}^{n} a_j^2 - \sum_{j=2}^{n} r_{1j} a_j = 0$$

Solving for a_1 gives

(4.16)
$$a_1 = \frac{\sum_{j=1}^{n} r_{1j} a_j}{\sum_{j=1}^{n} a_j^2}, \qquad j \neq 1$$

Differentiating $F(\mathbf{R}, \mathbf{A})$ in Eq. (4.14) partially with respect to a_2 and setting the result equal to zero gives an expression analogous to Eq. (4.15) which can be simplified similarly to

(4.17)
$$a_2 = \frac{\sum_{j=1}^{n} r_{2j} a_j}{\sum_{j=1}^{n} a_j^2}, \qquad j \neq 2$$

The same procedure may be followed for each element a_i resulting in a general expression

(4.18)
$$a_i = \frac{\sum_{j=1}^{n} r_{ij} a_j}{\sum_{j=1}^{n} a_j^2}, \qquad j \neq i$$

Substituting in turn each value from 1 to n for i in Eq. (4.18) results in n equations in n unknowns which would give the unknown factor loadings $a_1, a_2, \ldots, a_n$ if the equations could be solved. These equations are not linear equations, however; so the solution is not easy to obtain algebraically.

It is possible, however, to solve these equations by means of an iterative procedure. For example, if all the correct factor loadings were known except the first, these values could be substituted in Eq. (4.16) to find the correct a_1. In general, if all the correct a values are known except a_i, they may be substituted in Eq. (4.18) to find a_i.

In practice, intial estimates of the a values are made. These estimates are substituted in Eq. (4.18) to solve for each a_i value. Since the first estimates are usually incorrect, the a values derived by putting them in Eq. (4.18) are still not likely to be correct. These new values, however, can be substituted in turn into Eq. (4.18) to get still another estimate of the a values. These two sets of derived values may be averaged term by term to derive new estimated a values before starting the whole cycle anew. Such a procedure is helpful in finding the correct a values only if the values obtained from each iterative cycle are closer to the correct values than those with which the iterative cycle was begun, that is, if the iterative process converges. When the values put into the iterative cycle are the same as those that come out of it, then the iterative process is terminated.

The iterative process may be schematized for a four-variable example as follows:

(4.19)

$$\frac{1}{\sum_{j=1}^{4} a_j^2} \cdot \underset{\mathbf{R}}{\begin{bmatrix} 0 & r_{12} & r_{13} & r_{14} \\ r_{21} & 0 & r_{23} & r_{24} \\ r_{31} & r_{32} & 0 & r_{34} \\ r_{41} & r_{42} & r_{43} & 0 \end{bmatrix}} \times \underset{\mathbf{A}_0}{\begin{bmatrix} a_{10} \\ a_{20} \\ a_{30} \\ a_{40} \end{bmatrix}} = \underset{\mathbf{A}_1}{\begin{bmatrix} a_{11} \\ a_{21} \\ a_{31} \\ a_{41} \end{bmatrix}}, \qquad j \neq i$$

(4.20)

$$\frac{1}{\sum_{j=1}^{4} a_j^2} \cdot \mathbf{R} \cdot \mathbf{A}_1 = \mathbf{A}_2, \qquad j \neq i$$

(4.21)

$$\tfrac{1}{2}(\mathbf{A}_1 + \mathbf{A}_2) = \mathbf{A}_0$$

Reinsert the $\mathbf{A}_0$ from Eq. (4.21) into Eq. (4.19) to calculate a new $\mathbf{A}_1$. The new $\mathbf{A}_1$ is inserted into Eq. (4.20) to get a new $\mathbf{A}_2$, and so on, until $\mathbf{A}_1$ and $\mathbf{A}_2$ are as close to one another as desired. Three-decimal-place accuracy is ordinarily quite sufficient for empirical work.

Particular note must be taken of the fact that j may not equal i in Eq. (4.18)–(4.20). This means that in getting a_1 in Eq. (4.18), the first element of the sum is omitted in both numerator and denominator. In getting a_2, the second element of these sums is omitted. In general, in computing the value a_i, the ith elements of the sums in both numerator and denominator are omitted. The appropriate numerator elements are effectively eliminated from the sums by placing zeros in the diagonal cells in the matrix multiplication in Eq. (4.19). This does *not* mean that zeros are being used as communalities. This device merely eliminates the unwanted terms in the numerator summations. The denominator sums for a_i, however, must be computed separately by

(4.22)

$$\left(\sum_{j=1}^{n} a_j^2\right) - a_i^2$$

Thus instead of adding each time, omitting the term where $i = j$, it is more efficient to add once over all values of a_j and then to subtract out the unwanted term as in formula (4.22).

After a factor is obtained through convergence of the iterative process, it is removed from the matrix of correlations by the usual method; that is, $\mathbf{R}_1 = \mathbf{R} - \mathbf{A}_1 \mathbf{A}_1'$, where $\mathbf{A}_1$ in this case is the factor vector computed previously and referred to in Eqs. (4.12)–(4.14) as $\mathbf{A}$.

If enough factors are extracted, the point will be reached where convergence on a factor vector does not occur with this method of iteration. To understand this phenomenon, it is helpful to see the relationship between the minimum residual and the principal factor methods. Equations (4.1) and (4.2) lead to the schematic matrix result:

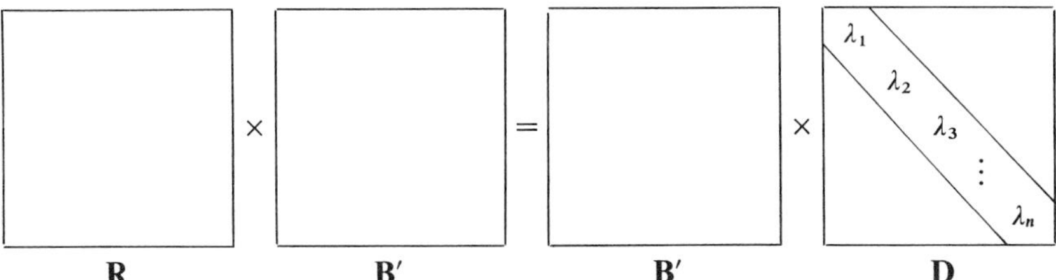

or, in matrix equation form:

(4.23) $$\mathbf{RB}' = \mathbf{B}'\mathbf{D}$$

The effect of the multiplication of $\mathbf{B}'$ on the right by $\mathbf{D}$ is to multiply each column of $\mathbf{B}'$ by a λ value. Every element of column 1 is multiplied by λ_1, every element of column 2 is multiplied by λ_2, and so on. If the first column of $\mathbf{B}'$ is isolated from the rest, thus considering only the first eigenvector, we have

(4.24) $$\underset{\mathbf{R}}{\square} \times \underset{\mathbf{B}_1'}{\square} = \lambda_1 \underset{\mathbf{B}_1'}{\square}$$

That is, multiplying $\mathbf{R}$ by the column matrix $\mathbf{B}_1'$, the first eigenvector, equals that same eigenvector multiplied by a constant, λ_1. It has already been indicated that λ_1 equals the sum of squares of the first factor loadings and that the vector $\mathbf{B}_1'$ can be converted from a normalized vector to the first factor vector $\mathbf{A}_1$ by multiplying each element of $\mathbf{B}_1'$ by $\sqrt{\lambda_1}$. Then since $\sqrt{\lambda_1}\,\mathbf{B}_1' = \mathbf{A}_1$, multiplying both sides of Eq. (4.24) by $\sqrt{\lambda_1}$ would give

(4.25) $$\mathbf{RA}_1 = \lambda_1 \mathbf{A}_1$$

Dividing both sides of Eq. (4.25) by λ_1 gives

(4.26) $$\mathbf{RA}_1/\lambda_1 = \mathbf{A}_1$$

For a particular element a_i in the product column matrix $\mathbf{A}_1$ on the right of Eq. (4.26), the formula for that equation in algebraic terms would be

$$a_i = \frac{\sum_{j=1}^{n} r_{ij} a_j}{\lambda_i} = \frac{\sum_{i=1}^{n} r_{ij} a_j}{\sum_{j=1}^{n} a_j^2} \tag{4.27}$$

Equation (4.27) is just like Eq. (4.18), the expression for the minimum residual solution, except that the restriction $j \neq i$ is removed. Thus, if the diagonal elements are used in the minimum residual method of factor extraction, instead of being omitted, the minimum residual method becomes equivalent to a principal factor method and yields a principal factor solution. The principal factor solution, then, is also a minimum residual solution for the particular communality estimates employed. Thus, if a principal factor solution is desired, the minimum residual method including the diagonal cells, that is, removing the restriction $i \neq j$ in the equations, may be used provided estimated communalities are inserted in the diagonal cells instead of the zeros that are normally used with this method. The ordinary minimum residual solution without communalities may or may not reduce the sum of squares of the off-diagonal residuals as well as the principal factor solution. This will depend on the communalities used for the principal factor solution.

Equation (4.25) expresses a rather remarkable theorem, namely, that multiplying the correlation matrix times a factor from a principal factor solution equals the result of multiplying every term of that factor by a constant, namely, the eigenvalue associated with that factor. In the minimum residual solution, while iterating to achieve convergence by a formula like (4.26), the **A** on the left-hand side and the **A** on the right-hand side are somewhat different. As the process converges, these two **A** vectors become more and more alike. If the factor involved is associated with a negative eigenvector, however, λ will be negative and

$$\frac{\mathbf{RA}}{\sum_{i=1}^{n} a_i^2} = -\mathbf{A}$$

That is, the successive iterated vectors will tend to be equal, term by term, in absolute value but opposite in sign. When Eq. (4.21) is applied to obtain the $\mathbf{A}_0$ for the next iteration, the approximately equal but oppositely signed elements of $\mathbf{A}_1$ and $\mathbf{A}_2$ will tend to add up to zero. This phenomenon results in a failure to achieve convergence. This does not occur, however, until all the major factors have been removed. When this failure to converge occurs, it is equivalent to encountering a negative eigenvalue with the principal factor solution which does not result in a legitimate factor.

An improved method of averaging the successive iterated vectors suggested by Ahumada (Comrey & Ahumada, 1964, 1965) avoids this problem. This improved method, called the "length adjustment" method, finds a vector that has a length equal to the geometric mean of the lengths of the two vectors to be averaged. Under this system of averaging, the signs of the loadings do not affect the averaging since the vector lengths are given by the square roots of the sum of squares of the loadings. The signs of the loadings are lost in the squaring process.

To illustrate this method, consider the vector **A** with elements $a_1, a_2, \ldots, a_n$ to be a correct factor vector with length L, that is, one satisfying Eq. (4.18). Derive two other vectors $\mathbf{A}_1$ and $\mathbf{A}_2$ as follows:

$$(4.28)\qquad \mathbf{A}_1 = \frac{1}{K}a_1, \frac{1}{K}a_2, \ldots, \frac{1}{K}a_n$$

$$\mathbf{A}_2 = Ka_1, Ka_2, \ldots, Ka_n$$

The lengths of vectors $\mathbf{A}_1$ and $\mathbf{A}_2$ in (4.28) are given by

$$(4.29)\qquad L_1 = \sqrt{\left(\frac{1}{K}a_1\right)^2 + \left(\frac{1}{K}a_2\right)^2 + \cdots + \left(\frac{1}{K}a_n\right)^2} = \frac{1}{K}L$$

$$L_2 = \sqrt{(Ka_1)^2 + (Ka_2)^2 + \cdots + (Ka_n)^2} = KL$$

The geometric mean of the lengths of L_1 and L_2 is given by

$$(4.30)\qquad \sqrt{L_1 L_2} = \sqrt{\frac{1}{K}L \cdot KL} = L$$

Thus, if two vectors differ only by a constant from the correct vector, the geometric mean of their lengths will produce the length of the correct vector. This fact can be used to derive a method of averaging successive iterated vectors with the minimum residual factor extraction procedure. If $a_1, a_2, \ldots, a_n$ are the elements of $\mathbf{A}_1$ and $a_1', a_2', \ldots, a_n'$ are the elements of $\mathbf{A}_2$ where $\mathbf{A}_1$ and $\mathbf{A}_2$ are successively derived vectors in the minimum residual iterative process, the correct averaged vector may be obtained by multiplying the elements of the second vector $\mathbf{A}_2$ by the fourth root of the ratio of the squared length of the vector $\mathbf{A}_1$ to the squared length of the vector $\mathbf{A}_2$ as follows:

$$(4.31)\qquad a_{i_0} = a_i' \sqrt[4]{\frac{L_1^2}{L_2^2}}$$

If L_0, the length of the vector $\mathbf{A}_0$ derived by Eq. (4.31), is computed, it will be the geometric mean of the lengths of vectors $\mathbf{A}_1$ and $\mathbf{A}_2$. This can be shown as follows:

$$L_0^2 = a_1'^2 \sqrt{\frac{L_1^2}{L_2^2}} + a_2'^2 \sqrt{\frac{L_1^2}{L_2^2}} + \cdots + a_n'^2 \sqrt{\frac{L_1^2}{L_2^2}}$$

$$L_0{}^2 = (a_1'^2 + a_2'^2 + \cdots + a_n'^2)\sqrt{\frac{L_1{}^2}{L_2{}^2}}$$

$$L_0{}^2 = L_2{}^2 \cdot \frac{L_1}{L_2} = L_2 \cdot L_1$$

$$L_0 = \sqrt{L_1 L_2}$$

Using this method of averaging successive vectors avoids the failure to converge at some point in the factor extraction process that characterized the averaging of vectors by taking the arithmetic mean of the two. Instead of reaching the point where convergence fails, the length-adjustment method of iteration will eventually reach convergence where the two successive vectors, $\mathbf{A}_1$ and $\mathbf{A}_2$, in the iterative cycle are equal in absolute value but opposite in sign for each element. For the principal factor solution, which requires communalities, this corresponds to the case where the factor being extracted has a negative eigenvalue. This terminology is perhaps inappropriate for the minimum residual solution without communalities, since the diagonal cells of the matrix are not involved, but an analogous situation occurs; that is, a factor is reached ultimately where the iterative process converges on vectors of opposite sign. The process of extraction is terminated at this point, with the factor that converges on vectors of opposite sign being dropped.

If the arithmetic average method of averaging vectors were applied after the length-adjustment method converged on vectors of opposite sign, additional factors might be extracted before reaching a point where this method would fail to converge. These extra factors are analogous to factors with positive eigenvalues that are smaller in absolute value than the largest negative eigenvalue. Such factors are considered to be too minor to be retained.

The steps in computing a minimum residual solution with the length adjustment method of averaging successive vectors are given below. This set of steps will yield either a minimum residual solution without using communality estimates, or it will permit the calculation of a principal factor solution using the diagonal cells with communality estimates inserted.

1. Place zeros in the diagonal cells of the correlation matrix for the minimum residual solution without communality estimates. For a principal factor solution by the minimum residual method of factor extraction, place communality estimates in the diagonal cells of the **R** matrix.

2. Estimate the factor loadings for the factor to be extracted. This is vector $\mathbf{A}_0$. The largest column value in each column with sign attached will serve adequately. If more accurate estimates are available, the iterative process will be shortened.

3. Inner multiply each row of the correlation matrix with vector $\mathbf{A}_0$. This is like multiplying $\mathbf{R}$ on the left by $\mathbf{A}_0$ on the right, that is, $\mathbf{RA}_0$.

4. Divide the inner products from step 3 by $(\sum_{j=1}^{n} a_j^2) - a_i^2$, as in formula (4.22), when communalities are *not* being used and the minimum residual solution is desired. Divide these inner products by $\sum_{j=1}^{n} a_j^2$ for the principal factor solution where communalities are being used. The result of step 4 is to obtain the vector $\mathbf{A}_1$, the first iterated vector.

5. Repeat steps 3 and 4, substituting $\mathbf{A}_1$ for $\mathbf{A}_0$ to obtain a second iterated vector $\mathbf{A}_2$.

6. Vectors $\mathbf{A}_2$ and $\mathbf{A}_1$ are averaged by the length-adjustment method to obtain a new $\mathbf{A}_0$. This is done by multiplying each element of $\mathbf{A}_2$ by the fourth root of the ratio of the squared length of $\mathbf{A}_1$ to the squared length of $\mathbf{A}_2$, as in Eq. (4.31).

7. The vectors $\mathbf{A}_1$ and $\mathbf{A}_2$ are compared term by term to see if they are sufficiently identical to permit termination of the iterative process. In practice, every such difference might be reduced to less than ε in absolute value where ε equals some suitably small number, for example, $\varepsilon = .0005$ for third place accuracy in the results.

8. If all differences are less than ε, terminate iteration and take $\mathbf{A}_2$ as the factor vector. If some differences are still larger than ε, use the new $\mathbf{A}_0$ to recycle through steps 3–7. Continue until convergence is reached.

9. Compute the residuals by the usual process. Compute the matrix $\mathbf{A}_i \mathbf{A}_i'$, where $\mathbf{A}_i$ is the second iterated vector $\mathbf{A}_2$, and subtract this matrix product from the matrix of correlation coefficients.

10. Repeat steps 2–9 for the second and subsequent factors to extract as many factors as desired, or until the iterative process converges on vectors of opposite sign. That is, each element of $\mathbf{A}_1$ will be equal in absolute value to the corresponding element of $\mathbf{A}_2$ but will be opposite in sign.

The results of applying the procedure just outlined to the four-variable sample problem are shown in Table 4.3. A minimum residual solution without communalities was obtained. The communalities derived as a result of the minimum residual solution by summing the squares of the factor loadings for each variable were .538, .469, .655, and .710. These values are considerably different from the "true" communalities of .65, .40, .65, and .82 which reduce the rank of the matrix to exactly 2 when a solution with communalities is obtained. The minimum residual method operating on the off-diagonal elements only, therefore, will not necessarily yield derived communalities that can be considered the "correct" communalities that reduce the rank of the correlation matrix to a minimum. This is particularly true where the matrix is small, in which case the number of unknown communalities is large compared with the number of known off-diagonal elements.

TABLE 4.3

Minimum Residual Solution to the Four-Variable Problem

Factor I

	Initial R matrix				Cycle 1			Cycle 2			Cycle 3		
	1	2	3	4	A_0	A_1	A_2	A_0	A_1	A_2	A_0	A_1	A_2
1	.00	.50	.16	.17	.500	.373	.381	.360	.342	.325	.326	.320	.315
2	.50	.00	.22	.24	.500	.445	.449	.425	.407	.388	.389	.383	.377
3	.16	.22	.00	.73	.730	.700	.801	.758	.784	.784	.787	.795	.794
4	.17	.24	.73	.00	.730	.714	.823	.779	.809	.811	.814	.825	.825

	Cycle 4			Cycle 5			Factor
	A_0	A_1	A_2	A_0	A_1	A_2	A_1
1	.315	.313	.312	.312	.311	.311	.311
2	.377	.375	.374	.374	.373	.373	.373
3	.795	.797	.796	.797	.797	.797	.797
4	.826	.829	.829	.830	.831	.831	.831

	Residual matrix I			
	1	2	3	4
1	.000	.384	−.088	−.088
2	.384	.000	−.077	−.070
3	−.088	−.077	.000	.068
4	−.088	−.070	.068	.000

Factor II

	Cycle 1			Cycle 2			Cycle 3			Cycle 4		
	A_0	A_1	A_2	A_0	A_1	A_2	A_0	A_1	A_2	A_0	A_1	A_2
1	.384	1.000	.390	.627	.631	.635	.635	.638	.641	.641	.644	.646
2	.384	.984	.379	.609	.605	.601	.601	.598	.595	.595	.593	.591
3	−.088	−.229	−.089	−.142	−.142	−.142	−.142	−.142	−.142	−.142	−.142	−.142
4	−.088	−.222	−.085	−.137	−.137	−.137	−.137	−.137	−.137	−.137	−.137	−.137

TABLE 4.3 continued

Factor II (continued)

	A_0	A_1	A_2	A_0	A_1	A_2	A_0	A_1	A_2	A_0	A_1	A_2
1	.646	.648	.650	.650	.652	.653	.653	.655	.656	.656	.657	.658
2	.591	.589	.587	.587	.586	.584	.584	.583	.581	.581	.581	.579
3	−.142	−.142	−.142	−.142	−.142	−.142	−.142	−.142	−.142	−.142	−.142	−.142
4	−.137	−.137	−.137	−.137	−.137	−.137	−.137	−.137	−.137	−.137	−.137	−.137
		Cycle 5			Cycle 6			Cycle 7			Cycle 8	

	A_0	A_1	A_2	A_0	A_1	A_2	A_0	A_1	A_2	A_0	A_1	A_2
1	.658	.660	.660	.660	.661	.662	.662	.663	.663	.663	.664	.664
2	.579	.579	.578	.578	.577	.577	.577	.576	.575	.575	.575	.575
3	−.142	−.142	−.142	−.142	−.142	−.142	−.142	−.142	−.142	−.142	−.142	−.142
4	−.137	−.137	−.137	−.137	−.137	−.137	−.137	−.137	−.137	−.137	−.137	−.137
		Cycle 9			Cycle 10			Cycle 11			Cycle 12	

	1	2	3	4
1	.000	.002	.006	.003
2	.002	.000	.004	.009
3	.006	.004	.000	.048
4	.003	.009	.048	.000

Residual matrix II

	I	II	h_2
1	.311	.664	.538
2	.373	.575	.469
3	.797	−.142	.655
4	.831	−.137	.710

Factor matrix

Rotating the minimum residual factor loadings and retaining only one-place accuracy gives .1, .2, .8, and .9 for factor I and .7, .7, .1, .1 for factor II. These values are within .1 in absolute value of the original coordinates in every case despite the fact that the minimum residual factor loadings were derived using only six data values out of a total of ten needed to fix these coordinates. The communality values represent the remaining four data values that were not used in obtaining the solution, It would be reasonable to expect greater accuracy as the proportion of known to unknown data values rises, for example, with larger matrices.

The minimum residual solution, without communalities, for the 12-variable sample problem is shown in Table 4.4. Five factors were extracted before the

TABLE 4.4

Minimum Residual Solution to the 12-Variable Problem

	I	II	III	IV	V	h^2
1	.080	.846	.160	−.046	.011	.749
2	.011	.707	.101	.048	−.058	.515
3	.037	.652	.221	−.100	−.045	.488
4	.544	.099	−.287	.261	.122	.471
5	.589	.157	−.180	.366	.092	.546
6	.435	.098	−.212	.482	.105	.487
7	.563	−.006	−.161	−.339	.310	.553
8	.661	.013	.076	−.205	.070	.489
9	.758	.023	−.210	−.230	.372	.811
10	.431	−.174	.506	.045	.019	.475
11	.537	−.197	.591	.152	.082	.706
12	.370	−.281	.522	.052	.043	.493
SSQ	2.779	1.833	1.198	.686	.288	(6.783)

iterative process converged on vectors of opposite sign, although the fifth factor was very small, accounting for only 4 percent of the total extracted variance. The principal factor solution gave only four positive eigenvalues, suggesting a maximum of four factors, but this was using the SMCs as communality estimates. The sum of these SMCs gave a total of 5.369, which is appreciably less than the sum of the communalities derived from the minimum residual solution, for example, 6.783. The communalities derived from the minimum residual solution suggest that the SMCs here were underestimates of the correct communalities. In any event, the smaller SMCs permitted the extraction of fewer factors, four instead of five.

The differences between the principal factor solution and the minimum residual solution are attributable primarily to the communalities selected for the principal factor solution. The smaller the matrix, the more pronounced this effect can be, other things being equal, since the number of diagonal values is a larger proportion of the total number of entries in the **R** matrix with smaller matrices.

There is nothing in the minimum residual procedure to prevent a factor loading from exceeding 1.0 as far as the mechanics of the procedure are concerned. This anomalous result will sometimes occur with small or otherwise unusual matrices. If this occurs, it is necessary to recompute with reasonable communality estimates, obtaining a minimum residual solution using these diagonal cell values. This situation has not been encountered by the author with any matrices of 15 or more variables. The intermediate loadings during iteration will often exceed 1.0, but by the time convergence has taken place, these values have been reduced below this upper limit.

[4.3] THE COMMUNALITIES

If 1's are placed in the diagonal cells of **R**, all the eigenvalues will be positive, unless **R** has a determinant equal to zero, which would almost never occur with real data matrices. Analyses with 1's in the diagonals are called "principal components" analyses to distinguish them from principal factor analyses where communalities are inserted in the diagonals. With 1's in the diagonals, all the eigenvalues will ordinarily be positive and then the matrix is said to be positive definite. If the determinant of the **R** matrix, with the diagonal cells filled, should happen to be zero, then one or more of the eigenvalues will be zero. If all the remaining eigenvalues are positive, however, the matrix is said to be positive semidefinite. With some zero eigenvalues, fewer than n factors are needed to reproduce the **R** matrix exactly. The factors with sums of squares equal to eigenvalues of zero are obviously factors with all zero entries; hence they may be dropped. If there are m nonzero eigenvalues, then the **B** matrix needed in the equation $\mathbf{BRB}' = \mathbf{D}$, which diagonalizes the **R** matrix, will have only m rows and n columns. The matrix $\mathbf{B}'$, the transpose of **B**, will have n rows and m columns as will the factor matrix **A** derived from $\mathbf{B}'$.

In the case where communality values less than 1 are put in the diagonal cells of **R**, it often occurs that the **R** matrix is no longer positive definite or even positive semidefinite. When this happens, negative eigenvalues occur in the **D** matrix. Since the sum of squares of the loadings on a factor is supposed to equal its eigenvalue, mathematicians tend to prefer the use of diagonal values, for example, ones, which will avoid this troublesome possibility. If 1's are placed in the diagonals, however, the total variance for each variable is input to the analysis. This includes common, specific, and error variance. In the limiting case where *n* factors are extracted, all the variance for each variable will be extracted so that both the entered and the derived communalities will equal 1.0 for every variable. This would occur even if the variable had a zero correlation with every other variable in the matrix. In this case, the specific and error variance would represent all the variance for that variable since there could be no common variance.

In most factor analytic applications, the interest is focused on studying the relationships between variables, that is, what they share in common. The extracted variance in such analyses should be common variance only, omitting the variance that should be properly assigned to specific factor and error factor sources. Inserting values in the diagonal cells that exceed the correct communalities results in extraction of extra specific and error variance that is then treated as common variance. The more the diagonal cells are inflated, the more pronounced is this distortion. The result of such distortions is to obtain factors that are not just common factors, but factors that mix up common and unique variance in an inextricable way that obscures the view of what the variables have in common with each other. Thus, while it is true that use of 1's in the diagonals avoids a troublesome mathematical problem, the treatment has unfortunate side effects which interfere with the proper interpretation of the data. Factor analytic workers with an empirical orientation, therefore, have tended to prefer working with communalities, rather than 1's, in the diagonals, coping with the negative eigenvalue problem by dropping any factors that do not have positive associated eigenvalues.

Some factor analysts use 1's in the diagonal cells and then retain only those factors with eigenvalues greater than 1.0. This procedure is based on the fact that since a single variable contributes 1.0 to the total extractable variance by being added to the matrix, any factor must have a total variance in excess of that added by a single variable. This seems reasonable enough *per se*, but in using this procedure there is no guarantee, or even a good reason to suppose, that only common factor variance will be extracted by this approach. The communalities that emerge from such an analysis, however, could be reinserted in the matrix to institute an iterative process that would terminate when the entered communalities prove to be essentially identical with the extracted communalities. For all practical purposes, 10 to 15 iterations will often suffice. In some cases, a very large number of iterations may be required. The same number of factors must be retained in each iteration. If the number of factors extracted is allowed to increase, the communalities will tend to increase and the number of factors extractable will rise until finally the process will stabilize with communalities of 1.0 and *n* factors. Iterating the communalities requires doing many factor analyses instead of just one, however; so the cost is much greater than that for a solution which does not involve this procedure.

Another starting point for such an iterative factor analytic process is to use the SMCs for the first factor analytic solution. The derived communalities from this solution are recycled, and the iterative process continues until the communalities stabilize sufficiently for the experimenter's purposes. Guertin and Bailey (1970) present an informative treatment of the distortions that can occur with the use of SMCs as communality estimates. Although they generally find the SMCs to be good estimates with a large N and a small n, which is the case with the present 12-variable sample problem, the SMCs are sometimes much too small. They recommend iterating the communalities with a fixed number of factors until the values derived from the solution are essentially identical with the communalities input to the solution (see also Tucker, Koopman, & Linn, 1969).

Still another starting point for an iterative cycle of factor analytic solutions is the set of communalities derived from the minimum residual solution. These values, derived without using the diagonal cells, are fed back into a minimum residual solution using the diagonal cells, which makes it a principal factor solution. With these derived communalities as the estimates for a second factor analysis, a new solution is obtained. The communalities derived from this solution are recycled, and so on, as before. After the first analysis, the remaining analyses are all principal factor solutions since they all use the diagonal cells filled with communality values derived from the previous

TABLE 4.5

Minimum Residual Solution to the 12-Variable Problem with Iterated Communalities[a]

	I	II	III	IV	h_{15}^2	$h_1{}^2$
1	.099	.835	.266	.044	.780	.749
2	.013	.678	.186	−.055	.498	.512
3	.045	.653	.316	.090	.536	.486
4	.549	.137	−.329	−.217	.475	.456
5	.603	.190	−.223	−.317	.550	.538
6	.464	.138	−.282	−.454	.520	.476
7	.585	.016	−.230	.448	.596	.457
8	.638	.013	.042	.214	.455	.485
9	.772	.053	−.279	.317	.777	.673
10	.466	−.227	.429	−.034	.454	.474
11	.628	−.289	.552	−.153	.806	.699
12	.414	−.352	.435	−.043	.486	.491
					(6.933)	(6.496)

[a] This solution is that obtained after 15 analyses using h^2 values from each analysis as h^2 estimates for the next trial. The $h_1{}^2$ values above are the values derived from the first minimum residual solution based only on off-diagonal correlation matrix entries.

solution. Table 4.5 gives the solution after 15 such cycles for the 12-variable problem. The communalities tend to rise as the process stabilizes because higher communalities in the right places will permit a greater reduction of the sum of squares of the off-diagonal elements.

Iteration of the communalities to stability is no guarantee that the "correct" communalities will be attained as an end result. For the four-variable sample problem, for example, the communalities after 15 iterations were .539, .472, .702, and .760, compared with the "correct" communalities of .65, .40, .65, and .82. These correct communalities reduce the rank of the matrix to 2, but the iterating of communalities does not find these rank-reducing diagonal values.

In the absence of a definitive solution to the communality problem, the author prefers to use the minimum residual solution without communalities, taking the derived h^2 values as the best indication of the amount of common factor variance. Iteration of the communalities runs the risk of capitalizing on chance errors, elevating the communalities in order to drive down the sum of squares of the off-diagonal residuals. For example, the sum of the derived communalities after 15 iterations with four factors in the 12-variable sample problem (see Table 4.5) is 6.933. This is only slightly larger than the corresponding sum for the five-factor minimum residual solution, 6.783. On the other hand, the communality for variable 11 is driven up to .806 in the iterated communality solution compared with .706 for the five-factor minimum residual solution. The value of .806 could be correct, but it appears to the author to be unrealistically high. With the uniterated minimum residual solution, at least the derived communalities are entirely a function of the known correlation coefficients and the factor loadings are independent of any errors in estimating communalities.

[4.4] DETERMINING HOW MANY FACTORS TO EXTRACT

It is easy to establish an upper bound on the number of factors to be extracted. With communalities specified, the maximum number of factors equals the number of factors with positive eigenvalues. Using the minimum residual method without communalities, it is the number of factors extracted before the iterative process converges on vectors of opposite sign. The author prefers the latter solution where the two disagree, because the number of factors with positive eigenvalues in the principal factor solution will vary with the particular communalities inserted. Since these are unknown, the solution is not exactly determined. With the centroid method, neither of these guidelines is available. By reestimating communalities, it is possible to continue extracting centroid factors almost indefinitely.

In actual practice with real data, the number of factors that merit retention for purposes of scientific description and explanation will be less than the upper bound, often considerably so. With such data, there are typically several prominent factors with a large amount of variance in each, followed by several factors of lesser importance trailing off into insignificance. Trying to decide precisely where to draw the line, considering the remaining factors to be too small to be retained, is somewhat like trying to decide how short someone must be to be called short. There is no precise solution to this problem in any practical sense, although a kind of precise solution may be imposed on the situation through appeal to mathematical criteria. For example, it may be determined that there is not another "significant" factor in the statistical sense to a certain level of confidence. Decisions of this kind ultimately are based on the sample size employed, which, after all, has nothing to do with the nature of the variables being studied. The sample size is determined by a decision of the investigator, not by the factor structure of a group of variables. The precision of such solutions to the problem of how many factors to extract, therefore, is more apparent than real.

Several signs are useful in trying to decide whether or not to stop extracting factors before the maximum number possible has been reached. For example, if the sums of squares of the loadings on the extracted factors are no longer dropping but are remaining at a low and rather uniform level, factor extraction may be reasonably terminated. If the maximum remaining residual correlation has dropped to a very low value, for example, less than .10, it is usually unnecessary to extract any more factors. Also, if the factors being extracted have no loadings as large as .30 or higher in absolute value, there is ordinarily little reason to extract more factors.

The main thing to consider in deciding when to stop factoring is that it is better to err on the side of extracting too many factors rather than too few. There are two main reasons for this rule. First, if there are ten strong factors, for example, in the matrix and only eight are extracted, most of the variance for all ten factors has been taken out with the first eight factors. Since the amount of variance extracted gets progressively smaller from the largest to the smallest factor extracted, variance is extracted on a single large factor that ultimately may be distributed to several different factors by the subsequent rotation process. In the rotation process, however, this variance for ten factors must be distributed on only eight factors. Not only are two factors lost, but the character of the remaining factors is distorted by the intrusion of variance from the two factors that are not retained in the final solution. Secondly, if too many factors have been extracted, proper management of the rotation process following factor extraction will permit discarding of the extra factors as residual factors of minimal importance without distorting the character of the major factors.

The recommended procedure, then, is to extract enough factors to be relatively certain that no more factors of any importance remain. If in doubt, continue until the maximum number has been extracted, that is, with the minimum residual method until convergence on vectors of opposite sign has taken place using the length-adjustment procedure. In the rotation process following factor extraction, eliminate the excess factors. Just how this can be done is explained later. It should be emphasized, however, at this point, that appropriate steps *must* be taken to eliminate these extra factors. If too many factors are rotated without thought, for example, with a canned computer program, distortions to the solution may occur which are just as serious as those that result from extracting and rotating too few factors.

[4.5] OTHER METHODS OF FACTOR EXTRACTION

There are many other methods of factor extraction besides the three discussed here. Thurstone (1967) describes the group centroid and multiple group methods as labor-saving alternatives to the centroid method described in this book. Harman and Jones (1966) present a method, called the "minres" method, which minimizes the residuals after a specified number of factors without using the diagonals. In their book on factor analysis, Lawley and Maxwell (1963) describe the maximum likelihood method of factor analysis which is computationally demanding but has the property of permitting a statistical test of significance for the number of factors to be extracted.

Horst (1965) presents an exhaustive treatment of factor extraction methods and their variations. Harman (1967) also presents a thorough treatment of factor extraction methods. The reader who aspires to achieve the status of an expert in factor analysis will, of course, wish to become familiar with these other available methods. The methods presented in this book, however, are probably adequate for most ordinary research purposes. The particular method of factor extraction used will usually have less effect on the final outcome than such considerations as the communalities chosen and whether or not they are used, the number of factors extracted, and how the rotations are carried out.

CHAPTER FIVE

Orthogonal Hand Rotations

[5.1]
INTRODUCTION

It has already been pointed out in previous chapters that the extracted factor matrix does not usually provide a suitable final solution to a factor analysis problem. Most extraction methods are designed to extract something approximating the maximum amount of variance at each step. The factors, therefore, get progressively smaller until finally they are too small for retention. At each step, variance from many different common factor sources is being extracted because the factor vector is placed in such a way that as many of the variables as possible have substantial projections on it. It is not atypical for entirely uncorrelated data vectors to have substantial projections on the same extracted factor vector. Such extracted factor vectors are, then, complex composites of partially overlapping and even unrelated data variables rather than narrowly defined factors represented by homogeneous collections of closely related data variables.

Suppose, for example, that several verbal abilities tests and several measures of physical size were included in the same matrix. A reasonable final factor solution might give two equally prominent factors, Verbal Ability and Size. The first extracted factor, however, would be a combination of Verbal Ability and Size, with all measures having positive loadings of substantial proportions. The second factor might have positive loadings for the size variables and negative loadings for the verbal ability variables, but the second factor loadings in general would be smaller in absolute size than those for the first factor. Only by rotating the extracted factors would it be possible to obtain the expected factors of Verbal Ability and Size.

Failure to understand how the first factor can pull variance from uncorrelated variables has led some investigators to infer mistakenly that a general factor exists in the data just because all the variables have positive loadings on the first factor. Consider the example in Fig. 5.1. Let variables 1, 2, and 3 represent three substantially intercorrelated verbal tests: 1. Vocabulary, 2. Verbal Fluency, and 3. Verbal Analogies. Variables 4, 5, and 6 represent size variables: 4. Height, 5. Weight, and 6. Chest Volume. The two groups of variables represent essentially uncorrelated clusters, yet the first extracted factor, a vector about equidistant between these two clusters, will have substantial positive projections for all these variables. These projections are indicated by the dotted lines from the data vector termini perpendicular to the factor vector I. The second factor will be bipolar, with positive loadings for variables 4, 5, and 6 and negative loadings for variables 1, 2, and 3. A rotation of the two extracted factors about 45° counterclockwise will bring the factors in line with the two independent clusters of homogeneous variables.

It is inappropriate to think of the unrotated factor I position as representing a scientifically meaningful, interpretable factor. Such a factor consists of unrelated components that do not go together. The rotated factor positions, identified with the Verbal Ability and Size factors, are much easier to interpret and use than the unrotated factors which would represent, respectively, a Size plus Verbal Ability factor and a Bipolar Size plus Lack of Verbal Ability factor.

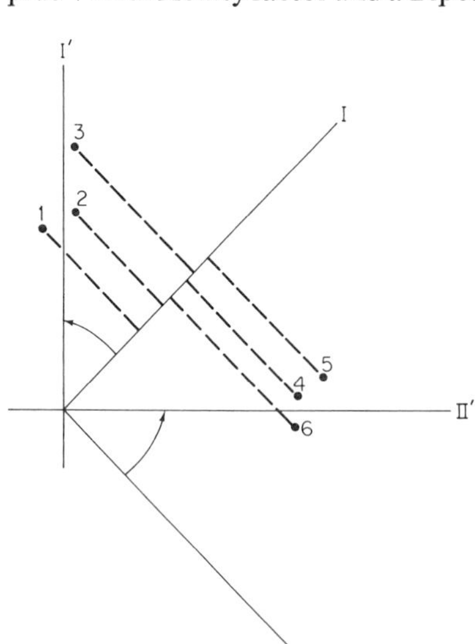

FIG. 5.1. Projection of data vectors on first extracted factor. 1: Vocabulary; 2: Verbal Fluency; 3: Verbal Analogies; 4: Height; 5: Weight; 6: Chest Volume.

In the event that all variables have substantial loadings on the first extracted factor and *in addition* all variables correlate positively to a substantial degree with each other, the case for a general factor would be much stronger, although still not conclusive. It could still be possible that all the variables included were themselves complex composites, making it impossible in the analysis to isolate the separate factor components. It can be stated, therefore, that having all variables in the matrix load substantially on the first extracted factor is a necessary but not sufficient condition to demonstrate the presence of a general factor.

The initial extracted factor matrix, then, usually must be rotated before the final factor solution is achieved. In Chapter 2 it was shown that a factor matrix may be transformed to a rotated factor matrix by the matrix operation $\mathbf{V} = \mathbf{A}\mathbf{\Lambda}$, where $\mathbf{V}$ is the rotated matrix, $\mathbf{A}$ is the unrotated matrix, and $\mathbf{\Lambda}$ is an orthogonal transformation matrix in which rows and columns have sums of squares equal to 1.0 and inner products of nonidentical rows or columns equal to zero. As shown in Chapter 2, such a transformation does not affect the capacity of the factor matrix to reproduce the original correlation matrix since

$$\mathbf{V}\mathbf{V}' = (\mathbf{A}\mathbf{\Lambda})(\mathbf{A}\mathbf{\Lambda})' = \mathbf{A}\mathbf{\Lambda}\mathbf{\Lambda}'\mathbf{A}' = \mathbf{A}\mathbf{I}\mathbf{A}' = \mathbf{A}\mathbf{A}' = \mathbf{R}$$

In other words, the transformed or rotated matrix $\mathbf{V}$ when multiplied by its transpose $\mathbf{V}$ will reproduce the $\mathbf{R}$ matrix just as well as $\mathbf{A}$ multiplied by its transpose $\mathbf{A}'$ does.

In order to carry out the rotation of the $\mathbf{A}$ matrix, it is necessary to find the appropriate $\mathbf{\Lambda}$ matrix to transform $\mathbf{A}$ to $\mathbf{V}$ by the matrix operation $\mathbf{V} = \mathbf{A}\mathbf{\Lambda}$. This chapter is concerned with methods of determining this $\mathbf{\Lambda}$ matrix through inspection of plots of factors two at a time. This traditional method of determining what rotations to perform has been losing popularity in recent years with the increasing availability of standard computer programs. Because of the availability of these easier and more convenient computer programs, the general sentiment seems to be growing that it is no longer important to teach the "old-fashioned" rotation procedures based on inspection of graphical plots of factors two at a time.

The author takes exception to this point of view. Computer methods of rotating factor axes by means of analytic criteria (as discussed in a later chapter) cannot replace intelligent judgment by the investigator as to what kind of rotational solution is appropriate. A computer program will carry out rotations according to some specified criterion which may be appropriate for certain situations, but it cannot be appropriate for all situations. The investigator must be prepared to ascertain what criteria are suitable for guiding the rotations in his particular situation and then carry out the rotations in such a way that the data are properly treated. The only way to prepare the investigator to meet all possible situations is to teach him how to position the rotated vectors wherever he pleases. He may then use any criterion he desires and place the axes accordingly. The method of orthogonal rotation described here permits the investigator this freedom of choice. If these procedures are not mastered, the student is left only with the alternative of choosing among available computer programs to carry out the rotation of factor axes, and none of these programs may be suitable.

To illustrate how orthogonal rotations are carried out to place the rotated factor axes in any position desired, a rotation of the minimum residual solution to the 12-variable sample problem (see Table 5.1) is shown in the next section.

TABLE 5.1

Unrotated Minimum Residual Solution to the 12-Variable Problem[a]

	Factors						
Variable	I	II	III	IV	V	Variable name	Factor
1	.080	.846	.160	−.046	.011	Law Enforcement	C
2	.011	.707	.101	.048	−.058	Acceptance of Social Order	C
3	.037	.652	.221	−.100	−.045	Intolerance of Nonconformity	C
4	.544	.099	−.287	.261	.122	Lack of Inferiority Feelings	S
5	.589	.157	−.180	.366	.092	Lack of Depression	S
6	.435	.098	−.212	.482	.105	Lack of Agitation	S
7	.563	−.006	−.161	−.339	.310	Lack of Reserve	E
8	.661	.013	.076	−.205	.070	Lack of Seclusiveness	E
9	.758	.023	−.210	−.230	.372	No Loss for Words	E
10	.431	−.174	.506	.045	.019	Sympathy	P
11	.537	−.197	.591	.152	.082	Helpfulness	P
12	.370	−.281	.522	.052	.043	Generosity	P

[a] Factor names are C: Social Conformity versus Rebelliousness; S: Emotional Stability versus Neuroticism; P: Empathy versus Egocentrism; E: Extraversion versus Introversion.

These rotations are carried out using the rotational criteria of "positive manifold" and "simple structure," criteria that have been traditional guides in carrying out the rotation process in factor analysis. Many of the computer rotational procedures are designed to achieve an approximation to this solution except that the computer does it without the benefit of any human judgment. Traditional application of these criteria has involved a series of judgments by the investigator as the rotation process unfolds.

Although simple structure and positive manifold are the guiding principles in carrying out the rotations for the 12-variable sample problem, the methods of executing the rotations are not limited to use with these criteria. The same methods of executing the rotations may be used by the investigator in applying any criterion to guide the rotations, however arbitrary. The only restriction is that the methods shown in this chapter limit the investigator to an orthogonal solution with uncorrelated factors. Oblique solutions are considered in the next chapter.

[5.1.1] POSITIVE MANIFOLD

Even if all correlations in the original matrix are nonnegative, i.e., zero or positive, except for a few chance correlations slightly below zero, the extracted factors past the first factor frequently exhibit many substantial negative as well as positive loadings. This is an arbitrary result, based on the requirement in the centroid method, for example, that the sum of the factor loadings be zero for the second and subsequent factors. Since all the original correlations are nonnegative, however, the final rotated factors should be positioned in such a way that there are no substantial negative loadings for any of the variables on any of the factors. In looking at a plot of two factors in Cartesian coordinates, therefore, the investigator looks for a rotation that will reduce or eliminate negative loadings as much as possible on the rotated factors. Trying to rotate to obtain nonnegative loadings is known as rotating to "positive manifold." The rotation of factors I and II about 45° counterclockwise in Fig. 5.1 to positions I′ and II′ is a good illustration of this principle. Before the rotation, variables 1, 2, and 3 had substantial negative loadings on factor II. After the rotation, however, there are no negative loadings of any consequence on either factor I′ or II′.

The idea behind positive manifold, then, is that if all the data variables in a matrix have intercorrelations that are either zero or positive, it is unreasonable to anticipate an underlying factor with substantial negative loadings for any of the data variables. Most factor analytic methods, however, will produce *extracted* factors after the first factor with substantial negative loadings for some variables, even if all the intercorrelations were nonnegative. The rotated factors which represent the meaningful constructs ordinarily should not exhibit these large negative loadings. Even with nonnegative correlations among the data variables, however, it would be theoretically possible to obtain one or more underlying rotated factor constructs with substantial negative loadings, since the extracted factors exhibit such loadings; on the other hand, such constructs would be contrary to reasonable expectation, probably complex in character, and difficult to interpret. In the absence of strong evidence suggesting the presence of such bipolar factors (that is, with both large positive and negative loadings), therefore, extracted factors should be rotated to eliminate substantial negative loadings. Of course, if a contemplated rotation would make all or most of the factor loadings negative, this would not necessarily represent a violation of the positive manifold concept. The direction of the factor can be reversed following the rotation, thereby changing all the signs. Positive loadings become negative and vice versa. As long as none of the negative loadings after reversal is large, reversal of the factor direction would bring the factor into agreement with positive manifold rotational principles.

[5.1.2] *SIMPLE STRUCTURE*

Thurstone (1947) developed the criterion of "simple structure" to guide the investigator in carrying out rotations of factor axes to positions of greater "psychological meaningfulness." A generation of factor analysts followed his lead and even today, most factor analyses are carried out using this method or some procedure that was developed to approximate it without the necessity for judgment by the investigator.

Thurstone reasoned that if we sample widely and more or less randomly from all the possible data variables which describe a complex domain such as the domain of human characteristics, it is inevitable that a large collection of such variables will measure many factors, not just one or two. He also supposed that any one variable will be related to only a few of these many factors. Furthermore, he reasoned that no one factor will be substantially related to more than a small proportion of the variables. Different factors, of course, will be defined by different data variables. If these conditions are met, Thurstone deduced that the rotated factor matrix representing these factors will take on a characteristic form.

1. Most of the loadings on any one factor will be small, that is, more or less randomly distributed about zero, with only a few of the loadings being of substantial size. This means that any column of the factor matrix should have mostly small values, as close to zero as possible.
2. A given row of the factor matrix, containing the loadings for a given variable on all the factors, should have nonzero entries in only a few columns. The fewer the columns in which the variable has loadings, the closer to the expected form is the solution.
3. Any two factor columns should exhibit a different pattern of high and low loadings. Otherwise, there would be no apparent difference between the two factors represented by these columns.

The situation represented by Fig. 5.1 does not conform to the ideal circumstances for applying simple structure since there are only two factors with three variables for each. Only half the variables on a given factor, therefore, can have low loadings. Within these limitations, it is clear that the rotation shown in Fig. 5.1 brings the rotated factors as close to a simple structure position as possible. Whereas the unrotated factors had no low loadings, the rotated factors both have 50 percent of the loadings at essentially zero level. Also, each variable has a major loading only in one of the two rotated factors. The pattern of loadings is markedly different in the two factors since those variables with high loadings on one have low loadings on the other.

The rotation in Fig. 5.1, then, is a good one from the standpoint of meeting both positive manifold and simple structure criteria. In practice, rotations are sought which will in fact do just this. A rotation which works against one of these criteria while improving the other would be less desirable and in fact might be actually undesirable. Positive manifold is less useful as a guide to rotations with factor matrices derived from correlation matrices with large negative correlations present. There is no large negative correlation in the 12-variable sample problem correlation matrix, however; so positive manifold is very useful with this problem.

Sometimes the negative correlations in a matrix can be reduced or eliminated by the reflection of data variables. For example, suppose one variable is represented by the number of mistakes in a certain task. All the other variables represent the number of correct answers on similar kinds of tasks. The variable based on the number of mistakes will very likely have negative correlations with the other variables. This variable can be reflected by changing all the signs for all the correlations between this variable and the other variables. This will eliminate these negative signs from the matrix, making it easier to use positive manifold to guide the rotations. However, the name of the reflected variable would have to be changed as, for example, from "Mistakes" to "Lack of Mistakes."

[5.2] ORTHOGONAL ROTATION OF THE 12-VARIABLE PROBLEM

The first step in the rotational process is to plot every pair of factors that are to be a part of the solution. In this case, only the first four factors from the minimum residual solution to the 12-variable problem, given in Table 4.4 and repeated in Table 5.1, will be rotated. The fifth factor could be included in all the rotation cycles, but it is a very small factor; so it will be omitted from consideration until after the main factors have been rotated. In plotting the factors, that is, I with II, I with III, and so on, until finally III with IV is plotted, the loadings for a variable in the two factors establish the location of a point on the plot in a Cartesian coordinate system. For example, if a variable had loadings of .60 and $-.40$ in factors I and II, respectively, there would be a single point at the location (+.60, −.40) on a two-dimensional plot of factor I with factor II. There would be a point on this plot for each variable. Each factor axis would extend from −1.0 to +1.0 to accommodate the largest possible loadings, positive or negative. These plots for the minimum residual factors are shown in Fig. 5.2.

From the first set of plots, two rotations were selected to be performed. Factor I is rotated 22°47′ away from factor IV, that is, counterclockwise in this plot, and to keep the rotations orthogonal, factor IV is rotated the same angle (22°47′) toward factor I. This keeps the angle between the new factor I and the new factor IV at 90°. Any angle approximately equal to 22°47′ would have

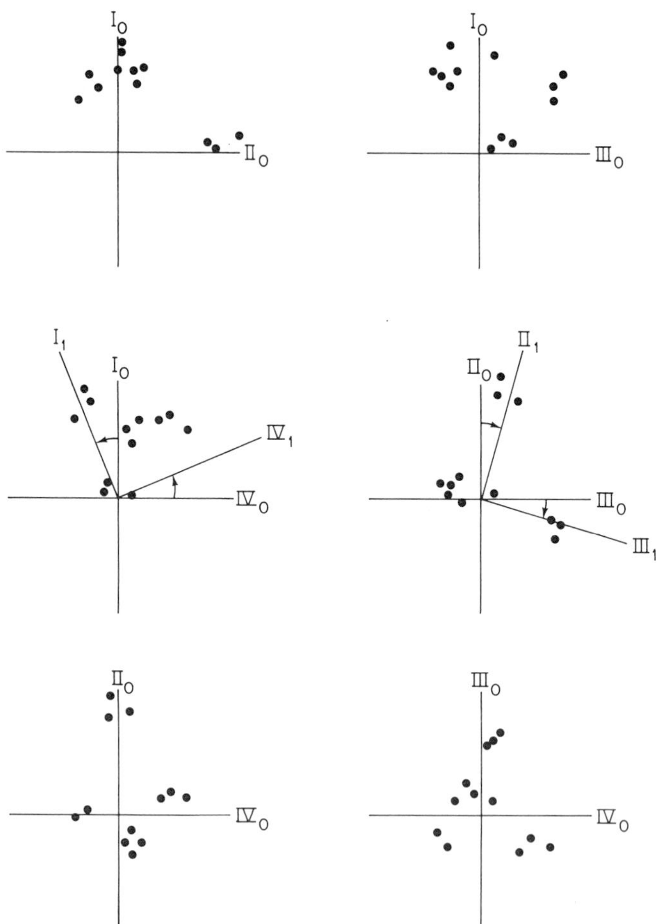

FIG. 5.2. First set of plots in orthogonal rotation of the 12-variable problem.

served equally well as the angle of rotation. The unrotated minimum residual factors are labeled with a zero subscript, that is, I_0 and IV_0. The positions after one rotation are labeled with a subscript of 1, that is, I_1, and after i rotations with a subscript i, for example, III_i.

This rotation of factor I with factor IV was selected because it eliminates the negative loadings on factor IV while bringing the formerly negatively loaded variables into the "hyperplane" of factor IV; that is, these variables have loadings close to zero on factor IV. The line perpendicular to the vector labeled IV is at once the hyperplane for factor IV and also the factor vector for factor I. Any points close to the factor axis I will have perpendicular projections onto the factor axis IV near the origin and hence will have near-zero loadings on factor IV. Any points close to factor axis IV will have perpendicular projections onto the factor axis I near the origin and hence will have near-zero loadings on factor I. The line perpendicular to the factor axis is referred to as a hyperplane because the termini of a collection of vectors in four-dimensional space have been projected onto a two-dimensional surface, that is, a plane, when we plot two factors at a time. For a three-dimensional space, we see the plane perpendicular to one of the factors as a line only because we are looking along the edge of the plane. If, for example, a square piece of plywood is held up before the eyes, it can be positioned in such a way that it will show only one edge; hence it appears as a line. If the board were clear plastic and points were embedded in the plastic, these points would all appear along the edge, but the distances between them in the direction of the line of sight would not be apparent. Only distances along the axis perpendicular to this line would be apparent. For a three-dimensional space, the hyperplane perpendicular to a factor axis is really a plane and as such has only two dimensions. If vectors project out into space of higher dimensions, such as the four-dimensional space with this sample problem, then the space perpendicular to a factor is no longer a plane of two dimensions. It is a "plane" of many dimensions and hence is referred to as a "hyperplane." In a two-dimensional plot, however, only the "edge" of this plane or hyperplane is seen. Any data vector that has no loading on a given factor will be wholly contained in the hyperplane somewhere. The point that represents the terminus of this vector, therefore, will show up in the plot of this factor with some other factor as a point along the line perpendicular to this factor (or close to it), and hence it is said to be in the hyperplane of this factor even though it appears only as a point close to a line. When the hyperplane has several dimensions, of course, it is impossible to visualize it, just as it is impossible to visualize a space of more than three dimensions.

In determining the loadings of the data vectors on the new factors I_1 and IV_1 in Fig. 5.2, perpendicular projections from the data points are dropped to the two axes I_1 and IV_1. The distances from the origin to the points where these perpendicular projections intersect with the factor axes give the factor loadings on the orthogonal factors. The loadings on the old axes I_0 and IV_0 were used to locate the points on the plot serving as the coordinates of the points with respect to the old axes I_0 and IV_0.

The rotation of factor I with factor IV also takes some of the variance away from factor I and puts it on factor IV. The first factor usually has maximum variance on it, representing a composite factor substantially related to as many data variables as possible. This means that the rotational process must remove variance from factor I (that is, reduce the sum of squares of loadings on factor I) and spread it around to other factors if the rotated factors are to be more nearly equal in the amount of variance they account for than were the original unrotated factors. Other early extracted factors may also have to give up some of their total variance to later factors in the rotational process.

To summarize, in Fig. 5.2 the rotation of factor I away from factor IV, and of IV toward I, improves positive manifold, moves closer to simple structure, and reduces some of the excess variance on factor I. The other rotation worth carrying out from the plots in Fig. 5.2 is that for factors II and III. Factor II is rotated toward factor III, that is, clockwise, by an angle of 15°39′ and factor III is rotated away from factor II by the same angle. This rotation improves both positive manifold and simple structure for both factors, making it an excellent choice for execution.

This series of plots (in Fig. 5.2) also offers another excellent rotational possibility. Factor I could have been rotated away from factor III about 22°, that is, counterclockwise, and factor III toward factor I by the same amount to maintain orthogonality. This rotation would be about as good as the actually performed rotation of I with IV. In a given series of rotations, however, each factor may be rotated only once without making new plots for that factor with the other factors. Factor I cannot be rotated both with IV and with III at the same time except through the use of techniques more complicated than those to be demonstrated in this text. After I has been rotated with IV, however, the new factor I, that is, I_1, can be rotated with the original, as yet unrotated, factor III, that is, III_0, if in fact factor III has not yet been rotated. In this example, factor III is rotated with factor II in the first series of rotations; so when the rotation of I with III is considered again, it will be I_1 plotted with III_1, as shown in Fig. 5.3.

Instead of rotating factor I counterclockwise away from factor III by an angle of about 22°, as considered in the preceding paragraph, we could achieve the same effect by rotating factor I clockwise toward factor III by an angle of $(90-22)°$. This would interchange factors I and III relative to those obtained with the 22° counterclockwise rotation. It would also reverse the direction of factor III. Since III would be rotated away from I in this clockwise rotation, the major loadings of the variables on the new factor III would be negative, whereas they were positive for the counterclockwise rotation. This clockwise rotation would not be considered a violation of the positive manifold principle, however, because the orientation of factor III can be reversed arbitrarily by changing all the signs of the factor loadings on the new III. Then, the major loadings would be positive as expected with the positive manifold principle. The signs of the third column entries in the transformation matrix would also have to be reversed to maintain consistency with the reversed signs in the third column of the rotated factor matrix. Thus, at any point in the rotation process, should $\mathbf{V} = \mathbf{A}\mathbf{\Lambda}$ and the signs of any column of $\mathbf{V}$ be predominately negative, they may be changed, reversing the signs of all column entries, if the signs in the corresponding column of $\mathbf{\Lambda}$ are also reversed. This may be done arbitrarily and at will. It merely reverses the direction of the factor. If the factor were Introversion–Extraversion before this reflection, after the reflection it would become Extraversion–Introversion. Many rotation problems will require this type of arbitrary reflection of factors at one or more points in the rotation process to eliminate a predominance of unwanted negative signs in the factor loadings for a given factor.

The remaining plots in Fig. 5.2 are less attractive as rotational possibilities than the three already given, although factor III could be rotated a small amount with factor IV in a clockwise direction with resulting improvements in both positive manifold and simple structure. This rotation was not chosen, however, since for these factor axes good rotations of greater magnitude were available with other factors than each other. The rotation of III with IV will actually be chosen as the best rotation in a later series of plots (see Fig. 5.3). Higher numbered factors are rarely to be rotated with each other before they have been rotated with lower numbered factors to build up their total variance. Initially, these factors have rather low sums of squares of their factor loadings before rotation.

The plots of I with II and II with IV in Fig. 5.2 offer little opportunity for improving positive manifold and simple structure simultaneously. To find three good rotations in a series of plots, however, is all that the most optimistic investigator could hope for. Even one good rotational possibility is enough to keep the process moving forward.

[5.2.1] *ALGEBRAIC EXECUTION OF THE FIRST ROTATION*

Computations for the first rotation are shown in Table 5.2. It is possible to do the rotations entirely graphically (Zimmerman, 1946), and in precomputer days this was the usual method. Today, however, the plots are used most commonly only as a means of deciding which factors to rotate, in what direction, and by how big an angle. The actual execution of the rotation is carried out algebraically rather than graphically; it is more accurate, provides a better

TABLE 5.2

First Orthogonal Rotation of the 12-Variable Problem

A

	I	II	III	IV
1	.080	.846	.160	−.046
2	.011	.707	.101	.048
3	.037	.652	.221	−.100
4	.544	.099	−.287	.261
5	.589	.157	−.180	.366
6	.435	.098	−.212	.482
7	.563	−.006	−.161	−.339
8	.661	.013	.076	−.205
9	.758	.023	−.210	−.230
10	.431	−.174	.506	.045
11	.537	−.197	.591	.152
12	.370	−.281	.522	.052

×

$\mathbf{\Lambda}_1$

I	II	III	IV
.922	0	0	.387
0	1	0	0
0	0	1	0
−.387	0	0	.922

=

$\mathbf{V}_1$

	I	II	III	IV
1	.092	.846	.160	−.011
2	−.008	.707	.101	.049
3	.073	.652	.221	−.078
4	.400	.099	−.287	.451
5	.401	.157	−.180	.566
6	.214	.098	−.212	.613
7	.650	−.006	−.161	−.095
8	.689	.013	.076	.067
9	.788	.023	−.210	.081
10	.380	−.174	.506	.208
11	.436	−.197	.591	.348
12	.321	−.281	.522	.191

mathematical record of what was done, and with the availability of computing equipment, is actually less work. A program for carrying out these rotations by computer is described in Chapter 12.

Each orthogonal rotation is carried out by multiplying the present matrix of factor loadings on the right (postmultiplying) by an orthogonal transformation matrix $\mathbf{\Lambda}_i$. For the first rotation, the matrix equation is

$$\mathbf{V}_1 = \mathbf{A}\mathbf{\Lambda}_1 \tag{5.1}$$

where **A** is the unrotated matrix of extracted factor loadings, $\mathbf{V}_1$ is the matrix of rotated loadings after the first rotation, and $\mathbf{\Lambda}_1$ is the orthogonal matrix that carries the matrix **A** into matrix $\mathbf{V}_1$.

Matrix $\mathbf{\Lambda}_i$ is determined by the rotation to be performed, which in turn is based on the plots of the factors, two at a time (see Fig. 5.2). If factor j is to be rotated with factor k, the elements of $\mathbf{\Lambda}_i$ are determined as follows: (1) $\lambda_{jj} = \lambda_{kk} = \cos\varphi$ where φ is the number of degrees in the angle of rotation to be performed; (2) $\lambda_{jk} = \sin\phi$ when the lower numbered factor is rotated away from the higher numbered factor and $-\sin\varphi$ when the lower numbered factor is rotated toward the higher numbered factor; (3) $\lambda_{kj} = -\sin\varphi$ when $\lambda_{jk} = \sin\varphi$, and $\lambda_{kj} = +\sin\varphi$ when $\lambda_{jk} = -\sin\varphi$; (4) other diagonal entries of $\mathbf{\Lambda}_i$ are equal to 1.0; (5) all other entries of $\mathbf{\Lambda}_i$ are zero.

In the first rotation, factor I is rotated away from factor IV by an angle of 22°47′. The cosine of this angle is .922, which appears as λ_{11} and λ_{44} of $\mathbf{\Lambda}_1$, Table 5.2. The sine of 22°47′ is .387 and since factor I is rotated *away* from factor IV, $\lambda_{14} = +.387$. Since $\lambda_{14} = +.387$, λ_{41} will equal $-.387$, also shown in Table 5.2. The other two diagonal cells of $\mathbf{\Lambda}_1$ equal 1 and all other entries of $\mathbf{\Lambda}_1$ are zero.

It should be noted in Table 5.2 that multiplying **A** by $\mathbf{\Lambda}_1$ has the effect of subtracting a proportion of the fourth factor loadings from the first factor loadings multiplied by .922. Thus, when factor I is rotated *away* from factor IV, the minus sign in the fourth row, first column of $\mathbf{\Lambda}_1$ causes the new factor I to be less like the old factor IV. The new factor IV, however, is being made more like the old factor I because it is being rotated toward factor I. This is reflected in the matrix multiplication by the plus sign in the first row, fourth column of $\mathbf{\Lambda}_1$ in Table 5.2. This causes a portion of the first factor loadings to be added to the fourth factor loadings multiplied by .922. The placement of the minus sign in the transformation matrix, therefore, is determined by whether the lower numbered factor is to be made more like the higher numbered factor (rotated toward it), which requires a plus sign in the λ_{ji} cell of $\mathbf{\Lambda}$, or whether it is to be made less like the higher numbered factor (rotated away from it), which requires a minus sign in the λ_{ji} cell of $\mathbf{\Lambda}$.

Multiplying **A** by $\mathbf{\Lambda}_1$ changes only columns I and IV, which is to be expected since only factors I and IV were being rotated. Factors II and III of $\mathbf{V}_1$ are seen in Table 5.2 to be identical with factors II and III of matrix **A**. The values in columns I and IV of matrix $\mathbf{V}_1$ could be obtained by taking the perpendicular projections of the test vectors on the rotated factor axes I_1 and IV_1 in Fig. 5.2. The reader may verify this to assure himself that the indicated matrix operations actually carry out the desired rotation of the factor axes.

The second rotation is performed by the matrix equation

$$\mathbf{V}_2 = \mathbf{V}_1 \mathbf{\Lambda}_2 \tag{5.2}$$

where $\mathbf{V}_2$ is the matrix of rotated factor loadings after two rotations, $\mathbf{V}_1$ is the matrix of rotated factor loadings after one rotation, and $\mathbf{\Lambda}_2$ is the orthogonal transformation matrix for the second rotation. For the 12-variable sample problem, the second rotation consisted of rotating factor II toward factor III by an angle of 15°39′. Factor III was rotated away from factor II by the same angle at the same time. The computations for this rotation are shown in Table 5.3. The cosine of φ, therefore, appears in λ_{22} and λ_{33} of $\mathbf{\Lambda}_2$. Ones appear in the other two diagonal cells, λ_{11} and λ_{44} of $\mathbf{\Lambda}_2$. Since factor II is rotated *toward* factor III, $\lambda_{23} = -\sin\varphi$ and $\lambda_{32} = +\sin\varphi$.

The matrix $\mathbf{V}_2$ will be altered by the next rotation as follows:

$$\mathbf{V}_3 = \mathbf{V}_2 \mathbf{\Lambda}_3 \tag{5.3}$$

Then

$$\mathbf{V}_4 = \mathbf{V}_3 \mathbf{\Lambda}_4 \tag{5.4}$$

and, in general,

$$\mathbf{V}_{i+1} = \mathbf{V}_i \mathbf{\Lambda}_{i+1} \tag{5.5}$$

TABLE 5.3

Second Orthogonal Rotation of the 12-Variable Problem

V_1:

	I	II	III	IV
1	.092	.846	.160	−.011
2	−.008	.707	.101	.049
3	.073	.652	.221	−.078
4	.400	.099	−.287	.451
5	.401	.157	−.180	.566
6	.214	.098	−.212	.613
7	.650	−.006	−.161	−.095
8	.689	.013	.076	.067
9	.788	.023	−.210	.081
10	.380	−.174	.506	.208
11	.436	−.197	.591	.348
12	.321	−.281	.522	.191

× Λ_2:

I	II	III	IV
1	0	0	0
0	.963	−.270	0
0	.270	.963	0
0	0	0	1

= V_2:

	I	II	III	IV
1	.092	.858	−.074	−.011
2	−.008	.708	−.093	.049
3	.073	.687	.037	−.078
4	.400	.018	−.303	.451
5	.401	.103	−.216	.566
6	.214	.037	−.231	.613
7	.650	−.049	−.153	−.095
8	.689	.033	.070	.067
9	.788	−.034	−.208	.081
10	.380	−.031	.534	.208
11	.436	−.030	.622	.348
12	.321	−.130	.578	.191

The effect of all the rotations performed by the many orthogonal matrices, one pair of factors at a time, can be achieved through one transformation matrix Λ, which is found as follows:

$$\Lambda = \Lambda_1 \Lambda_2 \cdots \Lambda_q \tag{5.6}$$

where Λ_1 is the first orthogonal transformation matrix that performs the first rotation, Λ_2 is the second orthogonal transformation matrix that performs the second rotation, and so on, with Λ_q being the last transformation matrix that effects the last orthogonal rotation. Then

$$\mathbf{V} = \mathbf{A}\Lambda \tag{5.7}$$

where $\mathbf{V}$ is the final orthogonal rotated matrix of factor loadings, $\mathbf{A}$ is the unrotated matrix of extracted factor loadings, and Λ is the orthogonal transformation matrix that does all the rotations at once. The matrix Λ is orthogonal itself because it is a product of orthogonal matrices [see Eq. (5.6)]. The product of $\mathbf{V}$ by its transpose gives

$$\begin{aligned} \mathbf{VV}' &= (\mathbf{A}\Lambda)(\mathbf{A}\Lambda)' = \mathbf{A}(\Lambda\Lambda')\mathbf{A}' \\ \mathbf{VV}' &= \mathbf{AIA}' = \mathbf{AA}' \end{aligned} \tag{5.8}$$

Equation (5.8) shows that the rotated matrix $\mathbf{VV}'$ reproduces the $\mathbf{R}$ matrix just as well as $\mathbf{AA}'$ does; hence the matrix $\mathbf{V}$ is mathematically equivalent to the matrix $\mathbf{A}$ in this sense.

The matrix $\mathbf{V}_2$ contains factors that have each been rotated once. New plots of these factors must be made to find more rotations to perform in the quest for better positive manifold and simple structure. Since two of the possible rotations have just been carried out, that is, I with IV and II with III, it is unnecessary to make these two plots in the next series of plots since these pairs will not be chosen for the next rotation. It is necessary, therefore, to make only four plots, I with II, I with III, II with IV, and III with IV. These plots are shown in Fig. 5.3. The loadings that provide the coordinates of the points for the plots in Fig. 5.3 are obtained from matrix $\mathbf{V}_2$ in Table 5.3.

Inspection of the second series of plots in Fig. 5.3 indicates that the best rotation is factor III toward factor IV by an angle of 21°36′. This rotation clearly improves both simple structure and positive manifold simultaneously. Factor I could be rotated away from factor III and factor III toward factor I, but this would involve factor III for which a more clear-cut rotation is available with factor IV. The matrix operations for this third rotation are shown in Table 5.4. The plots of factors I with II and II with IV in Fig. 5.3 offer no significant opportunities for improving either simple structure or positive manifold.

After the rotation of factor III with factor IV, new plots must be made for these two factors with factors I and II. The plot for factors I and II in Fig. 5.3 is still current and offers no rotational opportunities. The new plots are shown in

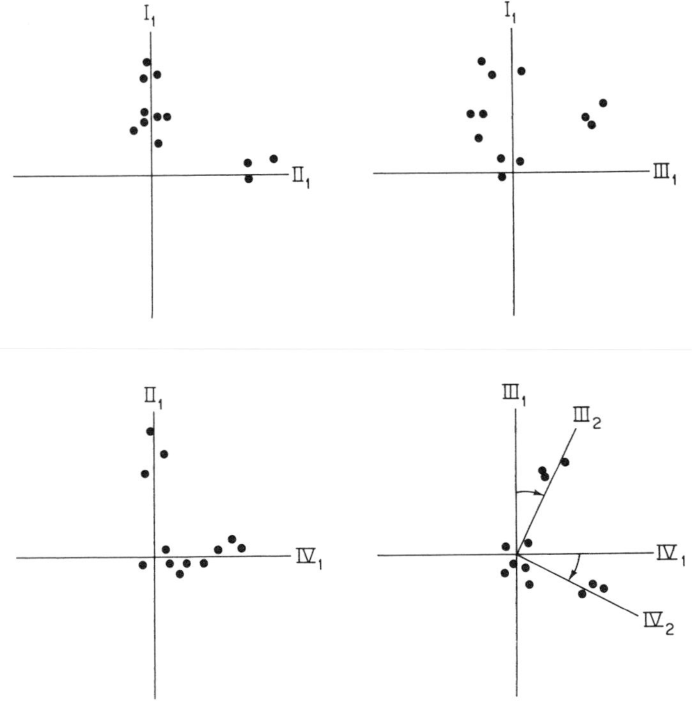

FIG. 5.3. Second set of plots in orthogonal rotation of the 12-variable problem.

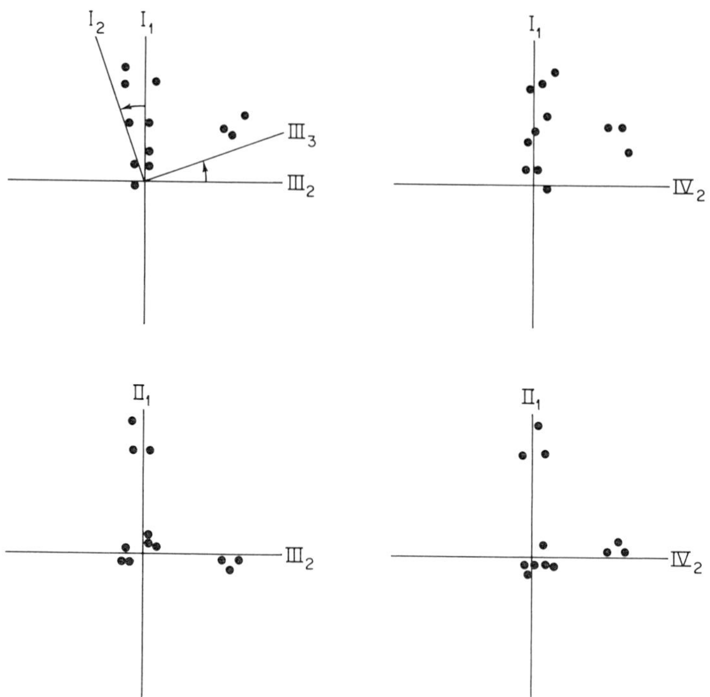

FIG. 5.4. Third set of plots in orthogonal rotation of the 12-variable problem.

Fig. 5.4. The loadings for these plots were taken from matrix $\mathbf{V}_3$ in Table 5.4. Pairs II, III and II, IV in Fig. 5.4 offer no rotational opportunities. The plot of I with III is better than the plot for I with IV from the standpoint of simultaneously improving both simple structure and positive manifold. In addition, I has not been rotated previously with III, whereas I and IV have already been rotated once. For these reasons, factor I is rotated away from factor III, by an angle of 19°17′, to make the fourth rotation.

Although the rotation of factor I away from factor III improves positive manifold by eliminating the small negative loadings on factor III, the improvement in simple structure is less marked. A position for factor I in between I_2 and I_1 would give a better hyperplane for factor III, but at the expense of making the three-point cluster to the right of the plot further from the hyperplane of factor I. On the other hand, if factor III were placed right through the three-point cluster just above the present location of III_3 in Fig. 5.4, the hyperplane for factor I would be improved, but since this would force I_2 even further counterclockwise than it is now, the hyperplane for III would be made worse.

TABLE 5.4

Third Orthogonal Rotation of the 12-Variable Problem

$\mathbf{V}_2$

	I	II	III	IV
1	.092	.858	−.074	−.011
2	−.008	.708	−.093	.049
3	.073	.687	.037	−.078
4	.400	.018	−.303	.451
5	.401	.103	−.216	.566
6	.214	.037	−.231	.613
7	.650	−.049	−.153	−.095
8	.689	.033	.070	.067
9	.788	−.034	−.208	.081
10	.380	−.031	.534	.208
11	.436	−.030	.622	.348
12	.321	−.130	.578	.191

×

Λ_3

I	II	III	IV
1	0	0	0
0	1	0	0
0	0	.930	−.368
0	0	.368	.930

=

$\mathbf{V}_3$

	I	II	III	IV
1	.092	.858	−.073	.017
2	−.008	.708	−.069	.080
3	.073	.687	.006	−.086
4	.400	.018	−.116	.531
5	.401	.103	.008	.605
6	.214	.037	.011	.655
7	.650	−.049	−.177	−.032
8	.689	.033	.089	.036
9	.788	−.034	−.164	.152
10	.380	−.031	.573	−.003
11	.436	−.030	.707	.094
12	.321	−.130	.608	−.035

The rotation actually chosen for I with III in Fig. 5.4 is a compromise solution, which tries to get both hyperplanes as close as possible to the lines of maximum point density. In rotating to simple structure, the object is to get as many points in the hyperplanes as possible rather than to put the factors through clusters of points. The factor locations are determined by the hyperplane positioning rather than vice versa. The failure to get the hyperplanes right through the lines of maximum point density with factors I and III, as shown in Fig. 5.4, is due to the fact that some correlation between factors I and III is indicated by the pattern of points. When factors are correlated, an orthogonal rotated solution can only approach simple structure; it cannot reach it. Thus, the present orthogonal solution will be taken as the best approximation to simple structure that can be achieved without going to an oblique solution. It is only by allowing the factor axes to depart from orthogonality, however, that the requirements for simple structure can ordinarily be completely satisfied. The methods of carrying out oblique rotations where the factors are allowed to depart from orthogonal positions are described and illustrated in the next chapter. There, further oblique rotations are performed on this 12-variable sample problem to bring it to a final oblique simple structure position.

Computations for the fourth rotation, factor I with III, are shown in Table 5.5. After this rotation, it is necessary to replot the remaining factors with the new factors I and III. Using the factor loadings in matrix $\mathbf{V}_4$ of Table 5.5 leads to the plots for the new I and III with the other factors shown in Fig. 5.5. Inspection of these plots reveals that no improvements in simple structure and positive manifold can be attained with plots I with II, II with III, and III with IV. The plot of I with IV, however, affords a good rotation. Factor I is rotated away from factor IV by an angle of 14°2′, thereby improving the

TABLE 5.5

Fourth Orthogonal Rotation of the 12-Variable Problem

	I	II	III	IV
1	.092	.858	−.073	.017
2	−.008	.708	−.069	.080
3	.073	.687	.006	−.086
4	.400	.018	−.116	.531
5	.401	.103	.008	.605
6	.214	.037	.011	.655
7	.650	−.049	−.177	−.032
8	.689	.033	.089	.036
9	.788	−.034	−.164	.152
10	.380	−.031	.573	−.003
11	.436	−.030	.707	.094
12	.321	−.130	.608	−.035

$\mathbf{V}_3$

×

I	II	III	IV
.944	0	.330	0
0	1	0	0
−.330	0	.944	0
0	0	0	1

$\mathbf{\Lambda}_4$

=

I	II	III	IV
.111	.858	−.039	.017
.015	.708	−.068	.080
.067	.687	.029	−.086
.416	.018	.023	.531
.377	.103	.139	.605
.199	.037	.081	.655
.672	−.049	.047	−.032
.621	.033	.312	.036
.798	−.034	.105	.152
.169	−.031	.667	−.003
.179	−.030	.811	.094
.102	−.130	.680	−.035

$\mathbf{V}_4$

balancing of the two hyperplanes. Prior to the rotation, the hyperplane for factor IV was well located, but the hyperplane for I was too far from the three points above it. By the indicated rotation, the hyperplane for IV is made worse to gain some improvement in the positioning of the hyperplane for I. Had the first rotation of I with IV been better chosen, this rerotation of I with IV would not have been necessary. In practice, however, such rerotations are frequently necessary before the final positions for the factors are located. The computations for this final orthogonal rotation are shown in Table 5.6.

In Fig. 5.5 the plot of factor I with IV shows that simple structure is not well achieved with this final orthogonal positioning of these two factors. Factors I and III also are not close to simple structure when plotted with each other, as shown in Fig. 5.4. This fifth orthogonal rotation of factor I with factor IV, however, takes the solution about as close to simple structure as it can get within the framework of orthogonal rotations. Oblique rotations of I and IV with each other and also I and III with each other in the next chapter will improve the extent to which simple structure is achieved in the solution.

The same final solution as that in matrix $\mathbf{V}_5$, in Table 5.6, may be obtained a different way. First, multiply all the orthogonal rotation matrices together as follows:

$$\mathbf{\Lambda} = \mathbf{\Lambda}_1 \mathbf{\Lambda}_2 \mathbf{\Lambda}_3 \mathbf{\Lambda}_4 \mathbf{\Lambda}_5 \tag{5.9}$$

Once this $\mathbf{\Lambda}$ matrix, which is the product of all the one-rotation $\mathbf{\Lambda}_i$ matrices, is obtained, the final rotated matrix $\mathbf{V}$ is obtained as follows:

$$\mathbf{V} = \mathbf{A}\mathbf{\Lambda} \tag{5.10}$$

The computations for this operation are shown in Table 5.7. The final $\mathbf{V}$

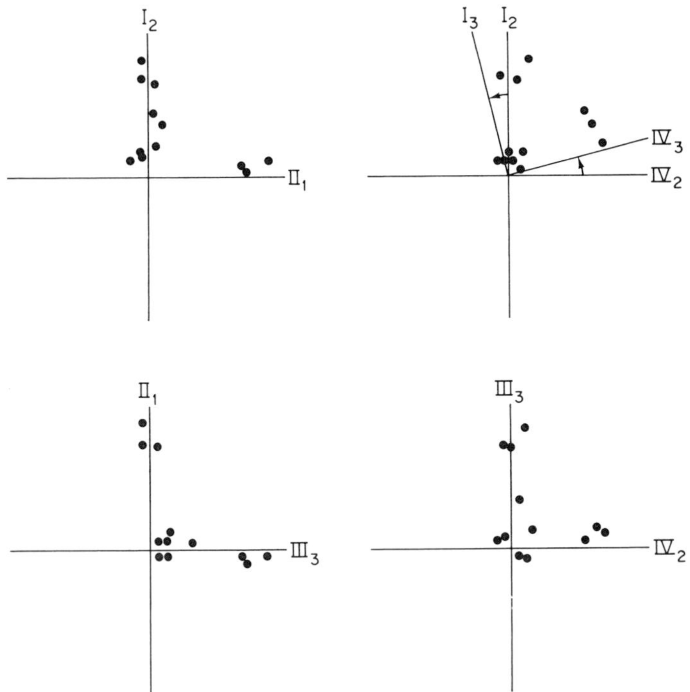

FIG. 5.5. Fourth set of plots in orthogonal rotation of the 12-variable problem.

matrix in Table 5.7 is seen to be identical with the matrix $\mathbf{V}_5$, in Table 5.6, which was the end result of the long series of rotations, modifying the matrix of rotated loadings by one rotation at a time. Equation (5.10) permits the execution of the rotations all at once and provides a matrix $\mathbf{\Lambda}$, which gives a final summarization of all the rotations that were performed.

In Fig. 5.6, the plots of the four rotated orthogonal factors are shown with the small fifth minimum residual factor, shown in Table 5.1. There is no way to build this factor up to an important separate factor except by splitting the variance for variables 7, 8, and 9 between factor I and this new factor, V. Factor V may be treated as a minor factor based on some surplus overlapping variance between variables 7 and 9 not shared as much with variable 8; or by rotating factor I slightly toward factor V and V slightly away from factor I, most of the variance on V can be shifted to factor I. This would reduce factor V to a residual factor with very little variance. A small rotation of IV toward V and V away from IV would complete the residualization of factor V by transferring further variance from factor V to factor IV. A four-factor solution, therefore, seems well justified for this 12-variable problem.

TABLE 5.6. Fifth Orthogonal Rotation of the 12-Variable Problem

	I	II	III	IV
1	.111	.858	−.039	.017
2	.015	.708	−.068	.080
3	.067	.687	.029	−.086
4	.416	.018	.023	.531
5	.377	.103	.139	.605
6	.199	.037	.081	.655
7	.672	−.049	.047	−.032
8	.621	.033	.312	.036
9	.798	−.034	.105	.152
10	.169	−.031	.667	−.003
11	.179	−.030	.811	.094
12	.102	−.130	.680	−.035

$\mathbf{V}_4$

×

I	I	III	IV
.970	0	0	.242
0	1	0	0
0	0	1	0
−.242	0	0	.970

$\mathbf{\Lambda}_5$

=

I	II	III	IV
.103	.858	−.039	.043
−.005	.708	−.068	.080
.086	.687	.029	−.067
.275	.018	.023	.616
.219	.103	.139	.678
.034	.037	.081	.683
.660	−.049	.047	.133
.593	.033	.312	.186
.737	−.034	.105	.341
.165	−.031	.667	.038
.150	−.030	.811	.135
.107	−.130	.680	−.010

$\mathbf{V}_5$

TABLE 5.7. Orthogonal Rotational Solution to the 12-Variable Problem

	I	II	III	IV
1	.080	.846	.160	−.046
2	.011	.707	.101	.048
3	.037	.652	.221	−.100
4	.544	.099	−.287	.261
5	.589	.157	−.180	.366
6	.435	.098	−.212	.482
7	.563	−.006	−.161	−.339
8	.661	.013	.076	−.205
9	.758	.023	−.210	−.230
10	.431	−.174	.506	.045
11	.537	−.197	.591	.152
12	.370	−.281	.522	.052

$\mathbf{A}$

×

I	II	III	IV
.711	.000	.439	.549
.057	.963	−.237	.116
−.201	.270	.845	−.416
−.671	.000	.192	.716

$\mathbf{\Lambda}$

=

I	II	III	IV
.103	.858	−.039	.043
−.005	.708	−.068	.080
.086	.687	.029	−.067
.275	.018	.023	.616
.219	.103	.139	.678
.034	.037	.081	.683
.660	−.049	.047	.133
.593	.033	.312	.186
.737	−.034	.105	.341
.165	−.031	.667	.038
.150	−.030	.811	.135
.107	−.130	.680	−.010

$\mathbf{V}$

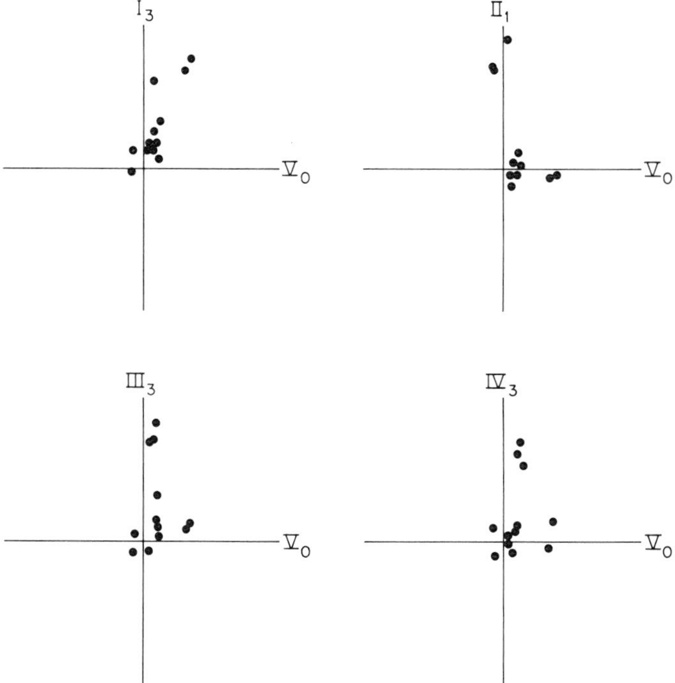

FIG. 5.6. Plots of final four factors with fifth unrotated factor.

In Table 5.1, the variable names and hypothetical factor identifications were given for each variable. The unrotated matrix in Table 5.1 shows some tendency for variables to cluster into factors according to the hypothesized factor structure, but the rotated factor matrix **V** in Table 5.7 shows this tendency to a much more marked degree. In the rotated solution in matrix **V** of Table 5.7, factor I corresponds to Extraversion versus Introversion (E), factor II is Social Conformity versus Rebelliousness (C), factor III is Empathy versus Egocentrism (P), and factor IV is Emotional Stability versus Neuroticism (S).

Matrix **V** represents a rotational solution based on "blind" rotations to the best orthogonal simple structure that can be achieved, using positive manifold as a supplementaıy aid to guide the rotations. The rotations are described as "blind" because the plots showed only points without identifying variable numbers. That is, the rotational decisions were made on the basis of the configuration of points in the plots and not on the basis of the experimenter's knowledge of the variables or his hypotheses regarding their factor structure. Cattell (1952) has been a prominent advocate of this particular method of carrying out rotations.

CHAPTER SIX

Oblique Hand Rotations

In the previous chapter, blind hand rotation to orthogonal simple structure was illustrated for the 12-variable sample problem through the inspection of orthogonal factor plots, two factors at a time. In this chapter, the process is extended to the case of nonorthogonal factors. Plots of factors are examined, two at a time, to determine if any oblique rotations are called for, that is, if the factor axes should depart from orthogonal positions.

[6.1] OBLIQUE FACTOR AXES

The extent to which simple structure could be achieved with orthogonal hand rotation was limited by the restriction that the factors had to be kept at right angles to each other. It was necessary, for example, in the rotation of factor I with factor III, Fig. 5.4, to position the factors so that the clusters of points lay near the hyperplanes, but not actually on the lines representing the hyperplanes. This type of situation is illustrated in Fig. 6.1. If the factor axes are kept orthogonal to one another, points 1, 2, and 3 cannot be brought into the hyperplane of factor II at the same time points 4, 5, and 6 are brought into the hyperplane of factor I. One cluster or the other may be brought into the

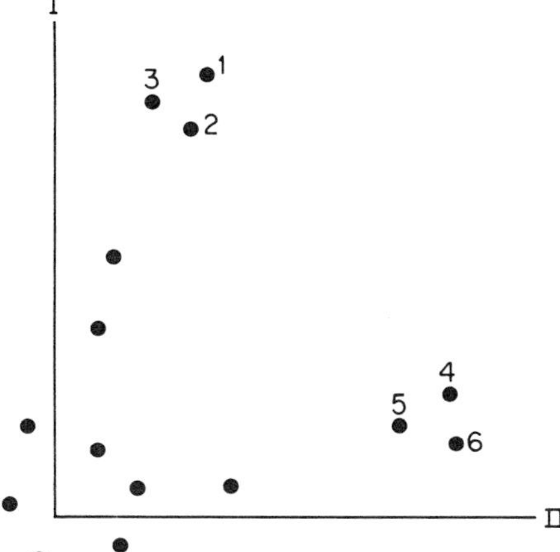

FIG. 6.1. Plot of factors requiring oblique rotation.

hyperplane of one of the factors, but only at the cost of putting the other cluster farther from the hyperplane nearest to it. If the factors are allowed to go to oblique (that is, nonorthogonal) positions, however, both clusters can be brought into the hyperplanes nearest to them. The proper oblique positions of these factors are shown in Fig. 6.2 with the angle φ between them. Now, the data vectors with termini in points 4, 5, and 6 have near-zero loadings on factor I′, and the points 1, 2, and 3 have near-zero loadings on factor II′. For the table of oblique factor loadings, variables 1, 2, and 3 will have essentially zero values on factor II, whereas in the table of orthogonal factor loadings, the loadings were not large but were definitely nonzero. The same situation applies for variables 4, 5, and 6 with respect to factor I.

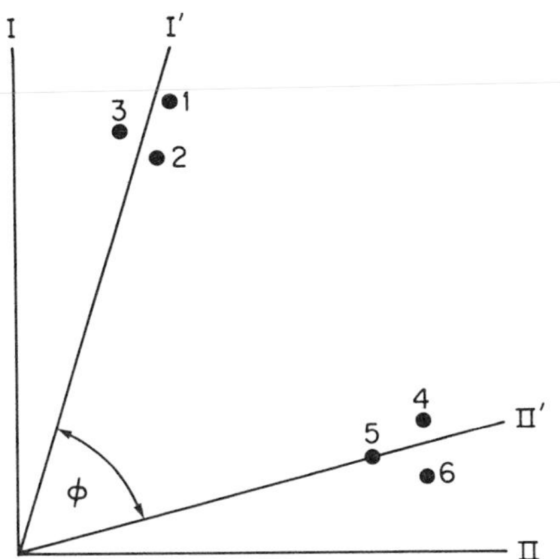

FIG. 6.2. Oblique factor positions.

Thus, by rotating factors to oblique positions, it is possible to place the factor axes so that simple structure is more closely approximated if the clusters of points are not orthogonal to one another. For this reason, those who place great faith in the value of blind rotation to simple structure prefer to rotate obliquely, thereby enabling them to obtain a closer adherence to simple structure criteria.

[6.1.1] *COORDINATES AND PROJECTIONS*

In Fig. 6.2, the loadings of the points 4, 5, and 6 are essentially zero on factor axis I′ because the loadings are the distances from the origin to the points where lines from points 4, 5, and 6 *parallel* to factor axis II intersect with factor axis I. The loadings are *not* determined by the distances from the origin to the points where lines from points 4, 5, and 6 *perpendicular* to factor axis I′ intersect with factor axis I′. These perpendicular projections of the vectors 4, 5, and 6 on factor axis I′ are referred to as "projections" rather than as "factor loadings" and are equal to the correlations of these data vectors with the factor vector I′. The factor loadings, on the other hand, are the coordinates of vectors 4, 5, and 6 with respect to factor axis I′, values which are not equal to the correlations between the data vectors and the factor axis.

These various relationships are illustrated by a single data vector P_1 in Fig. 6.3, a pair of orthogonal factor axes F_1 and F_2, and a pair of oblique factor axes F_1' and F_2'. The coordinates of the data vector P_1 with respect to the orthogonal factor axes F_1 and F_2, respectively, are .7 and .5, indicated in parentheses after P_1. The coordinates of the factor vectors with respect to vectors F_1 and F_2 as coordinate axes are also given in parentheses following F_1 and F_2 as well as F_1' and F_2'.

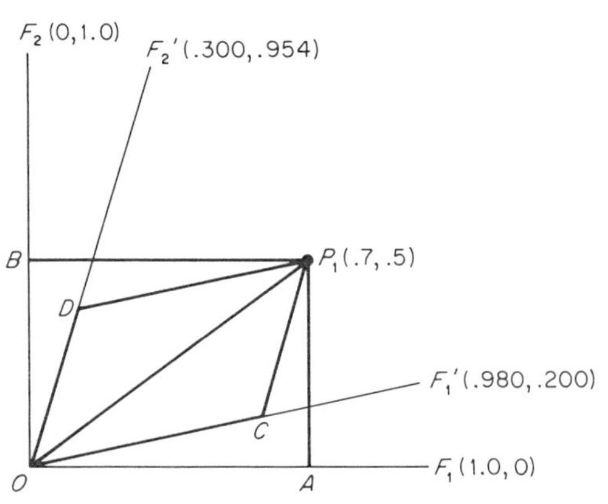

FIG. 6.3. Oblique coordinates.

When it is stated that the coordinates of the data vector P_1 are (.7, .5) with respect to the factor vectors F_1 and F_2, the following is meant:

$$P_1 = .7F_1 + .5F_2 = .7\begin{pmatrix}1\\0\end{pmatrix} + .5\begin{pmatrix}0\\1\end{pmatrix} = \begin{pmatrix}.7\\0\end{pmatrix} + \begin{pmatrix}0\\.5\end{pmatrix} = \begin{pmatrix}.7\\.5\end{pmatrix} \tag{6.1}$$

Thus, .7 is multiplied by the coordinates of the vector F_1 and added to .5 multiplied by the coordinates of the vector F_2 to get vector P_1. Thus, P_1 is expressed as a linear combination of the two factor vectors F_1 and F_2, that is, $P_1 = .7F_1 + .5F_2$. The factor vectors F_1 and F_2 are said to provide a "basis" for the two-dimensional space since any vector in the plane can be expressed as a linear combination of these two vectors. Even F_1 and F_2 can be expressed as a linear combination of the same basis vectors, that is, themselves:

$$\begin{aligned} F_1 &= 1\begin{pmatrix}1\\0\end{pmatrix} + 0\begin{pmatrix}0\\1\end{pmatrix} = \begin{pmatrix}1\\0\end{pmatrix} \\ F_2 &= 0\begin{pmatrix}1\\0\end{pmatrix} + 1\begin{pmatrix}0\\1\end{pmatrix} = \begin{pmatrix}0\\1\end{pmatrix} \end{aligned} \tag{6.2}$$

or, $F_1 = 1F_1 + 0F_2$ and $F_2 = 0F_1 + 1F_2$. The coordinates, therefore, for F_1 and F_2 are (1, 0) and (0, 1), respectively, with respect to basis vectors F_1 and F_2, as shown in Fig. 6.3. Since every vector in the plane can be expressed as a linear combination of the two factor vectors F_1 and F_2, the vectors F_1 and F_2 are said to "span" the space of two dimensions represented by this plane.

The coordinates of the vector P_1 with respect to the two orthogonal basis vectors F_1 and F_2 are equal to the perpendicular projections of the vector P_1 on the two factor vectors. Thus, vector coordinates and perpendicular projections of the vectors on the coordinate axes are the same as long as the coordinate axes are at right angles to one another.

It is not necessary, however, to give the coordinates of the data vector P_1 with respect to orthogonal factors such as F_1 and F_2. It is also possible to give the coordinates of vector P_1 with respect to vectors F_1' and F_2', which are not orthogonal. The angle between F_1' and F_2' is not equal to 90°; so these vectors are said to represent "oblique" factor axes.

Both F_1' and F_2' may be expressed as linear combinations of the orthogonal basis vectors F_1 and F_2 as follows:

$$\begin{aligned} F_1' &= .980\begin{pmatrix}1\\0\end{pmatrix} + .200\begin{pmatrix}0\\1\end{pmatrix} = .980F_1 + .200F_2 \\ F_2' &= .300\begin{pmatrix}1\\0\end{pmatrix} + .954\begin{pmatrix}0\\1\end{pmatrix} = .300F_1 + .954F_2 \end{aligned} \tag{6.3}$$

The vector P_1 may be expressed as a linear combination of the oblique factor vectors F_1' and F_2' by the equation $P_1 = .592F_1' + .400F_2'$ as follows:

$$P_1 = .592\begin{pmatrix} .980 \\ .200 \end{pmatrix} + .400\begin{pmatrix} .300 \\ .954 \end{pmatrix} = \begin{pmatrix} .120+.580 \\ .118+.382 \end{pmatrix} = \begin{pmatrix} .7 \\ .5 \end{pmatrix} \tag{6.4}$$

Thus, Eq. (6.4) shows that the coordinates of vector P_1 with respect to the basis vectors F_1 and F_2 can be expressed in terms of the coordinates of P_1 with respect to the factor vectors F_1' and F_2' *and* the coordinates of F_1' and F_2' with respect to the basis vectors F_1 and F_2. The values .592 and .400 are the coordinates (.592, .400) of the vector P_1 with respect to the two oblique factor axes F_1' and F_2'. These two values were computed readily by letting X and Y be the unknown coordinates. Then

$$X\begin{pmatrix} .980 \\ .200 \end{pmatrix} + Y\begin{pmatrix} .300 \\ .954 \end{pmatrix} = \begin{pmatrix} .7 \\ .5 \end{pmatrix} \tag{6.5}$$

This gives two equations in two unknowns:

$$\begin{aligned} .980X + .300Y &= .7 \\ .200X + .954Y &= .5 \end{aligned} \tag{6.6}$$

Solving these two linear equations simultaneously gives $X = .592$ and $Y = .400$, the coordinates of P_1 with respect to F_1' and F_2'. The coordinates of any other vector in the plane can also be determined with respect to factor axes F_1' and F_2' if their coordinates with respect to F_1 and F_2 are known. Thus, any vector in this space of two dimensions can be expressed as a linear combination of the vectors F_1' and F_2'; so these two oblique factor vectors also constitute an adequate set of basis vectors. They, too, span this two-dimensional space.

The coordinates of a data vector with respect to two oblique factor vectors cannot be obtained, however, by dropping perpendiculars to the factor axes and measuring the distance from the origin to these points of intersection. In Fig. 6.3, the line OC has a length equal to .592, the coordinate of vector P_1 with respect to the basis vector F_1'. The line OD has a length of .4 which is equal to the coordinate of P_1 with respect to the basis vector F_2'. The lines OD and P_1C are parallel as are the lines DP_1 and OC. This can be shown by demonstrating that the line CP_1 equals the line OD. This can be proved by showing that the angle COP_1 calculated from the assumption that $CP_1 = OD$ equals the angle COP_1 calculated from the scalar product [see Eq. (2.34)] of vectors P_1 and F_1'. Calculating the scalar product of vectors P_1 and F_1' gives

$$(.7\ .5)\begin{pmatrix} .980 \\ .200 \end{pmatrix} = l_1 l_2 \cos\varphi \tag{6.7}$$

where

$l_1 = \sqrt{(.7)^2 + (.5)^2} = \sqrt{.74}$ is the length of vector P_1,
$l_2 = \sqrt{(.980)^2 + (.200)^2} = 1.00$ is the length of vector F_1', and
$\cos\varphi$ is the cosine of the angle COP_1 in Fig. 6.3.

Substituting these values in Eq. (6.7) and dividing both sides by $l_1\, l_2$ gives

$$\cos\varphi = \frac{(.7)(.980)+(.5)(.200)}{l_1\, l_2} = \frac{.686+.100}{(\sqrt{.74})(1.00)} = \frac{.786}{\sqrt{.74}}$$

The cosine of φ can also be obtained using the law of cosines:

$$(6.8) \qquad \cos\varphi = \frac{OC^2+OP_1{}^2-CP_1{}^2}{2OC\cdot OP_1} = \frac{(.592)^2+.74-(.400)^2}{2(.592)\cdot\sqrt{.74}} = \frac{.786}{\sqrt{.74}}$$

Since the angle COP_1 has the same cosine computed by both Eqs. (6.7) and (6.8), the asserted proposition is established. This means that to obtain the coordinates of a vector P_1 with respect to two factor axes F_1' and F_2', oblique to one another, draw lines through the terminus of P_1 parallel to F_1' and F_2', respectively. The distances from the origin to the points where these parallels intersect the coordinate factor axes F_2' and F_1', respectively, are the coordinates of the vector P_1 with respect to these oblique factor axes.

Dropping a perpendicular line from the point P_1 in Fig. 6.3 to the vector F_1' does *not*, then, give the coordinate of vector P_1 with respect to factor F_1'. Instead, it gives the correlation of the data vector P_1 with the factor vector F_1'. From Eq. (2.35) the following relationship holds:

$$(6.9) \qquad r_{12} = h_1\, h_2 \cos\varphi$$

where r_{12} is the correlation between vectors 1 and 2, h_1 and h_2 are the lengths of vectors 1 and 2, and $\cos\varphi$ is the cosine of the angle between the vectors. In Fig. 6.3, the cosine of the angle between vector P_1 and factor F_1' is given by X/OP_1, where X is the distance from the origin to the point where a perpendicular projection from P_1 to the vector F_1' would intersect the vector F_1'. Therefore, $X = OP_1\cos\varphi$, where $\cos\varphi$ is the cosine of angle COP_1. The length of vector F_1' is $\sqrt{(.980)^2+(.200)^2} = 1.0$. Substituting these values in Eq. (6.9) gives the result that $X = r_{12}$, where vector 1 is factor vector F_1', vector 2 is data vector P_1, and r_{12} is the correlation between the two; X is, as before, the perpendicular projection of data vector P_1 on factor vector F_1'.

[6.1.2] *COORDINATES AND FACTOR LOADINGS*

The coordinates of a data vector with respect to oblique factor axes are referred to as the "factor loadings" for the data variable with respect to those factors. The term "factor saturations" is used by some authors. The matrix of these factor loadings for all the data variables with respect to the factors is called the factor "pattern" matrix by Harman (1967). The matrix of correlations of the data vectors with the oblique factor axes, the perpendicular projections of the data vectors onto the oblique factor axes, is called by Harman the factor "structure." The structure matrix contains the correlations of each data variable with each of the oblique factors. A complete oblique factor solution requires the determination of both the pattern matrix and the structure matrix.

In the orthogonal factor model, the factor axes are all at right angles to one another. In this case, the coordinates of the data vectors with respect to the factor axes and the perpendicular projections of the data vectors onto these factor vectors coincide. Thus, with orthogonal factors, the pattern matrix and the structure matrix are identical. This permits the interpretation of the factor loadings of a data variable with respect to orthogonal factors as though they were correlation coefficients. With oblique factors, the factor loadings cannot be interpreted in this manner. Only elements of the structure matrix may be interpreted as correlations with the oblique factors.

[6.2] REFERENCE VECTOR STRUCTURE

The ultimate objective in oblique rotations is to obtain the data-variable loadings on the oblique factors, that is, the pattern matrix **P**, after the factors have been rotated to a simple structure position. When oblique rotations are being carried out through inspection of plots of factors two at a time, however, it is more convenient to work initially with the "reference vector structure." Once the reference vector structure is found, the final solution matrices may be obtained: the pattern matrix **P**, the structure matrix **S**, and the matrix of correlations among the factors, **Φ**.

The reference vector structure consists of perpendicular projections, or correlations, of the data-variable vectors onto what Cattell (1952) calls "reference vectors." They are called "simple axes" by Thurstone (1947) and Fruchter (1954). The reference vector is illustrated in Fig. 6.4. This figure shows an oblique position for factor vectors I and II in which simple structure criteria are well met. In this plot, vector I is the trace of the hyperplane for factor II and vector II is the trace of the hyperplane for factor I.

If Fig. 6.4 were presented as a plot of orthogonal factors I and II′, leaving out of the picture the axes I′ and II, then vector I would be the trace of the hyperplane, or just "the hyperplane," for factor II′, and II′ would be the

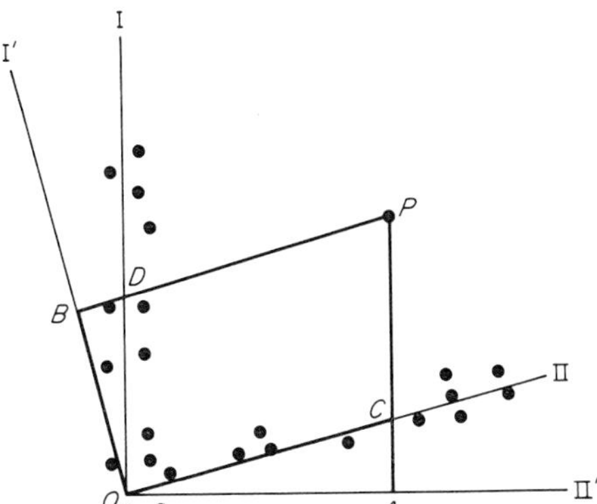

FIG. 6.4. Reference vectors: I′ and II′ are reference vectors; I and II are factor axes.

hyperplane for factor I. In this case factor II′ would be well positioned because its hyperplane lies along the path of maximum point density. Factor I, on the other hand, would be poorly positioned because its hyperplane, vector II′, does not lie along the line of maximum point density. Vector II would be a more appropriate hyperplane for factor I.

The reference vector associated with a given factor is the line from the origin perpendicular to the hyperplane for that factor. In the orthogonal case, with factors I and II′ in Fig. 6.4, the reference vector I and the factor vector I coincide, as do reference vector II′ and factor vector II′. This is because the factors are at right angles to one another and hence the hyperplanes are at right angles to each other in the orthogonal factor solution. With the oblique factor positions I and II in Fig. 6.4, however, vector I′ is the reference vector for factor I since it is perpendicular to the hyperplane for factor I, vector II. Vector II′ is the reference vector for factor II since it is perpendicular to the hyperplane for factor II, vector I.

If only an orthogonal rotation were to be done for the plot in Fig. 6.4, factor I would be rotated a bit away from factor II′, about half-way between the present I and I′, and factor II′ would be moved toward I′ an equivalent amount to maintain orthogonality. This would balance off the two hyperplanes, making the hyperplane for I a bit better and the hyperplane for II′ a bit worse.

By an oblique rotation in Fig. 6.4, however, it is possible to leave the hyperplane for II′ undisturbed, moving only the hyperplane for factor I. This is accomplished by rotating the reference vector for I to position I′, pulling the hyperplane at right angles to the reference vector to position II where it lies along the path of maximum point density. Vector II′ is now the reference vector for factor II, perpendicular to the hyperplane for factor II which is vector I.

The perpendicular projections of the data vectors on these reference vectors I′ and II′ are found by dropping perpendiculars from the data points to these axes and measuring the distances from the origin to the points of intersection. In Fig. 6.4, for example, OB is the projection of data vector OP on reference vector I′ and OA is the projection of data vector OP on reference vector II′. These projections are also the correlations of the data vectors with the reference vectors.

If the orthogonal factor positions I and II′ were being rotated in Fig. 6.4 to new orthogonal positions I′ and II, respectively, the rotation would be carried out as follows:

$$\underset{\mathbf{V}_i}{\overset{\text{I} \quad \text{II}'}{\begin{bmatrix} a_{11} & a_{12} \\ a_{21} & a_{22} \\ \vdots & \vdots \\ a_{n1} & a_{n2} \end{bmatrix}}} \times \underset{\mathbf{\Lambda}_{i+1}}{\begin{bmatrix} \cos\varphi & \sin\varphi \\ -\sin\varphi & \cos\varphi \end{bmatrix}} = \underset{\mathbf{V}_{i+1}}{\overset{\text{I}' \quad \text{II}}{\begin{bmatrix} a'_{11} & a'_{12} \\ a'_{21} & a'_{22} \\ \vdots & \vdots \\ a'_{n1} & a'_{n2} \end{bmatrix}}} \tag{6.10}$$

In this case, the column of values I′ in matrix $\mathbf{V}_{i+1}$ would be perpendicular projections of the data vectors on orthogonal factor vector I′, and the values in column II of matrix $\mathbf{V}_{i+1}$ would be the perpendicular projections of the data vectors on orthogonal factor vector II.

In doing an oblique rotation of reference vector I to position I′ in Fig. 6.4, however, the reference vector for factor II is not moved. The projections of the data vectors on reference vector II′ would remain unchanged. Only the projections on reference vector I would change. The projections of the data vectors on reference vector I′ are the same as those found in column I′ of the matrix $\mathbf{V}_{i+1}$ in Eq. (6.10). Rotating one reference vector, then, is like doing half an orthogonal rotation, using only half of the $\mathbf{\Lambda}_i$ transformation matrix. Multiplying matrix $\mathbf{V}_i$ in Eq. (6.10) by the left column of matrix $\mathbf{\Lambda}_{i+1}$ produces the new column I′ in matrix $\mathbf{V}_{i+1}$. If the right-hand column of $\mathbf{\Lambda}_{i+1}$ were 0 in λ_{12} and 1 in λ_{22}, then column II of $\mathbf{V}_{i+1}$ would be the same as column II′ of matrix $\mathbf{V}_i$.

Carrying out an oblique rotation of one reference vector, then, involves changing one column of the transformation matrix. To illustrate this operation, suppose that orthogonal rotations for a given problem had been carried as far as possible and had given the results $\mathbf{V}_0 = \mathbf{A}\mathbf{\Lambda}_0$, where $\mathbf{V}_0$ is the final orthogonal rotated matrix, $\mathbf{A}$ is the matrix of unrotated loadings, and $\mathbf{\Lambda}_0$ is the cumulated transformation matrix that carries out all the orthogonal rotations at once. Then, consider the rotation of reference vector I to position I′ in Fig. 6.4 to be the first oblique rotation to be performed. Assume that there are four factors. The rotation can be performed by obtaining a new $\mathbf{\Lambda}$ matrix with an altered column 1 to reflect the oblique rotation of reference vector I. This matrix operation is shown in matrix equation (6.11).

$$
(6.11)\qquad \underset{\mathbf{\Lambda}_1}{\begin{bmatrix} \lambda'_{11} & \lambda'_{12} & \lambda'_{13} & \lambda'_{14} \\ \lambda'_{21} & \lambda'_{22} & \lambda'_{23} & \lambda'_{24} \\ \lambda'_{31} & \lambda'_{32} & \lambda'_{33} & \lambda'_{34} \\ \lambda'_{41} & \lambda'_{42} & \lambda'_{43} & \lambda'_{44} \end{bmatrix}} = \underset{\mathbf{\Lambda}_0}{\begin{bmatrix} \lambda_{11} & \lambda_{12} & \lambda_{13} & \lambda_{14} \\ \lambda_{21} & \lambda_{22} & \lambda_{23} & \lambda_{24} \\ \lambda_{31} & \lambda_{32} & \lambda_{33} & \lambda_{34} \\ \lambda_{41} & \lambda_{42} & \lambda_{43} & \lambda_{44} \end{bmatrix}} \times \underset{\mathbf{\Theta}}{\begin{bmatrix} \cos\varphi & 0 & 0 & 0 \\ -\sin\varphi & 1 & 0 & 0 \\ 0 & 0 & 1 & 0 \\ 0 & 0 & 0 & 1 \end{bmatrix}}
$$

The angle φ in matrix $\mathbf{\Theta}$ is the angle of rotation of I away from II′. The sine of φ has a negative sign because I is being rotated away from II′. If it were being rotated toward II′, the sine of φ would be positive. The multiplication of $\mathbf{\Lambda}_0$ by $\mathbf{\Theta}$ changes only column 1, leaving $\mathbf{\Lambda}_1$ and $\mathbf{\Lambda}_0$ identical except for the elements of the first columns. The elements of column 1 of $\mathbf{\Lambda}_1$ are given by

$$
(6.12)\qquad \begin{aligned} \lambda'_{11} &= \lambda_{11}\cos\varphi - \lambda_{12}\sin\varphi \\ \lambda'_{21} &= \lambda_{21}\cos\varphi - \lambda_{22}\sin\varphi \\ \lambda'_{31} &= \lambda_{31}\cos\varphi - \lambda_{32}\sin\varphi \\ \lambda'_{41} &= \lambda_{41}\cos\varphi - \lambda_{42}\sin\varphi \end{aligned}
$$

Squaring both sides of Eq. (6.12) and adding them up gives

$$
\begin{aligned} &\lambda'^2_{11} + \lambda'^2_{21} + \lambda'^2_{31} + \lambda'^2_{41} \\ &\quad = \lambda^2_{11}\cos^2\varphi + \lambda^2_{12}\sin^2\varphi - 2\lambda_{11}\lambda_{12}\sin\varphi\cos\varphi \\ &\qquad + \lambda^2_{21}\cos^2\varphi + \lambda^2_{22}\sin^2\varphi - 2\lambda_{21}\lambda_{22}\sin\varphi\cos\varphi \\ &\qquad + \lambda^2_{31}\cos^2\varphi + \lambda^2_{32}\sin^2\varphi - 2\lambda_{31}\lambda_{32}\sin\varphi\cos\varphi \\ &\qquad + \lambda^2_{41}\cos^2\varphi + \lambda^2_{42}\sin^2\varphi - 2\lambda_{41}\lambda_{42}\sin\varphi\cos\varphi \end{aligned}
$$

which simplifies to

$$
\begin{aligned} &\lambda'^2_{11} + \lambda'^2_{21} + \lambda'^2_{31} + \lambda'^2_{41} \\ &\quad = (\lambda^2_{11}+\lambda^2_{21}+\lambda^2_{31}+\lambda^2_{41})\cos^2\varphi + (\lambda^2_{12}+\lambda^2_{22}+\lambda^2_{32}+\lambda^2_{42})\sin^2\varphi \\ &\qquad - 2(\lambda_{11}\lambda_{12}+\lambda_{21}\lambda_{22}+\lambda_{31}\lambda_{32}+\lambda_{41}\lambda_{42})\sin\varphi\cos\varphi \end{aligned}
$$

Since the sums of squares of the columns of $\mathbf{\Lambda}_0$ equal 1.0 and the inner products of the columns of $\mathbf{\Lambda}_0$ equal zero, the preceding equation reduces to

$$
\lambda'^2_{11} + \lambda'^2_{21} + \lambda'^2_{31} + \lambda'^2_{41} = \cos^2\varphi + \sin^2\varphi - 2(0)\sin\varphi\cos\varphi = 1.0
$$

The first column of values in the $\mathbf{\Lambda}_1$ matrix, therefore, has a sum of squares equal to 1.0; that is, the column is "normalized." The λ_{i1} values in this column are the direction cosines of the rotated reference vector with respect to the unrotated factor axes in matrix **A**. Any such set of direction cosines must have a sum of squares equal to 1.0.

That the columns of the transformation matrix **Λ** give the direction cosines of the rotated orthogonal factors with the original unrotated factors can be shown as follows. First, it has been shown in the last chapter that multiplying **A** by **Λ** gives the correlations of the data vectors with the rotated orthogonal factors. For data variable 1 and rotated factor I, this would be

$$r_{1\text{I}} = a_{11}\lambda_{11} + a_{12}\lambda_{21} + \cdots + a_{1m}\lambda_{m1} \tag{6.13}$$

But by Eq. (2.35),

$$r_{ij} = h_i h_j \cos\varphi_{ij}$$

and since factor vector I has length 1.0,

$$r_{1\text{I}} = h_1 \cos\varphi_{1\text{I}} \tag{6.14}$$

The a_{1i} values in Eq. (6.13) are projections of the data vector 1 on the unrotated factor vectors. If the angle of the data vector 1 with unrotated factor vector I is φ, then $\cos\varphi = a_{11}/h_1$ where h_1 is the length of data vector 1 and $a_{11} = h_1\cos\varphi$. Letting δ_{1i} be the cosine of the angle of the data vector 1 with unrotated factor i, Eq. (6.13) can be rewritten as

$$r_{1\text{I}} = h_1\delta_{11}\lambda_{11} + h_1\delta_{12}\lambda_{21} + h_1\delta_{13}\lambda_{31} + \cdots + h_1\delta_{1m}\lambda_{m1} \tag{6.15}$$

$$r_{1\text{I}} = h_1(\delta_{11}\lambda_{11} + \delta_{12}\lambda_{21} + \delta_{13}\lambda_{31} + \cdots + \delta_{1m}\lambda_{m1}) \tag{6.16}$$

Using Eqs. (6.14) and (6.16) and canceling h_1 on both sides gives

$$\cos\varphi_{1\text{I}} = \delta_{11}\lambda_{11} + \delta_{12}\lambda_{21} + \delta_{13}\lambda_{31} + \cdots + \delta_{1m}\lambda_{m1} \tag{6.17}$$

The cosine of the angle between data vector 1 and rotated factor vector I is the inner product of the δ_{1i} values and the column of λ_{i1} values from **Λ**. The δ_{1i} values are the direction cosines of the data-variable vector with the original unrotated factors. It was stated in Chapter 2 that the cosine of the angle between two vectors is the inner product of the direction cosines. Since the δ_{1i} values are the direction cosines of the data-variable vector with the original unrotated factors, the λ_{i1} values in Eq. (6.17) must be the direction cosines of the rotated factor vector I with respect to the same unrotated factor vectors.

With orthogonal factors, then, the **Λ** matrix that transforms unrotated factor loadings to rotated factor loadings has columns that represent direction cosines of the rotated factors with respect to the original unrotated factors. In oblique rotations of the matrix of reference vectors, columns of the matrix of projections on the axes are altered one at a time instead of two at a time, as in orthogonal rotations. This is accomplished through modification of the

$\mathbf{\Lambda}$ matrix one column at a time. As the ith column of $\mathbf{\Lambda}$ is altered, the ith column of the rotated matrix is changed in the equation $\mathbf{V}_i = \mathbf{A}\mathbf{\Lambda}_i$ where $\mathbf{\Lambda}_i$ includes all the rotations up to this one, orthogonal plus oblique. The matrix $\mathbf{\Lambda}_i$ is altered to $\mathbf{\Lambda}_{i+1}$ and $\mathbf{V}_i$ is altered to $\mathbf{V}_{i+1}$ to incorporate the current rotation.

[6.2.1] *OBLIQUE ROTATION OF A REFERENCE VECTOR*

A convenient way of changing one column of the $\mathbf{\Lambda}$ matrix to rotate one reference vector is a procedure described by Cattell (1952) which is equivalent to the multiplication of $\mathbf{\Lambda}$ by a matrix that is an identity matrix except for the column being changed. This process has already been illustrated in Eq. (6.11). To illustrate the equivalent process recommended by Cattell, we begin with the following equation:

$$\text{I}' = \text{I} \pm \tan\varphi \cdot \text{II} \tag{6.18}$$

where I represents the first column of $\mathbf{\Lambda}$ before the rotation and II is the second column of $\mathbf{\Lambda}$ before the rotation. In this case, vector I is being rotated with II by an angle φ. If I is rotated *toward* II, the plus sign is used in Eq. (6.18). If I is being rotated *away from* II, the minus sign will be used in Eq. (6.18). The term I′ represents the new column I of $\mathbf{\Lambda}$ *before* the column of values has been normalized. That is, the sum of squares of the values in I′ will not usually equal 1.0. Every column of $\mathbf{\Lambda}$ must have a sum of squares equal to 1.0, since these values are the direction cosines of the rotated factor with the original factors. These values are made to have a sum of squares equal to 1.0 by "normalizing" them. To normalize the I′ values, take the sum of squares of the values in column I′, take the square root of this total, and divide each element by I′ by this square root. The sum of squares of the resulting values will equal 1.0. These normalized values will replace the previous column 1 of the $\mathbf{\Lambda}$ matrix.

In Eq. (6.18), the roman numeral of whatever reference vector is being rotated is placed where the I is now placed. The roman numeral of the reference vector it is being rotated with goes where the II is now placed. The unnormalized new column of $\mathbf{\Lambda}$ goes where I′ now is placed. The normalized values will then replace the proper column of the $\mathbf{\Lambda}$ matrix. If reference vector II is being rotated, for example, then column 2 of the $\mathbf{\Lambda}$ matrix is changed. In multiplying the original unrotated factor matrix by this new $\mathbf{\Lambda}$ matrix, only the column of the rotated matrix that corresponds to the altered column of the $\mathbf{\Lambda}$ matrix will be altered.

That the procedure indicated in Eq. (6.18) is equivalent to the procedure described in Eq. (6.11) may be shown as follows. The procedure in Eq.

(6.18) is equivalent to altering the $\mathbf{\Lambda}$ matrix by the following matrix multiplication:

$$(6.19)\qquad \underset{\mathbf{\Lambda}}{\begin{bmatrix} \lambda_{11} & \lambda_{12} & \lambda_{13} & \lambda_{14} \\ \lambda_{21} & \lambda_{22} & \lambda_{23} & \lambda_{24} \\ \lambda_{31} & \lambda_{32} & \lambda_{33} & \lambda_{34} \\ \lambda_{41} & \lambda_{42} & \lambda_{43} & \lambda_{44} \end{bmatrix}} \times \underset{\mathbf{\Theta}'}{\begin{bmatrix} 1 & 0 & 0 & 0 \\ \pm\tan & 1 & 0 & 0 \\ 0 & 0 & 1 & 0 \\ 0 & 0 & 0 & 1 \end{bmatrix}}$$

If the matrix $\mathbf{\Theta}'$ on the right of Eq. (6.19) is normalized, only the values in the first column need to be changed since the other three columns already are normalized. Taking the sum of squares of the entries in the first column of $\mathbf{\Theta}'$, taking the square root of this total, and dividing the elements of the first column of $\mathbf{\Theta}'$ by this square root gives the values $1/\sqrt{1+\tan^2\varphi}$ and $\pm\tan\varphi/\sqrt{1+\tan^2\varphi}$ in the first two rows with zeros in the last two rows as before. These two values reduce trigonometrically, however, to the cosine and $\pm$sine of φ, respectively, and hence the normalized $\mathbf{\Theta}'$ matrix is equivalent to the $\mathbf{\Theta}$ matrix in Eq. (6.11). The more convenient procedure using Eq. (6.18) will be used, therefore, in making oblique rotations of the reference vectors.

[6.3] OBLIQUE ROTATION OF THE 12-VARIABLE PROBLEM

To illustrate these procedures, the 12-variable problem treated in the last chapter will be rotated to oblique simple structure. The point of departure for the oblique rotations is the final orthogonal solution shown in Table 5.7. It is best to rotate orthogonally as far as possible toward simple structure before attempting any oblique rotations. If oblique rotations are made early in the rotation process, a greater degree of obliquity may be introduced than is actually necessary to achieve simple structure. In the interest of keeping factor constructs as independent of one another as possible, angles between factors are kept as close to 90° as possible while bringing the factors to simple structure positions. As factors become more oblique to each other, they overlap each other more and more in what they represent, introducing redundancy into the system of constructs.

Results from orthogonal rotations of the factors in the 12-variable problem suggested that factor I might require oblique rotation with both factors III and IV. Plots of the final orthogonal factors I with III and IV are shown in Fig. 6.5. These plots are based on the factor loadings in Table 5.7. The plot of factor I

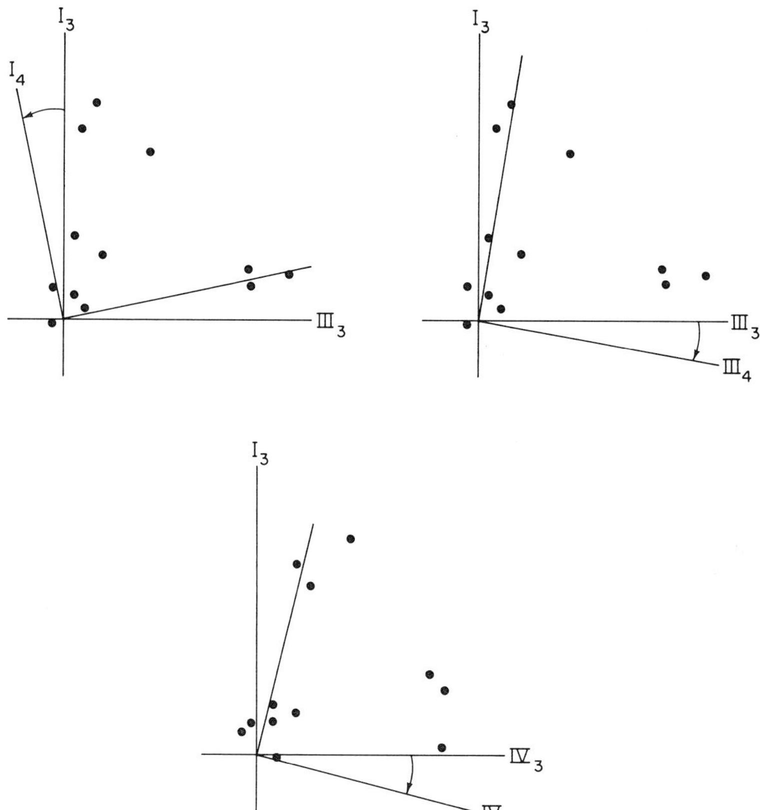

FIG. 6.5. First series of oblique rotations for the 12-variable problem.

with factor III is shown twice in Fig. 6.5 to allow presentation of one oblique rotation on each graph for the sake of clarity. Both oblique rotations could be shown on the same graph, but the profusion of lines would be confusing.

In the first plot of factor I with factor III in Fig. 6.5, reference vector I is moved to position I_4 from the original final orthogonal position at I_3. By rotating I away from III in this fashion, the three points to the right side of the graph are brought into the hyperplane of I_4. This is the whole purpose of this rotation. The reference vector I_4 is actually further from the points with high projections on this reference vector than was I_3, so the rotation has *not* been to bring reference vector I closer to a cluster of points high on I. Since this is the first oblique rotation, for the first time reference vector I and factor I will no longer coincide. Until a reference vector is rotated obliquely with some other reference vector, it will coincide with its associated factor vector.

The calculations for altering Λ, the final orthogonal transformation matrix, to include this first oblique rotation are as follows:

$$\text{I}' = \text{I} - \tan\varphi \cdot \text{III} \tag{6.20}$$

$$\begin{aligned}
\lambda'_{11} &= .711 - .2(.439) = .711 - .088 = .623\\
\lambda'_{21} &= .057 - .2(-.237) = .057 + .047 = .104\\
\lambda'_{31} &= -.201 - .2(.845) = -.201 - .169 = -.370\\
\lambda'_{41} &= -.671 - .2(.192) = -.671 - .038 = -.709
\end{aligned} \tag{6.21}$$

Summing the squares of the λ'_{i1} values in Eqs. (6.21) gives .3881+.0108+.1369+.5027 = 1.0385. The square root of 1.0385 is 1.019. Multiplying each λ'_{i1} value in Eqs. (6.21) by 1/1.019 gives the normalized λ_{i1} values in column 1 of the Λ_1 matrix in Table 6.1. The calculations for the oblique rotation in Fig. 6.5 are shown in Table 6.1.

In Eqs. (6.21) the first column of values, I′, the λ'_{i1} values, are the unnormalized first column transformation matrix values. The values in the I column are taken from the first column of the Λ matrix in Table 5.7, the final orthogonal transformation matrix, and the values in parentheses are the values from column 3 of the same Λ matrix in Table 5.7. The value .2 is the tangent of the angle of rotation of I with III in Fig. 6.5. The .2 value for the tangent has a minus sign before it because I is rotated away from rather than toward III. If I were rotated toward III by this same angle, the tangent would have a plus sign placed in front of it.

The second plot in Fig. 6.5 shows the rotation of reference vector III away from reference vector I by an angle whose tangent is .17. This rotation moves the reference vector away from the cluster of points with high loadings, which

TABLE 6.1

Calculations for the First Series of Oblique Rotations

A

	I	II	III	IV
1	.080	.846	.160	−.046
2	.011	.707	.101	.048
3	.037	.652	.221	−.100
4	.544	.099	−.287	.261
5	.589	.157	−.180	.366
6	.435	.098	−.212	.482
7	.563	−.006	−.161	−.339
8	.661	.013	.076	−.205
9	.758	.023	−.210	−.230
10	.431	−.174	.506	.045
11	.537	−.197	.591	.152
12	.370	−.281	.522	.052

× **Λ₁**

I	II	III	IV
.611	.000	.314	.360
.102	.963	−.243	.099
−.363	.270	.867	−.355
−.696	.000	.302	.857

= **V₁**

I	II	III	IV
.109	.858	−.056	.016
.009	.708	−.066	.079
.079	.687	.014	−.086
.265	.018	−.023	.531
.187	.103	.101	.605
.017	.037	.074	.655
.638	−.049	−.064	−.031
.520	.033	.208	.037
.702	−.035	−.019	.153
.031	−.031	.630	−.003
−.012	−.030	.775	.095
−.028	−.130	.653	−.035

again demonstrates that simple structure rotations are made to improve hyperplanes rather than to move reference vectors toward clusters. The hyperplane perpendicular to reference vector position III_4 is better located than it was in the original position perpendicular to III_3 (see Fig. 6.5). The calculations for getting the new column III of the $\mathbf{\Lambda}$ matrix to make this rotation are as follows:

$$\text{III}' = \text{III} - .17(\text{I})$$

(6.22)
$$\begin{aligned} \lambda'_{13} &= .439 - .17(.711) = .318 \\ \lambda'_{23} &= -.237 - .17(.057) = -.247 \\ \lambda'_{33} &= .845 - .17(-.201) = .879 \\ \lambda'_{43} &= .192 - .17(-.671) = .306 \end{aligned}$$

The sum of squares of the unnormalized λ'_{i3} values in Eqs. (6.22) is 1.0284 and the square root is 1.014. Multiplying the λ'_{i3} values by 1/1.014 gives the λ_{i3} values in Table 6.1, which shows the calculations for the oblique rotation of factor III. These normalized values are shown in the third column of the $\mathbf{\Lambda}_1$ matrix in Table 6.1.

Computations for the new column IV of the $\mathbf{\Lambda}_1$ matrix to rotate reference vector IV away from reference vector I are made in the same way, that is, $\text{IV}' = \text{IV} - .25(\text{I})$ to give the unnormalized column of values λ'_{i4}. These λ'_{i4} values are then normalized, giving the fourth column of $\mathbf{\Lambda}_1$ as shown in Table 6.1.

Multiplication of the original unrotated matrix by the new transformation matrix $\mathbf{\Lambda}_1$ gives the new matrix of projections on the oblique reference vectors, as shown in Table 6.1. Reference vectors I, III, and IV were rotated obliquely, while II was left unchanged. Thus, columns I, III, and IV of matrix $\mathbf{V}_1$ in Table 6.1 are different from the corresponding columns of the final orthogonal solution, matrix $\mathbf{V}$ of Table 5.7, while column II remains identical in the two matrices.

After the oblique reference vector projections in matrix $\mathbf{V}_1$ have been calculated by the equation $\mathbf{V}_1 = \mathbf{A}\mathbf{\Lambda}_1$, as shown in Table 6.1, the matrix of correlations among the reference vectors $\mathbf{C}$ must be calculated as follows:

(6.23)
$$\mathbf{C} = \mathbf{\Lambda}_1'\mathbf{\Lambda}_1$$

In Eq. (6.23), the transpose of the transformation matrix is multiplied by the transformation matrix itself to get the matrix of correlations among the reference vectors. Each row of $\mathbf{\Lambda}_1'$ in Eq. (6.23) is a row of direction cosines for a reference vector, as is each column of $\mathbf{\Lambda}_1$. Hence a row of $\mathbf{\Lambda}_1'$ inner multiplied with a column of $\mathbf{\Lambda}_1$ gives the inner product of direction cosines for two vectors which is a correlation between the two vectors. It is important to avoid the mistake of multiplying Λ_1 with Λ_1' because this reversed order of multiplying the matrices does not give a matrix of correlations among the reference vectors.

The **C** matrix for the 12-variable problem after the first series of oblique rotations of the reference vectors is shown in Table 6.2. When the reference vectors in matrix $\mathbf{V}_1$ of Table 6.1 are plotted with each other, the correlation between the two reference vectors is indicated on the plot to keep the investigator advised of the actual degree of obliquity present. This is kept in mind in planning further rotations, primarily to avoid allowing the axes to become too highly correlated with each other. If the axes are allowed to become too highly oblique with each other, two hyperplanes will collapse into one and a factor will be lost.

After the first series of oblique rotations, a new series of plots is made in order to look for additional oblique rotations. Even though the reference vectors are not at right angles to each other, they are plotted as if they were because it is difficult to make plots with axes at an angle with each other different from 90°. In practice, the angles of rotation chosen are little affected by plotting these oblique axes at right angles to each other.

Inspection of the available plots of the oblique reference vectors in matrix $\mathbf{V}_1$ of Table 6.1 with each other produces one additional oblique rotation that will make a noticeable improvement in the oblique simple structure. This last oblique rotation is shown in Fig. 6.6. Reference vector I is moved away from reference vector IV by an angle whose tangent is .2. Calculations for modifying column 1 of the transformation matrix Λ_1 to make this rotation are shown below:

$$\text{I}' = \text{I} - .2(\text{IV})$$

$$\begin{aligned} \lambda'_{11} &= .611 - .2(.360) &&= .611 - .072 = .539 \\ \lambda'_{21} &= .102 - .2(.099) &&= .102 - .020 = .082 \\ \lambda'_{31} &= -.363 - .2(-.355) &&= -.363 + .071 = -.292 \\ \lambda'_{41} &= -.696 - .2(.857) &&= -.696 - .171 = -.867 \end{aligned} \tag{6.24}$$

TABLE 6.2

Correlations between Reference Vectors After the First Series of Oblique Rotations

	I	II	III	IV
I	1.000	.001	−.358	−.237
II	.001	1.000	.000	−.001
III	−.358	.000	1.000	.040
IV	−.237	−.001	.040	1.000

$\Lambda_1'\Lambda_1$

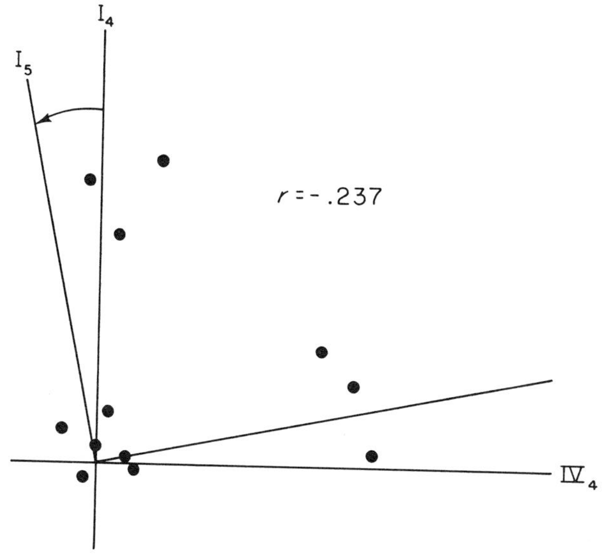

FIG. 6.6. Second series of oblique rotations for the 12-variable problem.

Normalizing these λ'_{i1} values gives the λ_{i1} values found in the first column of the final $\mathbf{\Lambda}$ matrix shown in Table 6.3. Only column 1 of $\mathbf{\Lambda}_1$ is changed to obtain the final $\mathbf{\Lambda}$; so only column I of $\mathbf{V}$, the final matrix of rotated oblique reference vector projections, is different from the corresponding column of matrix $\mathbf{V}_1$, Table 6.1.

In Table 6.3, the matrix $\mathbf{A}$ contains the original unrotated factor loadings. Matrix $\mathbf{\Lambda}$ is the transformation matrix that contains the effects of all

TABLE 6.3

Final Oblique Reference Vector Structure for the 12-Variable Problem

A

	I	II	III	IV
1	.080	.846	−.160	−.046
2	.011	.707	.101	.048
3	.037	.652	.221	−.100
4	.544	.099	−.287	.261
5	.589	.157	−.180	.366
6	.435	.098	−.212	.482
7	.563	−.006	−.161	−.339
8	.661	.013	.076	−.205
9	.758	.023	−.210	−.230
10	.431	−.174	.506	.045
11	.537	−.197	.591	.152
12	.370	−.281	.522	.052

×

Λ

I	II	III	IV
.506	.000	.314	.360
.077	.963	−.243	.099
−.274	.270	.867	−.355
−.814	.000	.302	.857

=

V

I	II	III	IV
.099	.858	−.056	.016
−.007	.708	−.066	.079
.090	.687	.015	−.086
.149	.018	−.023	.531
.061	.103	.101	.605
−.107	.037	.075	.655
.605	−.049	−.064	−.031
.482	.033	.208	.037
.630	−.035	−.019	.153
.029	−.031	.630	−.003
−.029	−.030	.775	.094
−.020	−.130	.653	−.035

orthogonal and oblique rotations performed. The columns of Λ are the direction cosines of the reference vectors with respect to the original unrotated factor axes. The **V** matrix in Table 6.3 contains the perpendicular projections of the data vectors on the oblique reference vectors. These values may be interpreted as correlations of the data-variable vectors with the oblique reference vectors.

Table 6.4

Final Matrix of Correlations among Reference Vectors for the 12-Variable Problem

	I	II	III	IV
I	1.000	.000	−.343	−.411
II	.000	1.000	.000	−.001
III	−.343	.000	1.000	.040
IV	−.411	−.001	.040	1.000

$\mathbf{C} = \Lambda'\Lambda$

Table 6.4 contains the matrix of correlations **C** among the final oblique reference vectors. The matrix is symmetric, since $r_{12} = r_{21}$, and has values of 1.0 in the diagonals, since the correlation of a reference vector with itself is 1.0. The correlations of reference vector **I** with reference vectors III and IV are negative, indicating an angle of greater than 90° between them. It should be noted that where the reference vectors are correlated negatively, the correlations between the factor axes associated with those reference vectors will be positive. If the reference vectors are correlated positively, the associated factor axes will, in general, be correlated negatively.

The factor axes coincide with the intersections of hyperplanes. If two such lines of intersection have an angle between them of less than 90°, then the angle between the reference vectors perpendicular to them must be greater than 90°, yielding a negative correlation. If the lines of intersecting hyperplanes are at an angle greater than 90°, however, their normals, the reference vectors, must have an angle between them of less than 90°.

The plots of the final reference vectors are shown in Fig. 6.7. Although some slight further improvements could be made, for example, moving II toward III by a small angle of rotation, the plots shown in Fig. 6.7 exhibit a reasonable approximation to simple structure. The fit to simple structure in Fig. 6.7 is clearly better than for the orthogonal factors in Fig. 5.4. No excessively large correlations between reference vectors were needed to reach this structure. The correlations between reference vectors should be kept as low as possible consistent with reaching simple structure positions. Values less than .5 should be preferred and rarely should a value larger than .6 be permitted.

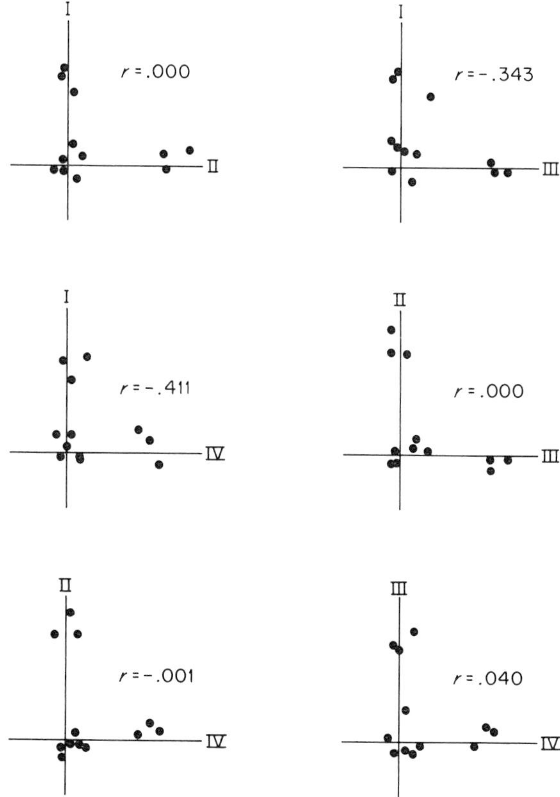

FIG. 6.7. Plots of final reference vectors.

When a correlation between reference vectors is negative, rotating one vector away from the other will merely increase the size of the negative correlation. A rotation of one vector toward the other under these circumstances will reduce the size of the negative correlation. With a positive correlation between vectors, rotating one reference vector toward the other makes the correlation higher and rotating one vector away from the other reduces the size of the correlation. When reference vectors have substantial correlations between them, rotations that reduce the absolute size of the correlation, while at the same time improving simple structure, are particularly welcome.

All the oblique rotations executed for the 12-variable problem involved moving one reference vector away from the other. This was an accidental phenomenon caused by the fact that the factors represented by these 12 variables are either correlated positively or not at all. This means that the reference vectors will be correlated negatively where the factors are correlated positively. If the small rotation of reference vector II toward reference vector III were carried out, the new column II of the $\mathbf{\Lambda}$ matrix would be obtained by using the equation $\text{II}' = \text{II} + .05(\text{III})$ and normalizing the λ'_{2i} values from II′. Notice that the tangent of φ, .05, multiplied by λ_{i3} is *added* rather than subtracted since the rotation of II is toward rather than away from III.

The reference vector projections in matrix **V** of Table 6.3 are not the factor loadings required for a complete factor analytic solution, although they are proportional to the factor loadings and hence are perfectly correlated with them. This can be seen in Fig. 6.8 where F_1 and F_2 are factor vectors which are the hyperplanes for the reference vectors RV_1 and RV_2, respectively. The horizontal lines from the data points are perpendiculars that intersect RV_1 at distances from the origin which equal the reference vector projections. The factor loadings on factor F_1 are the distances from the origin to the points where these horizontal lines intersect the axis F_1. Point 1 has the highest value in both. Points 2, 3, and 4 are tied for the next highest reference vector projections on RV_1 and factor loadings on F_1. Points 5 and 6 are tied for the next position in the rank order, and so on, down to point 19 which has the largest negative reference vector projection on RV_1 and factor loading on F_1. The rank order is identical for the points, whether reference vector projections on RV_1 are considered, or factor loadings on F_1.

Because of the identical rank order in size of reference vector projections and factor loadings, investigators have often presented the matrices **R**, **A**, $\mathbf{\Lambda}$, **V**,

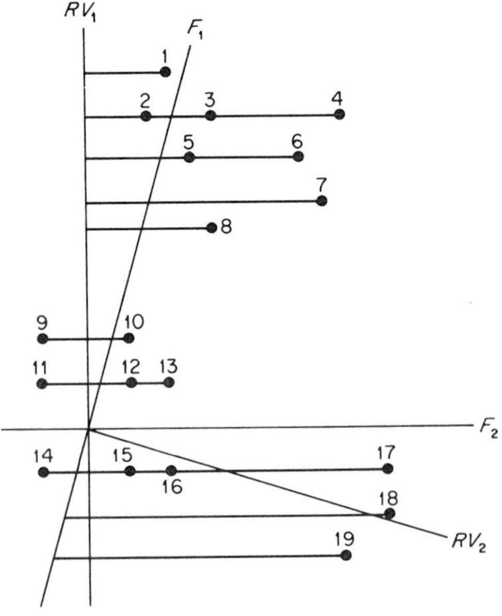

FIG. 6.8. Correlations between factor loadings and reference vector projections.

and **C** as the solution to an oblique factor analysis, failing to compute the factor pattern, factor structure, and matrix of correlations among the factors. This was particularly apt to be the case prior to the widespread availability of computers since the extra calculations needed to obtain these additional matrices represent a significant burden if the work is done by desk calculator. With computers readily available, however, there is less reason to omit these additional matrices from the final factor solution.

[6.4] OBTAINING THE ADDITIONAL MATRICES Φ, S, AND P

Consider the schematic matrix operation in Fig. 6.9, an adaptation of a similar figure presented by Thurstone (1947, p. 349). Ignoring, for the moment, the bottom part **T** of the matrix on the left and the bottom part **D** of the matrix on the right, this operation is just the **AΛ** = **V** of Table 6.3 that produced the final reference vector structure **V** by multiplying the matrix of unrotated loadings **A** by the oblique transformation matrix **Λ**. The columns of **Λ** are the direction cosines of the reference vectors with respect to the unrotated factors in matrix **A**. The sums of squares of each column of **Λ** equal 1.0.

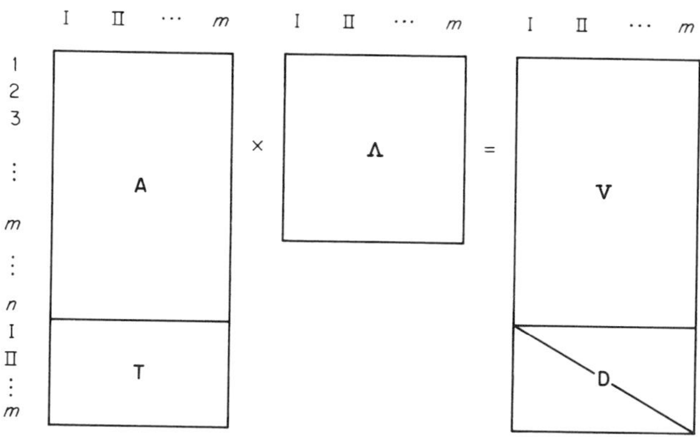

Fɪɢ. 6.9. Projections on oblique reference vectors: Rows of **A** are projections of data vectors on orthogonal unrotated factors; rows of **T** are direction cosines of oblique factor vectors with the unrotated factors; columns of **Λ** are direction cosines of oblique reference vectors with the unrotated factors; rows of **V** are projections of data vectors on the oblique reference vectors; rows of **D** are projections of oblique factor vectors on the oblique reference vectors. [Adapted from *Multiple factor analysis*; by L. L. Thurstone, Chicago, 1947, by permission of The University of Chicago Press.]

The rows of **A** are projections of the data vectors on the unrotated factor vectors, that is, correlations with the unrotated factors. Since $r_{ij} = h_i h_j \cos\varphi_{ij}$ where r_{ij} is the correlation between vectors i and j, h_i and h_j are the lengths of vectors i and j, and $\cos\varphi_{ij}$ is the cosine of the angle between vectors i and j, the values in a row of **A** equal $h_i \cos\varphi_{ij}$ where h_i is the length of the data vector i, and j is the factor vector with length equal to 1.0. The sums of squares of the rows of **A** are the communalities h_i^2, the squared lengths of the data vectors.

Let the rows of **T** represent the projections of the oblique rotated factor vectors on the original unrotated factor vectors, that is, the correlations between the oblique rotated factor vectors and the orthogonal unrotated factor vectors. These rows in **T** of Fig. 6.9 may be considered to represent additional data vectors added to the matrix of unrotated loadings **A**. The rows of **T** have sums of squares equal to 1.0 rather than communalities less than 1.0, however, because the rotated factor vectors have unit length, as do both the unrotated factor vectors and the reference vectors.

Since the rotated factor vectors that are represented by the rows of **T** in Fig. 6.9 have unit length and the unrotated factor vectors have unit length, the correlations between these two sets of vectors are given by

$$r_{ij} = h_i h_j \cos\varphi_{ij} = (1.0)(1.0)\cos\varphi_{ij} = \cos\varphi_{ij}$$

The values in each row of matrix **T**, therefore, are also the direction cosines of the oblique rotated factor vectors with respect to the orthogonal unrotated factor vectors. Row i of matrix **T** and column j of matrix **Λ**, therefore, represent direction cosines of factor vector i and reference vector j, respectively. Where $i = j$, they are the factor vector and reference vector that correspond to each other in the oblique solution.

Multiplying **A** by **Λ** gives a matrix in which the columns consist of the projections of the data vectors on the reference vectors. Since the rows of matrix **T** in Fig. 6.9 consist of projections of rotated factor vectors on the unrotated factor vectors in **A**, multiplying **T** by **Λ** will produce a matrix **D** for which the columns consist of projections of the rotated factor vectors on the same reference vectors.

The matrix **D** will be a diagonal matrix, however, since a factor vector has zero projections on every reference vector except the one with which it is associated. Thus, oblique factor vector I projects only on reference vector I, factor II projects only on reference vector II, and so on. This leaves nonzero entries only in the diagonal cells of **D**. These relationships can be seen more clearly by reference to Fig. 6.10 in which one reference vector has been rotated obliquely, from OG toward OA to position OH, by an angle whose tangent is .2. The hyperplane perpendicular to reference vector OH, which is really only a plane when three factors are represented, is the plane $OCJI$. This plane

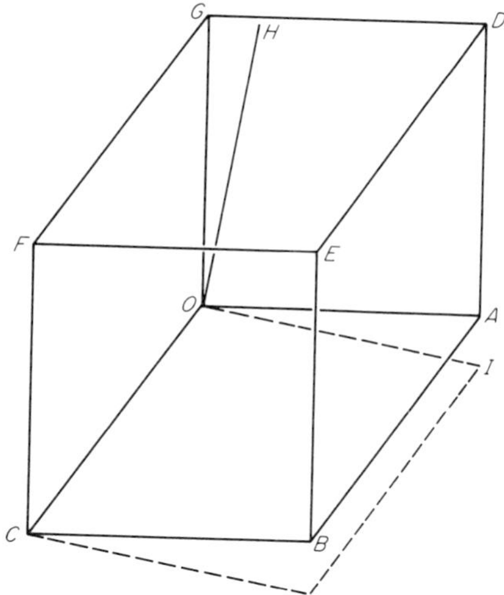

FIG. 6.10. Oblique hyperplanes and reference vectors.

is shown in dotted lines except for the line *OC* where plane *OCJI* intersects plane *OCBA*.

The factors are the intersections of the hyperplanes. Before the oblique rotation of the reference vector, the three hyperplanes in Fig. 6.10 were *OCBA*, *OADG*, and *OCFG*. The intersections of these hyperplanes, that is, the factors, were the lines *OA*, *OC*, and *OG*. In the orthogonal system, the reference vectors perpendicular to the hyperplanes were coincident with the factors, that is, *OA*, *OC*, and *OG*.

After the rotation of the reference vector *OG* toward reference vector *OA*, however, arriving at position *OH*, the hyperplane perpendicular to *OH* has been tipped below plane *OCBA*, rotating around axis *OC*. The line *OI* is in the plane *OADG* still because *OH* is in the plane of *OADG*. The rotation was performed in the plot of factor *OG* with factor *OA*; *OA* is the trace of the hyperplane for *OG* in the plane *OADG* and *OI* is the trace of the hyperplane for reference vector *OH* in plane *OADG*.

After the rotation of *OG* to *OH*, then, the intersections of the hyperplanes, and hence the factors, are the lines *OI*, *OC*, and *OG*. The reference vectors corresponding to these factors are *OA*, *OC*, and *OH*. The reference vector *OC* still coincides with factor *OC* because the oblique rotation took place only in the plane perpendicular to *OC*. The hyperplanes perpendicular to the reference vectors, then, are *OCFG*, *OADG*, and *OCJI*.

Now, it is apparent that a factor projects only on its own reference vector. It is orthogonal to all other reference vectors. Factor *OI* projects on reference vector *OA*, but it is orthogonal to reference vectors *OC* and *OH*. Factor *OG* projects on reference vector *OH* but is orthogonal to reference vectors *OA* and

OC. Finally, reference vector OC projects on factor OC (actually coincides with it) but is at right angles to reference vectors OH and OA.

Note that the angle between reference vectors OH and OA is less than 90°, giving a positive correlation between these reference vectors, but that the angle between their associated factors, OI and OG, is greater than 90°, leading to a negative correlation between the factors. The diagonal cells of the **D** matrix in Fig. 6.9, then, contain the projections of the factors on their associated reference vectors, and the off-diagonal cells are zero since they represent projections of these factor vectors on reference vectors other than their own.

[6.4.1] *CORRELATIONS BETWEEN THE FACTORS*

The matrix of correlations between the oblique factors **Φ** can be represented as follows:

$$\mathbf{\Phi} = \mathbf{TT'} \tag{6.25}$$

where **T**, as before, is a matrix whose rows are direction cosines of the oblique rotated unit-length factor vectors with respect to the original orthogonal unrotated factor vectors, as shown in Fig. 6.9. The matrix **TT′** gives the inner products of the direction cosines for all possible pairs of oblique rotated factor vectors. Since these rotated factor vectors are of unit length, these inner products equal the correlations between the rotated factor vectors. The matrix **Φ** is symmetric, since $r_{ij} = r_{ji}$, and has 1.0 in the diagonals since the correlation of each factor with itself is 1.0. Since the matrix **T** is not known from the results of the reference vector rotations, however, Eq. (6.25) must be developed further to provide a workable equation for obtaining **Φ**.

From Fig. 6.9 and the accompanying discussion, the following holds:

$$\mathbf{T\Lambda} = \mathbf{D} \tag{6.26}$$

Multiplying both sides of Eq. (6.26) by $\mathbf{\Lambda}^{-1}$, the inverse of **Λ**, gives

$$\mathbf{T\Lambda\Lambda}^{-1} = \mathbf{D\Lambda}^{-1} \tag{6.27}$$

Equation (6.27) simplifies to

$$\mathbf{T} = \mathbf{D\Lambda}^{-1} \tag{6.28}$$

Substituting Eq. (6.28) in Eq. (6.25) gives

$$\mathbf{\Phi} = \mathbf{D\Lambda}^{-1}(\mathbf{D\Lambda}^{-1})' \tag{6.29}$$

Since the transpose of a product of matrices equals the product of the transposes in reverse order, Eq. (6.29) simplifies to

$$\mathbf{\Phi} = \mathbf{D\Lambda}^{-1}(\mathbf{\Lambda}^{-1})'\mathbf{D}' \tag{6.30}$$

The transpose of a diagonal matrix is identical to the matrix itself; hence $\mathbf{D}' = \mathbf{D}$. Also, the inverse of a transpose equals the transpose of an inverse; so Eq. (6.30) may be rewritten as

(6.31) $$\mathbf{\Phi} = \mathbf{D}(\mathbf{\Lambda}'\mathbf{\Lambda})^{-1}\mathbf{D}$$

Since $\mathbf{\Lambda}'\mathbf{\Lambda}$ is the matrix $\mathbf{C}$, the matrix of correlations between reference vectors already referred to, Eq. (6.31) reduces to

(6.32) $$\mathbf{\Phi} = \mathbf{D}\mathbf{C}^{-1}\mathbf{D}$$

Equation (6.32) would permit the calculation of the matrix of correlations among factors $\mathbf{\Phi}$ if $\mathbf{C}^{-1}$ and $\mathbf{D}$ were available. A method for calculating the inverse of $\mathbf{C}$ is described in the next section. The unknown matrix $\mathbf{D}$ can be determined if $\mathbf{C}^{-1}$ is known.

The matrix $\mathbf{\Phi}$ must have the value 1.0 in all diagonal cells. For oblique solutions, the diagonal values of $\mathbf{C}^{-1}$ will generally be different from 1.0 even though the diagonal values of the $\mathbf{C}$ matrix itself are equal to 1.0. Selecting the unknown diagonal values of $\mathbf{D}$ in such a way that the product matrix $\mathbf{D}\mathbf{C}^{-1}\mathbf{D}$ gives diagonal values of 1.0 provides a unique solution for the diagonal values of $\mathbf{D}$, that is, d_{ii}.

The values of the diagonal cells in $\mathbf{D}$ which will make $\mathbf{D}\mathbf{C}^{-1}\mathbf{D}$ have unities in the diagonal cells are

(6.33) $$d_{ii} = \frac{1}{\sqrt{c_{ii}^{-1}}}$$

where c_{ii}^{-1} is the diagonal element of the matrix $\mathbf{C}^{-1}$ that corresponds to the diagonal element of $\mathbf{D}$, d_{ii}. Multiplying $\mathbf{C}^{-1}$ on the left by $\mathbf{D}$ has the effect of multiplying column i of $\mathbf{C}^{-1}$ by d_{ii}. Multiplying $\mathbf{C}^{-1}$ on the right (that is, postmultiplying) by $\mathbf{D}$ has the effect of multiplying the elements of row i by d_{ii}. Using Eq. (6.33) to obtain the d_{ii} values, an element in the diagonal of $\mathbf{\Phi} = \mathbf{D}\mathbf{C}^{-1}\mathbf{D}$ will be

(6.34) $$\varphi_{ii} = \frac{1}{\sqrt{c_{ii}^{-1}}}\frac{1}{\sqrt{c_{ii}^{-1}}}[c_{ii}^{-1}] = 1.0$$

[6.4.2] *COMPUTING THE INVERSE OF* **C**

There are many satisfactory methods of computing an inverse. Most computer centers have standard programs for this purpose. If at all possible, the inverse of the $\mathbf{C}$ matrix should be obtained by computer, using one of these available standard programs. If these facilities are not available so that the inverse must be calculated with desk computing equipment, the following method described by Fruchter (1954) will serve. Several decimal places should be carried in the computations even though only four are reported in the following example.

Begin by placing the **C** matrix to be inverted on the left and the identity matrix on the right as shown in (6.35). The general procedure is to use row operations to convert the **C** matrix on the left into an identity matrix. The same operations carried out on the rows of the **C** matrix are also carried out on the right-hand matrix. As the **C** matrix is carried by these row operations into the identity matrix, the identity matrix on the right is carried into the inverse matrix $\mathbf{C}^{-1}$.

The **C** matrix to be inverted in the following example is the matrix of correlations among reference vectors for the 12-variable problem shown in Table 6.4:

(6.35)

$$\underset{\mathbf{C}}{\begin{bmatrix} 1.0000 & .0000 & -.3430 & -.4111 \\ .0000 & 1.0000 & .0000 & -.0010 \\ -.3430 & .0000 & 1.0000 & .0400 \\ -.4111 & -.0010 & .0400 & 1.0000 \end{bmatrix}}, \quad \underset{\mathbf{I}}{\begin{bmatrix} 1.0000 & .0000 & .0000 & .0000 \\ .0000 & 1.0000 & .0000 & .0000 \\ .0000 & .0000 & 1.0000 & .0000 \\ .0000 & .0000 & .0000 & 1.0000 \end{bmatrix}}$$

If the entry c_{11} in matrix **C** above were not 1.0000, it would be necessary first to get a value of 1.0000 in that position by dividing every entry in row 1 of both matrices by the value in c_{11}. Since $c_{11} = 1.0000$ already, however, it is possible to proceed to the next step. The remaining values in column 1 of the matrix on the left, that is, c_{21}, c_{31}, and c_{41}, must be made equal to zero by multiplying row 1 by a constant such that when the product is added to the row in question, a zero appears in the first column position for that row. The entry c_{21} is already zero; so there is no need to eliminate this value. The element c_{31} is $-.3430$, however; so it must be made equal to zero. Multiplying the row 1 values by .3430 and adding the results to row 3 will give a zero in row 3, column 1 of the matrix on the left. Correspondingly, multiplying row 1 by .4111 and adding to row 4 will eliminate the $-.4111$ in row 4, column 1 of the matrix on the left, making it zero. These same operations are also carried out on the matrix at right. The results of these operations are shown below:

(6.36)

$$\begin{bmatrix} 1.0000 & .0000 & -.3430 & -.4110 \\ .0000 & 1.0000 & .0000 & -.0010 \\ .0000 & .0000 & .8824 & -.1010 \\ .0000 & -.0010 & -.1010 & .8311 \end{bmatrix}, \quad \begin{bmatrix} 1.0000 & .0000 & .0000 & .0000 \\ .0000 & 1.0000 & .0000 & .0000 \\ .3430 & .0000 & 1.0000 & .0000 \\ .4110 & .0000 & .0000 & 1.0000 \end{bmatrix}$$

Zeros will now be induced in the second column elements of the matrix on the left, except for the diagonal element. The process calls for making the entry

$c_{22} = 1.0000$ by dividing all row 2 entries of both matrices by c_{22} and then adding multiples of the second row to the other rows to eliminate the nondiagonal second column values of the matrix on the left. The element c_{22} is already equal to 1.0000; so that step is unnecessary. The entries c_{12} and c_{32} are already zero; so only the entry c_{42}, $-.0010$, needs to be eliminated. Multiply row 2 by .0010 and add to row 4, obtaining

(6.37)

$$\begin{bmatrix} 1.0000 & .0000 & -.3430 & -.4110 \\ .0000 & 1.0000 & .0000 & -.0010 \\ .0000 & .0000 & .8824 & -.1010 \\ .0000 & .0000 & -.1010 & .8311 \end{bmatrix}, \quad \begin{bmatrix} 1.0000 & .0000 & .0000 & .0000 \\ .0000 & 1.0000 & .0000 & .0000 \\ .3430 & .0000 & 1.0000 & .0000 \\ .4110 & .0010 & .0000 & 1.0000 \end{bmatrix}$$

To eliminate the nondiagonal elements of the third column of the matrix on the left, divide the third row by the element in the third column and third row, that is, .8824, to get a value of 1.0000 in the cell c_{33}. This is the third row in the matrix (6.38). Multiplication of this third row by .3430 and adding the result to row 1 of (6.37) gives the new row 1 of (6.38) with a zero in the third column position. In (6.37) the second row value is zero in the third column of the matrix on the left so nothing needs to be done about the second row. The second row of (6.37) is carried down unchanged to the second row of (6.38). Multiplying row 3 of (6.38) by .1010 and adding to row 4 of (6.37) will give the new row 4 of (6.38) with a zero in the third column position. Again, these operations are performed on both matrices. The results are

(6.38)

$$\begin{bmatrix} 1.0000 & .0000 & .0000 & -.4503 \\ .0000 & 1.0000 & .0000 & -.0010 \\ .0000 & .0000 & 1.0000 & -.1145 \\ .0000 & .0000 & .0000 & .8195 \end{bmatrix}, \quad \begin{bmatrix} 1.1333 & .0000 & .3887 & .0000 \\ .0000 & 1.0000 & .0000 & .0000 \\ .3887 & .0000 & 1.1333 & .0000 \\ .4503 & .0010 & .1145 & 1.0000 \end{bmatrix}$$

Only the last column of the matrix on the left now needs further treatment. Divide row 4 of both matrices by .8195 to get 1.0000 in the cell c_{44} and place the result in row 4 of (6.39) below. Then, multiply row 4 in (6.39) by .4503 and add to row 1 in (6.38) to get the new row 1 in (6.39). This places a zero in the cell c_{14}. Multiplying row 4 of (6.39) by .0010 and adding to row 2 in (6.38) gives the new row 2 in (6.39). This places a zero in the cell c_{24}. Finally, multiplying row 4 in (6.39) by .1145 and adding to row 3 in (6.38) gives the new row 3 in (6.39). This places a zero in the cell c_{34}:

(6.39)

$$\underset{\mathbf{I}}{\begin{bmatrix} 1.0000 & .0000 & .0000 & .0000 \\ .0000 & 1.0000 & .0000 & .0000 \\ .0000 & .0000 & 1.0000 & .0000 \\ .0000 & .0000 & .0000 & 1.0000 \end{bmatrix}}, \quad \underset{\mathbf{C}^{-1}}{\begin{bmatrix} 1.3807 & .0005 & .4516 & .5495 \\ .0005 & 1.0000 & .0001 & .0012 \\ .4516 & .0001 & 1.1493 & .1397 \\ .5495 & .0012 & .1397 & 1.2203 \end{bmatrix}}$$

Now, the matrix on the left in (6.39) is the identity matrix, and the matrix on the right is the inverse of **C**, that is, $\mathbf{C}^{-1}$. By multiplying **C** by $\mathbf{C}^{-1}$, and vice versa, it can be verified that the identity matrix is obtained as required by the definition of the inverse; for example, $\mathbf{CC}^{-1} = \mathbf{C}^{-1}\mathbf{C} = \mathbf{I}$.

[6.4.3] *FORMULAS FOR THE STRUCTURE AND PATTERN MATRICES*

The oblique rotation of reference vectors ends up with the formula

$$\mathbf{A\Lambda} = \mathbf{V} \tag{6.40}$$

where **A** is the matrix of unrotated factor loadings, **Λ** is the matrix with columns of direction cosines of the reference vectors with respect to the unrotated factor vectors, and **V** is the matrix of correlations of the data variables with the reference vectors. By analogy,

$$\mathbf{AT}' = \mathbf{S} \tag{6.41}$$

where the columns of $\mathbf{T}'$ are direction cosines of the oblique factor vectors with respect to the unrotated factor vectors and **S** is the matrix of correlations of the data variables with the oblique factor vectors, that is, the factor structure. The rows of matrix **T** gave the projections of the unit-length oblique factor vectors on the unrotated factor vectors. These are also direction cosines. The transpose $\mathbf{T}'$ merely puts these direction cosines in columns rather than rows. Equation (6.41) is not used in practice to obtain **S** because **T** is ordinarily not known. This equation will be used, however, in the derivation of a formula for the factor pattern **P**. The formula that is actually used to obtain **S** in practice requires knowing **P**, the pattern.

Equation (2.2) gave the matrix of standard scores as a product of the matrix of factor loadings multiplied by the orthogonal, unrotated factor scores:

$$\mathbf{Z} = \mathbf{A}_u \mathbf{F}_u \tag{6.42}$$

Each row of **Z** is a data-variable vector and each row of matrix $\mathbf{F}_u$ is a factor score vector. Each row of $\mathbf{A}_u$ is a set of weights for the factor scores for a given data variable. The data-variable vectors, then, are being expressed as linear combinations of the orthogonal unrotated factor score vectors. These factor vectors constitute a basis for the space.

It was explained earlier in this chapter that any set of vectors that can be expressed as a linear combination of one set of basis vectors can also be expressed as a linear combination of a different set of basis vectors spanning the same space. The basis vectors do not even have to be orthogonal. A new set of basis vectors requires new weights, or coordinates, however. Using a new set of basis vectors, Eq. (6.42) may be written as

$$\mathbf{Z} = \mathbf{P}_u \mathbf{W}_u \tag{6.43}$$

where **Z** is the matrix of data-variable scores, $\mathbf{W}_u$ is the matrix of basis vectors, and $\mathbf{P}_u$ is the matrix of weights, or coordinates, of the data variables with respect to the factor vectors in $\mathbf{W}_u$ that constitute the new basis for the space. Let the first m rows of $\mathbf{W}_u$ be the oblique rotated factor scores with the next $2n$ rows the same specific and error factor scores that appear in matrix $\mathbf{F}_u$ in Eq. (6.42). The first m columns of $\mathbf{P}_u$ in Eq. (6.43) are the coordinates of the data-variable vectors with respect to the common factors, and the next $2n$ columns are the coordinates of the data-variable vectors with respect to the specific and error factors. The portions of $\mathbf{P}_u$ and $\mathbf{W}_u$ that are concerned only with common factors will be designated as matrices **P** and **W**, respectively.

The relationship in Eq. (6.43) will be used now in developing another equation for the factor structure **S**. Since the factor structure is the matrix of correlations between the n data variables and the m oblique factors, it may be represented in schematic matrix form as follows:

(6.44)

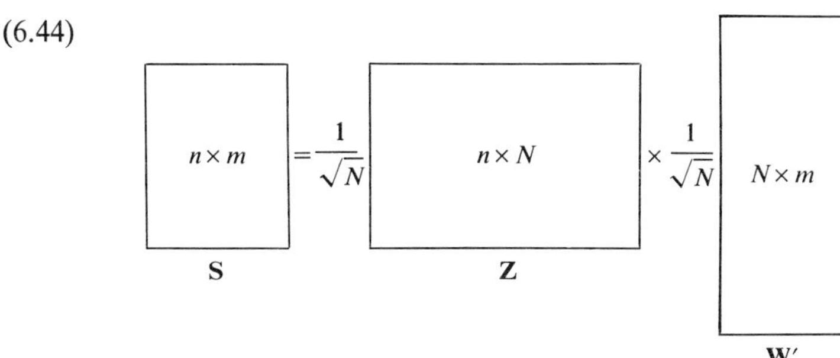

where **S** is the common factor structure, **Z** is the matrix of data-variable scores reproducible with common factors, and **W**′ is the transpose of the matrix of oblique common factor scores. An element of the matrix on the left, s_{ik}, is given by

$$s_{ik} = \frac{\sum_{j=1}^{n} z_{ij} w'_{jk}}{N}$$

Since this is an average cross product of standard scores, it represents a correlation coefficient for data variable i with oblique factor k.

If the specific and error factor correlations were added on as additional columns of **S** and the specific and error factor scores as additional columns of **W**′, Eq. (6.44) would become

$$\mathbf{S}_u = \frac{1}{\sqrt{N}} \mathbf{Z} \frac{1}{\sqrt{N}} \mathbf{W}_u' \tag{6.45}$$

By Eq. (6.43), $\mathbf{Z} = \mathbf{P}_u \mathbf{W}_u$. Substituting this value for $\mathbf{Z}$ in Eq. (6.45) gives

$$\mathbf{S}_u = \mathbf{P}_u \left[\frac{\mathbf{W}_u \mathbf{W}_u'}{N} \right] \tag{6.46}$$

The expression in brackets in Eq. (6.46) is the matrix of correlations among all the factors, common, specific, and error. Except for the self-correlations and the intercorrelations among the oblique common factors, these values are all zero since the specific and error factors are uncorrelated with any other factors. Using only the part of Eq. (6.46) relevant to the oblique common factors gives the equation

$$\mathbf{S} = \mathbf{P\Phi} \tag{6.47}$$

where $\mathbf{S}$ is the factor structure for the oblique common factors, $\mathbf{P}$ is the factor pattern for the oblique common factors, and $\mathbf{\Phi}$ is the matrix of correlations among the oblique common factors. Equation (6.47) is the formula actually used in practice to compute the factor structure $\mathbf{S}$. The pattern matrix $\mathbf{P}$ must be obtained first. A formula for finding $\mathbf{P}$ is developed next.

By Eq. (6.41), $\mathbf{S} = \mathbf{AT}'$; by Eq. (6.47), $\mathbf{S} = \mathbf{P\Phi}$. Equating these two expressions for $\mathbf{S}$ gives

$$\mathbf{P\Phi} = \mathbf{AT}' \tag{6.48}$$

Substituting $\mathbf{T} = \mathbf{D\Lambda}^{-1}$ from Eq. (6.28) in Eq. (6.48) gives

$$\mathbf{P\Phi} = \mathbf{A}(\mathbf{D\Lambda}^{-1})' \tag{6.49}$$

But, since $\mathbf{A\Lambda} = \mathbf{V}$ (see Fig. 6.9), and therefore $\mathbf{A} = \mathbf{V\Lambda}^{-1}$, Eq. (6.49) becomes

$$\mathbf{P\Phi} = \mathbf{V\Lambda}^{-1}(\mathbf{D\Lambda}^{-1})' \tag{6.50}$$

Equation (6.50) simplifies to

$$\mathbf{P\Phi} = \mathbf{V}(\mathbf{\Lambda}'\mathbf{\Lambda})^{-1}\mathbf{D} \tag{6.51}$$

By Eq. (6.31), however, $\mathbf{\Phi} = \mathbf{D}(\mathbf{\Lambda}'\mathbf{\Lambda})^{-1}\mathbf{D}$; hence, Eq. (6.51) can be expressed as

$$\mathbf{PD}(\mathbf{\Lambda}'\mathbf{\Lambda})^{-1}\mathbf{D} = \mathbf{V}(\mathbf{\Lambda}'\mathbf{\Lambda})^{-1}\mathbf{D} \tag{6.52}$$

Multiplying both sides of Eq. (6.52) by $\mathbf{D}^{-1}$ and substituting $\mathbf{C} = \mathbf{\Lambda}'\mathbf{\Lambda}$ gives

$$\mathbf{PDC}^{-1}\mathbf{DD}^{-1} = \mathbf{VC}^{-1}\mathbf{DD}^{-1} \tag{6.53}$$

Dropping $\mathbf{DD}^{-1}$ from both sides and multiplying both sides by $\mathbf{C}$ gives

$$\mathbf{PDC}^{-1}\mathbf{C} = \mathbf{VC}^{-1}\mathbf{C} \tag{6.54}$$

Dropping $\mathbf{C}^{-1}\mathbf{C}$ from both sides of (6.54) and multiplying both sides by $\mathbf{D}^{-1}$ and simplifying produces the final equation for $\mathbf{P}$, the pattern matrix:

$$\mathbf{P} = \mathbf{VD}^{-1} \tag{6.55}$$

The effect of multiplying the reference vector structure matrix **V** by $\mathbf{D}^{-1}$ is merely to multiply column i of matrix **V** by element d_{ii}^{-1} of $\mathbf{D}^{-1}$. This is consistent with the previously demonstrated fact that the factor loadings in **P** are perfectly correlated with and proportional to the corresponding reference vector projections in **V**. The matrix $\mathbf{D}^{-1}$ is also a diagonal matrix with elements equal to the reciprocals of the corresponding elements of matrix **D**.

A complete oblique solution, then, requires the following matrices: (1) **R**, the correlation matrix; (2) **A**, the matrix of unrotated factor loadings; (3) **V**, the reference vector structure; (4) **Λ**, the transformation matrix that carries **A** into **V**; (5) **C**, the matrix of correlations among the oblique reference vectors; (6) **P**, the pattern matrix of factor loadings of the data variables on the oblique common factors; (7) **S**, the structure matrix of correlations between the data variables and the oblique common factors; and (8) **Φ**, the matrix of correlations among the oblique common factors.

Some investigators would expect to carry out a second-order factor analysis of the matrix of correlations among the factors, **Φ**, usually with 1.0 values in the diagonal cells as communalities. This analysis is carried out to find the factors among the oblique factors at the next level of generality. Such factors are referred to as "second-order factors." Eysenck in particular has been an advocate of this approach, even doing third- and higher order factor analyses until the higher order factors are uncorrelated (Eysenck & Eysenck, 1969).

The **P**, **S**, and **Φ** matrices for the 12-variable sample problem are given in Table 6.5. Simple structure holds only for the pattern matrix **P**. The structure matrix **S** shows the extent to which data variables correlate with the oblique factors, whereas the pattern matrix does not. Both pattern and structure matrices are needed to present a clear picture of the interrelationships among the data variables and the oblique common factors.

As solutions become highly oblique, the values in the reference vector structure often become rather small. The factor loadings in the factor pattern, however, will be large in these cases, even exceeding 1.0. The oblique factor loadings are not to be interpreted as correlations, however; so these values over 1.0 are entirely appropriate as weights in the specification equation.

[6.4.4] *REPRODUCING* **R** *FROM THE OBLIQUE FACTOR LOADINGS*

Equation (2.2) stated that $\mathbf{Z} = \mathbf{A}_u\mathbf{F}_u$ where **Z** is the matrix of data-variable scores, $\mathbf{A}_u$ is the matrix of unrotated factor loadings, common plus specific and error factors, and $\mathbf{F}_u$ is the matrix of factor scores for the orthogonal unrotated factors. The correlation matrix can also be represented by the following equation:

$$\mathbf{R} = \frac{1}{N}(\mathbf{ZZ}') \tag{6.56}$$

where $\mathbf{Z}$ is the matrix of data-variable scores. Each row of $\mathbf{Z}$ inner multiplied with a column of $\mathbf{Z}'$, the transpose of $\mathbf{Z}$, gives a sum of cross products of standard scores which when divided by N gives the correlation coefficient for the

TABLE 6.5

Pattern, Structure, and $\mathbf{\Phi}$ Matrices for the 12-Variable Problem

	Pattern matrix (P) I	II	III	IV	Structure matrix (S) I	II	III	IV
1	.117	.858	−.060	.018	.103	.858	−.016	.061
2	−.008	.708	−.071	.088	.004	.708	−.064	.076
3	.105	.687	.016	−.095	.071	.687	.042	−.049
4	.175	.018	−.025	.587	.414	.018	.107	.658
5	.072	.103	.109	.668	.394	.103	.213	.712
6	−.125	.037	.080	.723	.209	.037	.120	.679
7	.710	−.049	−.068	−.035	.671	−.049	.182	.258
8	.566	.033	.223	.040	.663	.033	.431	.306
9	.740	−.035	−.020	.169	.804	−.034	.265	.479
10	.035	−.031	.675	−.003	.275	−.031	.687	.091
11	−.034	−.030	.830	.104	.308	−.030	.830	.188
12	−.023	−.130	.700	−.039	.211	−.130	.687	.033

	I	II	III	IV
I	1.000	.000	.359	.423
II	.000	1.000	.000	.000
III	.359	.000	1.000	.118
IV	.423	.000	.118	1.000

Correlations among factors
$\mathbf{\Phi}$ matrix

pair of data variables involved. Substituting $\mathbf{Z} = \mathbf{A}_u\mathbf{F}_u$ in Eq. (6.56) gives

$$\mathbf{R} = \frac{1}{N}(\mathbf{A}_u\mathbf{F}_u)(\mathbf{A}_u\mathbf{F}_u)' \tag{6.57}$$

Equation (6.57) simplifies to

$$\mathbf{R} = \mathbf{A}_u\left[\frac{\mathbf{F}_u\mathbf{F}_u'}{N}\right]\mathbf{A}_u' \tag{6.58}$$

The expression in brackets in Eq. (6.58) is the matrix of correlations among all the factors, common, specific, and error. Since specific and error factors do not contribute to the correlations between data variables, if communalities are used in the diagonals of **R** and only the common factor portions of the matrices in Eq. (6.58) are used, Eq. (6.58) simplifies to

$$\mathbf{R} = \mathbf{A}\boldsymbol{\Phi}\mathbf{A}' \tag{6.59}$$

where **A** is the matrix of unrotated factor loadings and **Φ** is the matrix of correlations among these factors. With the unrotated, orthogonal extracted factors, no correlations exist among the factors, so the matrix **Φ** would be the identity matrix, leading to the equation $\mathbf{R} = \mathbf{AA}'$, as shown in Eq. (2.11). With oblique factors, however, the correlations among the factors are not equal to zero; so Eq. (6.59) becomes

$$\mathbf{R} = \mathbf{P}\boldsymbol{\Phi}\mathbf{P}' \tag{6.60}$$

where **P** is the pattern matrix of oblique factor loadings and **Φ** is the matrix of correlations among the oblique factors. To reproduce the **R** matrix with oblique factor loadings, therefore, it is necessary to use the matrix of correlations among the oblique factors.

[6.4.5] *OBLIQUE VERSUS ORTHOGONAL SOLUTIONS*

The complexities of using oblique solutions and the added burden of computation involved have led many investigators to settle for the much simpler orthogonal solutions. Some investigators, for example, Cattell (1952), have insisted on using oblique solutions because they feel that factors in nature are not apt to be orthogonal and that the use of orthogonal solutions is therefore unjustified. Other investigators, such as Guilford, have generally preferred to stick with orthogonal factor solutions on the grounds that they wish to develop constructs as independent of one another as possible. The too-easy acceptance of oblique solutions may inhibit the search for truly independent constructs.

The author can sympathize with both these points of view. There seems to be little real doubt about the fact that the most satisfactory constructs in a given domain will usually exhibit some degree of obliquity. Height and weight, for example, are excellent constructs for anthropometric studies, but they are obviously correlated. On the other hand, just because constructs are apt to be oblique in nature does not mean necessarily that oblique factor solutions must always be employed. Often, an orthogonal solution may give for all practical purposes a perfectly adequate impression of the nature of the constructs in a domain and a reasonably adequate impression of the degree of obliquity that is involved from inspection of the orthogonal plots without actually executing an oblique solution. Furthermore, the data variables that define the oblique simple structure factors most clearly are usually the same ones that have the highest loadings on the corresponding orthogonal factors.

Using the best orthogonal approximation to a simple structure solution has the advantage of avoiding the extra computation involved in an oblique solution and perhaps more important, it simplifies the task of reporting results. The only feasible way of reporting all the matrices for a complete oblique solution is through the use of auxiliary publication facilities. Unless it is important to obtain a precise oblique solution for the particular study involved, it is often more practical to settle for the best orthogonal approximation to the simple structure solution. This is particularly true with crude, exploratory investigations. The latter stages of a programmatic series of factor analytic investigations will often be soon enough to expend the additional effort to obtain and report the more refined oblique solutions. If, however, the particular investigation reveals factors of great obliquity such that an orthogonal solution would seriously distort the true picture, it may be necessary to go to an oblique solution even in an exploratory study.

CHAPTER SEVEN

Simple Structure and Other Rotational Criteria

[7.1]
INTRODUCTION

In the preceding two chapters, rotation of factor axes according to simple structure criteria has been described, with positive manifold used as a supplementary aid in achieving this goal. Thurstone (1947, pp. 140ff.) attempted to prove the value of simple structure by presenting a solution to his famous "box" problem. He collected a sample of 20 boxes and took various mathematical functions of the three basic dimensions to obtain a total of 20 variables. With x, y, and z being the basic dimensions, Thurstone used as variables x^2, $\log x$, x^2+y^2, x^2+z^2, $2x+2y$, $x^2+y^2+z^2$, and e^x with similar functions for y and z. From the intercorrelations of these variables he obtained three centroid factors, none of which matched the basic dimension variables of length, width, or height, the presumed underlying factors in the system. Rotating to simple structure, however, produced factors that were readily identifiable with the three basic box dimensions of length, width, and height.

Cattell and Dickman (1962) published a similar kind of study, called the "ball" problem, designed to present a more obviously oblique set of factor constructs. In fact, two of the rotated factors ultimately derived from the solution to this problem correlated .76 with each other. Thirty-two properties were measured for 80 balls varying in size, weight, and elasticity. The 32 ball measures were intercorrelated over the 80 "cases" and factor analyzed. Oblique simple structure rotations produced four oblique factors corresponding to the physical influences expected in this situation. The orthogonally rotated factors, however, did not match the physical influences nearly as well.

Examples like the Thurstone "box" problem and the Cattell and Dickman "ball" problem are supposed to establish the proposition that simple structure, and oblique simple structure in particular, leads to satisfactory solutions. Cattell (1952, 1966) in particular is prone to reject as unsuitable rotational solutions which do not satisfy oblique simple structure criteria.

All that these examples demonstrate, however, is that there *are* instances in which oblique simple structure *does* work. Pointing out two such instances, or indeed any number of examples in which a method works, is not sufficient to establish the proposition that the method *always* works or even that it will work in the next situation where it is applied.

It is the author's view that although simple structure can often be a very valuable guide in the rotation process, it has definite limitations. In fact, the uncritical use of oblique simple structure as a criterion can lead to grossly distorted solutions. This conclusion was reached by the author as a result of his experimentation with a computer program that he developed (described in a later chapter) that searches systematically for good hyperplane positions. The general conclusion from these unpublished investigations was that with many variables and many factors in a factor matrix, hyperplanes can be "improved" to extremely high proportions of near-zero loadings if the factors are allowed to become sufficiently oblique to one another. Furthermore, frequently found with good hyperplanes were factor positions which made no sense whatsoever. For example, unrelated and uncorrelated variables in some cases had high loadings on the same factor. This was true for matrices which had alternate rotational solutions that were entirely satisfactory and meaningful as well as exhibiting a reasonable approximation to simple structure. Earlier experimentation (Comrey, 1959) with Thurstone's oblique analytic rotation method gave additional evidence questioning the universal dependability of oblique simple structure.

Thurstone's method starts with a vector coincident with some selected data variable. This vector is adjusted to a position not too far from the original position but with an improved hyperplane. Experiments with large matrices using this procedure showed that with the data analyzed, a good hyperplane could be found starting with *any* of the data variables, invariably winding up with that variable having the highest or at least a high loading on the adjusted factor vector.

The author's conclusion from these studies is that locating good hyperplanes is not always sufficient to establish good factor positions. In fact, it will be asserted that rotating to obtain the best oblique hyperplanes is neither necessary nor sufficient in general for obtaining the best factor solution, although it may have that effect in many cases.

One reason that the anomaly of a good hyperplane with a bad factor position is not encountered more often is that systematic searches for good oblique hyperplanes are not generally conducted starting immediately with the unrotated factors. Typically, orthogonal rotations are carried along until the hyperplanes have been fairly well established before going to oblique positions to improve them. Under this procedure, uninterpretable factor positions with good hyperplanes are not likely to be reached.

Some problems with oblique simple structure may still remain, however, even following the procedure of carrying the solution as far as possible orthogonally before going to oblique factor positions. The difficulty is that with many variables and many factors available for rotation, it is possible to continue reducing the size of the reference vector projections by allowing the vectors to go more and more oblique relative to each other.

At a given point, for example, rotating reference vector I away from reference vector III an additional 25° will bring two more variables into the hyperplane. But, should they be brought into the hyperplane? Is the resulting factor position representative of a factor construct that is superior to the factor construct represented by the position before this rotation? It is the author's view that the answer may sometimes be "yes"; but at other times it will be "no." The mere fact that an additional point has been brought into the hyperplane is, *per se*, no guarantee that the factor is better than before.

The number of zero or near-zero loadings that a factor will have in the "best" scientific solution cannot be stated independently of knowledge about the variables that are included in the analysis. It is possible to select a group of variables that would be best represented by one general factor with substantial loadings for all the data variables plus several additional smaller factors that would have a high proportion of near-zero loadings. A simple structure solution would be inappropriate here, although one might approximate such a solution by rotating the big general factor obliquely with the other factors. In less spectacular fashion, the best oblique position for a given factor might be one that would leave variable 3, for instance, with a moderate loading, although by oblique rotation it would be possible to bring variable 3 into the hyperplane. In many instances, it may be necessary to make a solution move away from simple structure rather than toward it to achieve the optimum positioning of factors.

How is it possible to know how oblique to let the factors go? Because this question is difficult to answer, investigators have often settled for a belief in the proposition that "simple structure" would guide them correctly. This author has limited faith in simple structure as a guide in making such decisions. This faith is strained more and more as the criterion of simple structure calls for more and more obliquity in the solution. Where factor axes are to be placed in very oblique positions, the author believes that not only should simple structure indicate the need for such positioning but that supporting evidence outside the geometric configuration of the data points themselves should be available to justify this type of solution. The fact that a better simple structure can be achieved by going to a more highly oblique position does not constitute a necessary and sufficient reason for doing so. If, however, apart from the configuration of data points, there are other good reasons that strongly suggest the need for a highly oblique solution, then there is more reason to accept the guidance of simple structure as being essentially valid in this instance.

Cattell and those who share his views are so convinced of the value of simple structure as a guide to rotations that they prefer to do their rotations "blindly," that is, without knowledge of the variables represented by the data points, to be certain that simple structure and that alone will influence their decisions in the rotation process.

This author cannot place that much faith in manipulations on a configuration of points as a royal road to truth. Few factor analyses indeed conform to the ideal situation Thurstone had in mind when he formulated the principles of simple structure. He felt that simple structure would be a good guide where a large random selection of variables was taken from the entire universe. In such a case, many factors would be represented among the variables, no factor would be related to very many variables, and no one variable would be related to very many factors. To the extent that this ideal is achieved, simple structure undoubtedly provides an excellent aid in the rotation process, but even here the best possible solution may diverge from the best simple structure solution because of random effects resulting in atypical variable selection. The best position for a given factor may call for certain variables to have loadings in the region of .2 to .3; a simple structure solution could be found where these variables would be brought down into the hyperplane by making the factors more oblique.

In the real-life situation, collections of variables chosen for a factor analysis rarely if ever represent a random selection of variables from the total universe. The investigator typically has something in mind in selecting the variables, or he is limited by the available data in the variables he has to work with. All this adds up to making it probable that the conditions optimum for applying oblique simple structure may not be met in his study. To obtain the best solution for his data, therefore, he may have to depart from the goal of obtaining the best oblique simple structure.

A rotational solution is nothing more than one interpretation of the data. It has been shown in earlier chapters that there are infinitely many possible such interpretations for most real-life correlation matrices. The best oblique simple structure solution frequently provides an interpretation that the investigator finds useful for his purposes. It has not been proved and cannot be proved, however, that the best oblique simple structure will always constitute the most meaningful solution for scientific purposes. This author feels, therefore, that it is a mistake to depend upon blind rotation to oblique simple structure as though this procedure takes precedence over all others.

The correctness of a factor matrix as an interpretation of the data cannot be validated by reference to anything within the analysis itself, including the conformity of the solution to oblique simple structure. To attempt to do so would be circular. The validity of an interpretation of the data must be assessed with reference to evidence that is gathered independently, that is, outside the analysis itself. This evidence may come from many sources, for example, knowledge about the characteristics of the variables that are being analyzed, previous factor analytic investigations, experiments conducted using these variables, experiments conducted using factor scores derived from this investigation, additional factor analytic studies in which variables have been added, dropped, or modified according to certain hypotheses, and so on.

Since the correctness of a factor analytic interpretation of the data can only be established by appeal to information independent of the analysis itself, it would seem to be reasonable to use all available knowledge in selecting the best possible factor interpretation to begin with. It would be absurd to select a factor solution which contradicts available knowledge when an alternate solution is available that is consistent with that information. The investigator is urged, therefore, to use, in addition to oblique simple structure, any available well-established knowledge that he has to guide him in arriving at a final rotational solution.

Rotation "with knowledge," therefore, is recommended by this author in preference to "blind" rotations. Where outside knowledge and simple structure criteria point to the same path, the investigator is fortunate. When several lines of evidence point to the same conclusion, confidence in that conclusion is increased. Where outside evidence and oblique simple structure call for traveling along divergent paths, the investigator is faced with a dilemma that must be resolved. A resolution in favor of the oblique simple structure may be the correct one, but there can be no assurance that this will always be the case. The investigator must carefully assess the situation and make the most intelligent decision that circumstances permit. Knowledge about the correctness of his decision, whatever it is, may be forthcoming only after a great deal of additional research.

Adherents of the blind-rotation approach warn that the use of knowledge about the variables will make it possible for the investigator to allow his preconceived notions to bias his observations. This is true, of course, and the investigator must work very hard to avoid selective perception favoring pet hypotheses. Closing his eyes, however, is not going to increase the probability that he will see the world as it is rather than as he would like it to be. If there were some guarantee that blind rotations to oblique simple structure would enable him to see the world as it is, this would be an excellent means of controlling this type of bias. Unfortunately, no such guarantee exists.

[7.2] ALTERNATIVE CRITERIA OF ROTATION

Although hand rotation to oblique simple structure, or to the best orthogonal approximation to it, has been the most traditionally favored method of rotation, alternate approaches have enjoyed some popularity. In fact, at the present time, the traditional simple structure approach described in previous chapters has been supplanted in popularity by computer methods primarily because of their convenience for the investigator. Some of the alternative approaches to simple structure are considered in the remainder of this chapter.

[7.2.1] *ROTATION TO PLACE FACTORS THROUGH CLUSTERS*

It will be recalled that simple structure rotations are carried out to improve the hyperplane for a factor, not to bring the factor closer to a cluster of vectors with large projections or loadings on the factor. Advocates of simple structure insist that clusters merely represent "surface traits," to use Cattell's terminology, rather than "real" factors; hence to rotate toward them is an error as a general policy although it might work out in a given instance. Suppose, for example, that several psychological tests were constructed that consisted of items drawing upon two distinct mental abilities factors, A and B. Some of the tests lean more on ability A and others more on B. These tests would correlate substantially with each other since they all depend on both factors A and B. They would form a well-defined cluster. To rotate toward this cluster, however, would be to put a factor about half-way between factor position A and factor position B, while no factor would represent either A or B alone. For this reason, simple structure advocates reject the cluster as a guide to factor positioning and rely on the hyperplane instead.

The late Professor Tryon has been one of the most prominent supporters of the cluster approach, developing with the collaboration of his students theory and procedures for carrying out cluster analyses (Tryon & Bailey, 1970). Tryon certainly did not accept the idea that the only way to locate useful constructs through analysis of correlations among variables is by the application of simple structure rotations to an unrotated factor matrix. The author agrees with Tryon and his collaborators that rotating toward clusters in certain analyses can produce excellent solutions with good scientific utility. It is also undoubtedly true, however, that rotation toward clusters in some instances may position factors in such a way that the constructs represented by those factor position are not univocal. This was the case in the hypothetical example previously described where the factor through the cluster of tests represented a composite of factors A and B and where neither univocal factor A nor B was represented separately by any factor in the analysis.

The investigator thoroughly familiar with the characteristics of these particular tests would know that they are factorially complex, that is, involving more than one factor, and would not choose or accept a rotational solution that would place a factor through the cluster of such related test variables.

It should be noted in passing that the application of simple structure is no guarantee that a factor will not be placed through a cluster of complex data variables. Suppose, for instance, that in the matrix there are no variables that measure either factor A or B alone, and the only variables measuring either factor A or B are the complex variables that measure both. Suppose further that the other 80 percent of the variables in this matrix clearly are uncorrelated with these tests and define other unrelated factors. Under these circumstances, a simple structure solution will produce a factor with substantial loadings for all these complex variables on a single factor. Thus, a composite A+B factor will be produced rather than two separate univocal factors, A and B. In this case, blind rotation to simple structure might give the same result as rotating toward clusters; both methods would then imply that these complex tests measure a single univocal factor when in fact they do not. Only by using his knowledge of the variables involved can the investigator avoid making the mistake of accepting this type of solution. Of course, the knowledgeable investigator would design his study properly to avoid this kind of situation if he had sufficient information and freedom in planning the study to do so. Proper design of factor analytic investigations is discussed in a later chapter.

In summary, the position taken here is that under certain circumstances, clusters can be used as guides in rotations. Factors should only be rotated toward clusters, however, when confirmed knowledge about the variables in the cluster suggests strongly that they are univocal in character. By the same token, knowledge about the variables in a cluster may clearly suggest a positioning of a given factor in some other relationship to one or more clusters. That is, it might be concluded that two factors should be positioned so that a given cluster lies in the plane of two factors and half-way between them rather than collinear with either. Clusters, then, can be helpful to the investigator in making rotations, but knowing how to use them requires information and intelligent judgment on his part.

[7.2.2] *ROTATION TO ACHIEVE CONSISTENCY WITH PAST RESULTS*

Significant scientific progress using factor analytic results is most likely to come from a series of programmatic investigations rather than from a single isolated study. Investigators involved in programmatic research naturally seek findings that will generalize across studies. They wish to be able to replicate findings in order to gain confidence in their conclusions. In studying human abilities Guilford (1967) and his collaborators have generally been more concerned in their rotations to relate present results to past findings than to satisfy simple structure criteria. They have also adhered generally to orthogonal solutions.

Cattell has also extended criteria for rotation from simple structure in one single study to results from several related studies in his "parallel proportional profiles" concept (Cattell, 1944a; Cattell & Cattell, 1955). Consider two different factor analytic studies involving the same or nearly the same variables, some of which were designed to give a factor of general intelligence. One of the studies is carried out with college students, and the other uses a random sample from the general population. Since the college students would show a restriction of range in intelligence, due to the elimination of people with low IQs from the sample, the correlations among the intelligence measures would be lower in this group than they would be in the sample from the general population. This would result in lower factor loadings for these variables on the Intelligence factor in the college sample. Cattell would position the Intelligence factor vector in the two samples in such a way that the loadings for the variables defining the factor would be proportional to each other, although they would all be higher in the general population sample than they would be in the college sample. To attain a solution of this kind, it may be necessary to depart from the best simple structure position in each of the two samples considered alone.

The method of "orthogonal additions" is perhaps best classified under the heading of rotating to achieve consistency with past results. An investigator may start out with a group of variables that define one well-established factor. Then he places a second factor at right angles to the first, using existing variables and attempting to develop new ones until the second factor is well defined. Then he goes on to develop a third factor, orthogonal to the first two, and so on. At each new step, results are built upon and made to exhibit consistency with previous findings.

Past factor analytic investigations of course do not constitute the only kind of past results that may be used to guide present rotations. Information from other kinds of investigations, knowledge about the data variables, and generally known facts in a given field are all examples of past results that have been used to guide rotations. The position has already been taken that any well-established information that the investigator has at his disposal and that has implications for factor positioning can and should be used as circumstances permit in settling upon a rotational interpretation of the data. This is not to say that it is justifiable to position a factor in a particular manner because it was so positioned in one or more previous studies. After all, it may have been improperly positioned in the first place. It would seem desirable, however, to strive for rotational solutions which reveal the extent to which continuity of results does exist over a series of investigations. This may occasionally require a rerotation of all the studies in a series to make allowances for recently discovered information that is relevant to all these factor solutions. Again, it must be emphasized that any rotational solution is merely a hypothesis about nature that can only be verified by appeal to evidence that goes beyond the results of any single factor analysis *per se*. If the investigator is to *prove* that his rotational interpretation is correct in some absolute sense, he must go beyond describing the mathematical procedures he used to obtain the solution. This is not often done in published descriptions of factor analytic investigations.

[7.2.3] *ROTATION TO AGREE WITH HYPOTHESES*

This approach in some of its applications is very similar to the last one in that by rotating present results to be consistent with known facts, including findings from previous investigations, there is an attempt to achieve congruence between present results and a complex hypothesis about their nature. In other instances, the hypothesis about present results may be generated strictly by theory rather than by past results. In such cases, factor analytic results are used as a test of the adequacy of the theory, much as a set of data points is related to a theoretically generated curve. In the latter case, a statistical test may be used to determine the adequacy of the fit between the data points and the theoretical curve. In the case of the factor analysis, the rotational solution may be manipulated to achieve the maximum possible agreement with theoretical expectations consistent with the reality of the data matrix of correlations. In such instances, the number of factors rotated is usually an important part of the test of congruence between theoretical expectations and outcome.

Rotations of this kind have been criticized by some individuals, apparently on the basis of the mistaken belief that any desired degree of congruence with theoretical expectations can be achieved through the occult art of rotation by unscrupulous but clever scientific charlatans. These same critics, in their own work, may well take advantage of generous error margins in asserting that their experimentally observed data points were not statistically inconsistent with their theoretical expectations. The truth is that the unrotated factor matrix admits only a certain class of transformations. Any investigator with much experience in carrying out factor analytic rotations is painfully aware of how intractable a configuration of data points can be for him who would achieve a preconceived result.

Rotation to agree with hypothesis may take the form of establishing a "target" matrix. An attempt is made then to rotate the unrotated matrix into a form that agrees as closely as possible with this target matrix. There has been considerable interest in recent years in developing computerized methods of achieving such results. These solutions have come to be known as "Procrustes"-type solutions. Some of these are mentioned in the next section.

[7.3] ANALYTIC METHODS OF ROTATION

The basic distinction between so-called "analytic" and "nonanalytic" methods of rotation is that analytic methods can be reduced to clearly specified rules that require no judgment on the part of the investigator. Nonanalytic methods, such as hand rotation to oblique simple structure, require that the investigator exercise judgment to determine what is to be done at various stages in the rotation process.

Analytic methods of rotation have an enormous inherent appeal for many reasons. First, since they require no investigator judgment, they can be applied with equal facility by novice and expert alike, giving the same results no matter who does it. Second, since the experimenter exercises no judgment, the process is more objective, giving the experimenter no opportunity to influence the results as a function of his preconceived notions. Third, since such methods are reducible to a set of well-defined rules, they can be programmmed for computer application. Since the authors of these methods ordinarily provide such programs for the use of the scientific community, the user is even spared the task of having to develop his own computer program to carry out the calculations.

Most computer centers today have available standard programs that carry out factor extractions and analytic rotations by one or more methods. Many investigators, therefore, are able to prepare data for such "package" computer treatment and to obtain factor analytic results for interpretation with little understanding of what goes on in between or of the extent to which the program may be appropriate as a form of treatment for their data. Such applications of factor analysis often tend to be afterthoughts of the kind, "Well, what do we do with the data now that we have collected it?" Data collected with no advance plans for data analysis are apt to be poor data. Factor analysis, unfortunately, has no magical power to extract good results from poor data, especially when the investigator is making his choice of the form of factor analytic treatment with no knowledge of the relevance of that treatment for his particular data.

Factor analysis as a tool is not responsible for the poor quality of such research investigations any more than the hammer is at fault for the imperfections in the work of an amateur carpenter. He who would effectively exploit the potential of factor analysis as a research tool must acquire an understanding of the method and how it is to be properly applied. Proper planning of the factor analytic investigation before data collection starts is just as important in factor analytic work as it is in experimental work that is to be followed by data treatment using analysis of variance. In the remainder of the chapter, attention is given to the properties of some of the more useful analytic methods of rotation. It should be made very clear, however, that no existing analytic method of rotation can be correct or appropriate for all data situations. For any set of rotation rules, it is possible to devise a data situation that makes this set of rules inappropriate. Analytic methods of rotation do eliminate judgment of the investigator in carrying out the rotations after the method of analytic rotation has been selected. They do not eliminate, however, the necessity of judgment on the part of the investigator in determining whether the selected method of analytic rotation is appropriate for his data. The elimination of experimenter judgment with analytic methods of rotation, therefore, is more apparent than real.

There can be no doubt that the use of analytic criteria of rotation reduces discrepancies between the solutions obtained by different investigators. Barring errors in applying the rules, all investigators should end up with the same results starting from the same unrotated matrix. This is a desirable outcome, other things being equal, but such unanimity among investigators is of limited scientific value if though agreed, they are all wrong. This is a distinct possibility when a randomly chosen analytic criterion of rotation is applied to a randomly chosen unrotated matrix of factor loadings. The author has yet to find an experienced factor analyst who is willing to assert that he has been uniformly satisfied with the results of any analytic method of rotation when it is applied to a variety of problems. Marks, Michael, and Kaiser (1960), for example, found little agreement between Varimax analytic rotation factors and factors rotated visually by the Zimmerman method. The reader is cautioned, therefore, to avoid the uncritical acceptance of computer-produced factor analytic results. They may be appropriate and they may not be. Experimenter judgment may be necessary to decide.

The fact that analytic rotation methods can be programmed for computer application provides a tremendous labor-saving advantage over traditional hand rotation using 2×2 plots. What is not well understood, however, is the fact that the drudgery of plotting and calculation for the traditional hand-rotation procedures can be programmed for computer, too, so that the investigator is left only with the necessity of inspecting the plots and making decisions about what factors to rotate and where (Cattell & Foster, 1963). Once this decision is made, the computer can be used to carry out the numerical operations and to produce the new plots. A program, prepared by the author, that will carry out these operations is described in a later chapter. With computer assistance, then, traditional hand-rotation procedures are only slightly less convenient than analytic methods as far as the amount of effort involved. Certainly the difference in labor would not be sufficient to be of any practical significance as far as the amount of effort involved in the total research project is concerned. Of far greater importance is the choice of those procedures that will lead to the best possible solution.

[7.3.1] *COMPUTERIZED SIMPLE STRUCTURE SOLUTIONS*

Although rotation to oblique simple structure is not classified as an analytic method, it is possible to formulate a set of rules to apply by computer for the purpose of searching for solutions that appear to meet the demands of oblique simple structure. Since such applications involve an explicit set of rules, they will be classified as analytic methods.

Cattell and Muerle (1960) developed the "Maxplane" program for factor rotation to oblique simple structure. This program counts the number of points that fall "in the hyperplane" for each factor, for example, within the range of +.10 to −.10. Factors are rotated systematically by computer to increase the number of points falling in the hyperplanes. Cattell (1966) states that Maxplane gives higher hyperplane counts than analytic rotation programs based on satisfying some mathematical criterion. Eber (1966) has produced a more recent version of the Maxplane rotation program that starts off by spinning the factor axes to positions at 45° to the unrotated factor positions and normalizing all data vectors to unit communalities. Then, the search for hyperplane improvement begins. A weight function is used, however, to adjust the hyperplanes rather than an all-or-none decision which merely labels the variable "in" or "out" of the hyperplane. Guertin and Bailey (1970, pp. 135ff.) have had some success with the Maxplane method, although they report in general that oblique analytic solutions tend to put too many points in the hyperplanes and to give angles between factors that are more oblique than necessary. The author's experience is in agreement with that reported by Guertin and Bailey.

The author has also developed a program for obtaining oblique simple structure solutions by computer. This program operates on a principle similar to that used with the original Maxplane program. The number of points in the hyperplane for each reference vector is counted at the beginning. Then, trial rotations a specified number of degrees toward and away from each other reference vector are tried to see if the number of points in the hyperplane is increased. If so, the new reference vector position is adopted. If not, the old one is retained. After every reference vector has been tried out in comparison with every other reference vector, the size of the trial angle of rotation is reduced and the process is repeated. At each step, the size of the correlation between reference vectors is checked. No rotation is permitted that would result in a correlation between reference vectors that exceeds a predetermined level.

Experience with this program, which is described in a later chapter in more detail, indicates that it can "clean up" a solution that is already nearing simple structure. A maximum correlation between reference vectors that is near the largest value currently found among the oblique reference vectors is selected, with the assumption that oblique hand rotations have already been completed or nearly completed. A search angle that is not too large to begin with is preferred. These constraints prevent the search for a "better" simple structure from getting out of hand. Allowing larger search angles and greater obliquity between reference vectors with the program tends to result in an increased number of points in the hyperplanes but at the expense of higher correlations among reference vectors. In applying this type of program, the best strategy appears to be to come as close as possible to simple structure positions that are acceptable to the investigator using hand-rotation methods. Then, apply the computer search procedure with restrictions on the size of search angle and the maximum permissible correlation between reference vectors to prevent any radical departure from the solution already achieved at the beginning of the computer run. Even with such restrictions, the computer search will usually result in significant hyperplane improvements. It is often convenient to start this process with the best orthogonal approximation to simple structure rather than with an oblique hand-rotation solution. Under these circumstances, it is best to start off with modest search angles and permissible correlations among reference vectors. These may be increased on a rerun later if this is strongly indicated. The greatest difficulty with oblique solutions is that of determining how oblique the reference vectors may be permitted to go. The greater the freedom allowed in this respect, the more points there will be in the hyperplanes and the greater will be the correlations between the reference vectors as a general rule. There is virtually no assurance that the angles will stabilize at reasonable values if greater latitude is permitted.

[7.3.2] *"PROCRUSTES" SOLUTIONS*

The Procrustes-type solutions are classified here as analytic in character because a specified set of rules is laid down governing how rotations are to be carried out to transform the unrotated factor matrix into a form maximally similar in some sense to a predetermined target matrix. The target matrix may have all the values specified or only some of them. Mosier (1939) derived equations to provide a least-squares fit to a target matrix, but these equations could not be solved algebraically. More recently, Browne (1967) has developed an iterative method for solving these equations, giving rise to an oblique Procrustes-type solution. Hurley and Cattell (1962) have also developed a program for this type of solution. Green (1952) presented procedures for a Procrustes-type solution in which the final factors are orthogonal to one another. See also Schönemann (1966) and Schönemann and Carroll (1970).

Guilford has found helpful in his work the Procrustes-type method of Cliff (1966). This method permits either the orthogonal rotation of two matrices to maximum agreement with each other or the orthogonal rotation of one matrix to maximum agreement with a specified target matrix. The target matrix may be formulated on the basis of theoretical expectation, knowledge of past results, or inspection of a preliminary rotational solution to the problem under investigation by other rotational methods.

Guilford has found Cliff's Procrustes-type solution to his liking because it enables him to maintain maximum continuity between past results and new findings. His extensive programmatic researches and exhaustive knowledge of the variables he is working with provide a wealth of constraints that can be reasonably applied in rotating the factors for a new study designed to verify or extend previous findings. He may have knowledge, for example, that a given factor has no pure-factor tests in the battery to identify it. It may be necessary to locate a factor in a position that would never be attained by usual rotational criteria. If the hypothesis is consistent with the data in the new matrix, a properly formulated "target" matrix and a Procrustes-type solution will produce the expected factor results. For investigators with less reliable knowledge about the probably appropriate form of solution, however, this approach should be applied with extreme caution.

[7.3.3] *ORTHOGONAL ANALYTIC ROTATION METHODS*

Dissatisfaction with the subjectivity and vagueness about simple structure criteria led to an effort to provide more precise mathematical statements of the criteria which a rotational solution should satisfy. Several methods were published in the fifties that proved to be rather similar to each other (Carroll, 1953; Ferguson, 1954; Neuhaus & Wrigley, 1954; Saunders, 1960) according to Harman (1967, p. 298). Harman refers to these similar methods collectively as the "Quartimax" method. The essential idea underlying this approach is to rotate in such a way that ideally each variable would have a major loading in one and only one factor. That is, each variable would appear to be a pure-factor measure. In practice, of course, it is not possible ordinarily to obtain such an idealized solution. It is only approached as closely as possible in the Quartimax solution.

Experience with the Quartimax approach has shown that it tends to give a general factor, that is, a factor with which all or most of the variables are substantially correlated. This undesirable property of the Quartimax class of solutions coupled with the appearance of the superior Varimax method of Kaiser has led to the virtual abandonment of these earlier methods.

The Kaiser Varimax method (Kaiser, 1958) has become perhaps the most popular rotational procedure in use today. It is a method that gives excellent results in certain situations. It is efficient and readily available at most computer centers, for example, as part of the BMD package of programs (Dixon, 1970a, b). The Varimax method does not suffer from the tendency to give a general factor, as the Quartimax method does, although Cattell (1966) states that it still does not pull enough variance away from the largest extracted factor. The Varimax method is of such great importance that it is considered in a separate section later in this chapter (Section 7.4). At this point, however, it can be compared briefly with the Quartimax method.

In a sense, the Quartimax method attempts to maximize the variance of squared factor loadings by rows, since if all the variance for a given variable is concentrated on one factor. leaving the other loadings essentially equal to zero, this result will be achieved. The Varimax method, by contrast, maximizes the variance of the squared factor loadings by columns. Thus, on any given factor, a pattern is desired such that there are some high loadings and lots of low loadings with few intermediate-sized loadings. This type of solution offers easy interpretability because variables with high loadings are similar to the factor in character, whereas variables with low loadings are not. A variable with an intermediate loading is neither fish nor fowl. It is of little use in telling what the factor is or is not like. Varimax rotations, then, tend to push high loadings higher and low loadings lower, to the extent that this is possible within the constraints maintained by an orthogonal reference frame. Factor positions are determined, therefore, by the locations of clusters as well as by the location of hyperplanes. A Varimax factor will tend to be drawn toward a neighboring cluster because this elevates the variance of the squared factor loadings for the factor (for further discussion of the Varimax method see Section 7.4).

Whereas most analytic methods of rotation represent an attempt to specify mathematically in some sense the essential characteristics of a simple structure solution, the Tandem Criteria for analytic rotation (Comrey, 1967) are based on a somewhat different idea. This method is a two-stage affair. The first stage carries out rotations on the principle that variables that are correlated with each other should appear on the same factor. This first stage has the effect of distributing the variance among the factors more evenly than in the unrotated factor solution, but less so than with the Varimax. Factors of minor importance are dropped out before applying the second stage of rotation, which is based on the principle that variables which are uncorrelated with each other should not appear on the same factor. The Tandem Criteria method of rotation has some advantages over the Varimax method, as described later.

[7.3.4] *OBLIQUE ANALYTIC METHODS OF ROTATION*

Computerized simple structure search methods and Procrustes-type approaches that give oblique solutions have already been mentioned. Well before the development of these methods, however, Thurstone (1947) presented his Single-Plane method which is a semianalytic procedure for obtaining oblique solutions. The investigator starts with a given variable thought to be a likely defining variable for a particular factor. A trial vector is placed through this variable and projections of the data-variable vectors on this trial vector are obtained. These values are plotted against all the unrotated factor vectors and an adjustment is made in the position of the trial vector simultaneously with respect to all unrotated factor axes to improve simple structure. Projections of the data-variable vectors on the new trial vector are computed and new plots are made. A new adjustment in the trial vector position is made if further adjustments are indicated. The process is continued until no further adjustments are needed. This positions the first reference vector. A second trial reference vector is placed through a variable that has a near-zero correlation with the first trial vector but still has a substantial communality. All the oblique reference vectors are positioned separately in this way, one after the other. Fruchter (1954) has a clear exposition of this process.

Thurstone (1954) developed a more sophisticated version of the Single-Plane method that uses a weighting function permitting calculation of the corrections to the trial vector position simultaneously with respect to all factors without plotting and without making judgments. It is more analytic in character, therefore, than the Single-Plane method. As previously mentioned, however, the author's experience with this method suggests that it can give misleading results if not used with considerable judgment and skill (Comrey, 1959).

More recently, several strictly analytic methods of carrying out oblique rotations have been suggested. Of these, the best known at the present time appear to be the Biquartimin criterion of Carroll (1957), the Promax method of Hendrickson and White (1964), and a procedure developed by Jennrich and Sampson (1966) to get factor loadings on the oblique factors directly without first obtaining projections on the reference vectors. Computer programs in the BMD series (Dixon, 1970a, b) are available for the Carroll and the Jennrich and Sampson methods. A good discussion of several of these oblique analytic rotation methods is to be found in Harman (1967). Gorsuch (1970) found the Promax method to give results comparable to those with other analytic methods but with considerably less computer time.

Guertin and Bailey (1970, p. 135) have found, however, that oblique analytic solutions tend to put too many points in the hyperplanes. There is little evidence available at the present time to suggest that any existing oblique analytic solution can perform as well with nonorthogonal situations as Varimax does with cases known to require orthogonal solutions. The lack of success in developing totally satisfactory analytic oblique methods of rotation probably can be attributed, at least in part, to the fact that it is difficult to decide how much obliquity to tolerate in oblique hand rotations. Simple structure fails to provide an entirely satisfactory guide because it is possible to continue to improve hyperplanes by allowing the axes to go more oblique. The point at which the investigator calls a halt to this process is to some extent arbitrary. Because of the present uncertainty about the value of these completely analytic rotation methods in actual practice, none of them will be described in detail. More experience with these methods or perhaps new methods will be required before satisfactory oblique analytic rotation becomes a practical reality, if in fact it ever does.

[7.4] THE KAISER VARIMAX METHOD

Because of its great popularity and general usefulness, the Varimax method of Kaiser (1958, 1959) is described here in some detail. It has already been stated that the Varimax method is based on the idea that the most interpretable factor has high and low but few intermediate-sized loadings. Such a factor would have a large variance of the squared loadings since the values are maximally spread out. Using the square of the formula for the standard deviation, the variance of the squared factor loadings on factor j may be symbolized as follows:

$$\sigma_j^2 = \frac{1}{n}\sum_{i=1}^{n}(a_{ij}^2)^2 - \frac{1}{n^2}\left(\sum_{i=1}^{n} a_{ij}^2\right)^2 \tag{7.1}$$

The variance should be large for all factors; so an orthogonal solution is sought where V is a maximum, V being defined as follows:

$$V = \sum_{j=1}^{m} \sigma_j^2 \tag{7.2}$$

In practice, V is not maximized in one operation. Rather, factors are rotated with each other systematically, two at a time in all possible pairs, each time maximizing $\sigma_i^2 + \sigma_j^2$. After each factor has been rotated with each of the other factors, completing a cycle, V is computed, and another cycle is begun. These cycles are repeated until V fails to get any larger. The fundamental problem, then, is to find an orthogonal transformation matrix $\mathbf{\Lambda}$ that will rotate two factors such that $\sigma_i^2 + \sigma_j^2$ for the rotated factors will be as large as possible. Consider the following:

$$\underset{\mathbf{V}_1}{\begin{bmatrix} x_1 & y_1 \\ x_2 & y_2 \\ x_3 & y_3 \\ \vdots & \vdots \\ x_n & y_n \end{bmatrix}} \times \underset{\mathbf{\Lambda}}{\begin{bmatrix} \cos\varphi & -\sin\varphi \\ \sin\varphi & \cos\varphi \end{bmatrix}} = \underset{\mathbf{V}_2}{\begin{bmatrix} X_1 & Y_1 \\ X_2 & Y_2 \\ X_3 & Y_3 \\ \vdots & \vdots \\ X_n & Y_n \end{bmatrix}} \tag{7.3}$$

where $\mathbf{V}_1$ is a matrix of factor loadings to be rotated to maximize $\sigma_i^2 + \sigma_j^2$, $\mathbf{V}_2$ is the matrix of factor loadings for which $\sigma_i^2 + \sigma_j^2$ is a maximum, and $\mathbf{\Lambda}$ is the orthogonal transformation matrix that will accomplish this desired rotation. The values in $\mathbf{V}_1$ are known. The values in $\mathbf{V}_2$ are not. The values in $\mathbf{V}_2$, however, are functions of the angle φ and the values in $\mathbf{V}_1$ as follows:

$$\begin{aligned} X_j &= x_j \cos\varphi + y_j \sin\varphi \\ Y_j &= -x_j \sin\varphi + y_j \cos\varphi \end{aligned} \tag{7.4}$$

The sum of the variances to be maximized is given by

$$\sigma_i^2 + \sigma_j^2 = \frac{1}{n}\sum (X^2)^2 - \frac{1}{n^2}(\sum X^2)^2 + \frac{1}{n}\sum (Y^2)^2 - \frac{1}{n^2}(\sum Y^2)^2 \tag{7.5}$$

The development from this point requires some knowledge of calculus. The reader may wish to skip to Eq. (7.8) if he is unfamiliar with this mathematical tool.

By substituting Eqs. (7.4) in Eq. (7.5), the unknown values of X and Y can be removed from Eq. (7.5), leaving φ as the only unknown. The resulting expression can be differentiated with respect to φ, set equal to zero, and a solution for φ determined that maximizes the expression in (7.5). It is easier, however, to differentiate (7.5) with respect to φ before making the substitutions of Eqs. (7.4) into (7.5). Doing it this way requires a knowledge of values for $dX/d\varphi$ and $dY/d\varphi$. Dropping the subscripts from Eqs. (7.4) and differentiating with respect to φ gives

(7.6)
$$\frac{dX}{d\varphi} = x(-\sin\varphi) + y(\cos\varphi) = Y$$
$$\frac{dY}{d\varphi} = -x(\cos\varphi) + y(-\sin\varphi) = -X$$

Differentiating (7.5) with respect to φ, substituting Eqs. (7.6) in the resulting equation, setting it equal to zero, and multiplying both sides by n^2 gives

(7.7)
$$n\sum XY(X^2-Y^2) - \sum XY\left(\sum X^2 - \sum Y^2\right) = 0$$

Substitution of Eq. (7.4) into Eq. (7.7) gives an expression involving only the unknown φ and the known values of x and y, the loadings before rotation. Using a table of trigonometric identities, the resulting expression can be simplified and solved for φ to give

(7.8)
$$\tan 4\varphi = \frac{2[n\sum(x^2-y^2)(2xy) - \sum(x^2-y^2)\sum(2xy)]}{n\{\sum[(x^2-y^2)^2-(2xy)^2]\} - \{[\sum(x^2-y^2)]^2 - [\sum(2xy)]^2\}}$$

Any value of φ that maximizes Eq. (7.7) must satisfy Eq. (7.8). To ensure that the value of φ gives a maximum rather than a minimum or point of inflection, however, requires that the second derivative of Eq. (7.7) with respect to φ shall be negative when evaluated at φ. The angles of rotation that will accomplish this result may be determined as follows:

First, $\tan 4\varphi = \sin 4\varphi/\cos 4\varphi$; so Eq. (7.8) may be rewritten as

(7.9)
$$\tan 4\varphi = \frac{\sin 4\varphi}{\cos 4\varphi} = \frac{\text{num}}{\text{denom}}$$

The angle 4φ will be in the first quadrant if both numerator and denominator of Eq. (7.8) are positive, and the angle of rotation will be φ itself. If both numerator and denominator are negative, the tangent will still be positive, but the required angle of rotation is $-\frac{1}{4}(180° - 4\varphi) = -(45° - \varphi)$ since the sine and cosine are both negative in the third quadrant. If the numerator is negative and the denominator is positive, the angle 4φ will be in the fourth quadrant and the angle of rotation will be $-\varphi$. Finally, if the numerator of Eq. (7.8) is positive and the denominator is negative, 4φ will be in the second quadrant and the angle of rotation will be $\frac{1}{4}(180° - 4\varphi) = (45° - \varphi)$.

Then, to determine the desired rotation transformation matrix in Eq. (7.3), compute the numerator (num) in Eq. (7.8) and the denominator (denom), keeping their signs. Ignoring these signs for the moment, find a positive angle 4φ in the first quadrant such that the tangent of 4φ equals the absolute value of num/denom. Find the angle φ as one-fourth the angle with this tangent. This angle must be $22\frac{1}{2}°$ or less. The transformation matrix Λ that maximizes the function in Eq. (7.5) will then be one of the following, depending on the combination of signs in the numerator (num) and denominator (denom) of Eq. (7.8):

$$\underset{\Lambda\,(\text{num}\,+,\ \text{denom}\,+)}{\begin{bmatrix} \cos\varphi & -\sin\varphi \\ \sin\varphi & \cos\varphi \end{bmatrix}}, \qquad \underset{\Lambda\,(\text{num}\,-,\ \text{denom}\,+)}{\begin{bmatrix} \cos\varphi & \sin\varphi \\ -\sin\varphi & \cos\varphi \end{bmatrix}} \tag{7.10}$$

$$\underset{\Lambda\,(\text{num}\,+,\ \text{denom}\,-)}{\begin{bmatrix} \cos(45°-\varphi) & -\sin(45°-\varphi) \\ \sin(45°-\varphi) & \cos(45°-\varphi) \end{bmatrix}}, \qquad \underset{\Lambda\,(\text{num}\,-,\ \text{denom}\,-)}{\begin{bmatrix} \cos(45°-\varphi) & \sin(45°-\varphi) \\ -\sin(45°-\varphi) & \cos(45°-\varphi) \end{bmatrix}}$$

As an illustration, the Λ matrix will be determined that will give a Varimax rotation of the two centroid factors in matrix **A** of Table 2.2, with the results hopefully approximating those in matrix **V** of Table 2.2. These calculations are shown in Table 7.1. The tangent of 4φ is $+.4737$, with both numerator and denominator negative. The angle 4φ is 25°21′; so the angle φ is 6°20′. Since the denominator is negative, the angle of rotation is $45°-\varphi$, or 38°41′. This angle was rounded to 39° to obtain the Λ matrix. Multiplying the unrotated centroid matrix **A** by the matrix Λ does give the expected **V** matrix as shown in Table 7.1. In this example, therefore, the Varimax recovered the original data-variable configuration very well.

With the Varimax method it is important to rotate the proper number of factors. Rotation of too few factors with the Varimax criterion, as with any method of rotation, crowds the variance for $n+k$ factors into a space of only n dimensions, thereby losing some factors and distorting the others. Rotating too many factors, however, can also bring about distortions with the Varimax criterion. The Varimax method will tend to build up the variance on extra small factors to increase the overall variance function V, thereby splitting up major factors and "robbing" larger factors of their variance for certain data variables. The best indication that this has happened is the appearance of a Varimax factor with only one large loading and all the rest of the loadings at much lower levels. The solution should be rerotated, dropping out any factor that is artificially overinflated in this way.

A more subtle type of distortion occurs where several variables define a major factor, but two or three of these variables are more highly correlated among themselves than with the other variables defining the factor. This often results in a small unrotated factor with modest loadings for these more highly

TABLE 7.1

Varimax Rotation of Two Centroid Factors from the Four-Variable Problem

x	y	x^2-y^2	$2xy$	$(x^2-y^2)^2$	$(2xy)^2$	$(x^2-y^2)(2xy)$	$(x^2-y^2)^2-(2xy)^2$
.578	.562	.0182	.6497	.0003	.4221	.0118	−.4218
.531	.344	.1636	.3653	.0268	.1334	.0598	−.1066
.687	−.422	.2939	−.5798	.0864	.3362	−.1704	−.2498
.765	−.484	.3510	−.7405	.1232	.5483	−.2599	−.4251
	Sum	.8267	−.3053	.2367	1.4400	−.3587	−1.2033

$$\begin{aligned}\operatorname{Tan} 4\varphi &= \frac{2[4(-.3587)-(.8267)(-.3053)]}{4\{(-1.2033)\}-\{(.8267)^2-(-.3053)^2\}}\\ &= \frac{-2.8696+.5048}{-.48132-(.6834-.0932)}\\ &= \frac{-2.4580}{-4.8132-.5902}\\ &= \frac{-2.4580}{-5.4034}\\ &= .47374\end{aligned}$$

The absolute value of $\tan 4\varphi$ is .47374 giving an angle of 25°21′ for 4φ. The angle φ is 6°20′. With a negative denominator, the angle of rotation will be $(45°-\varphi)$ or 38°41′ which is rounded off to 39°. The cosine and sine of 39°, respectively, are .7771 and .6293. With both numerator and denominator negative, the transformation matrix chosen from (7.10) is as follows:

$$\underset{\mathbf{A}}{\begin{bmatrix} .578 & .562 \\ .531 & .344 \\ .687 & -.422 \\ .765 & -.484 \end{bmatrix}} \times \underset{\mathbf{\Lambda}}{\begin{bmatrix} .7771 & .6293 \\ -.6293 & .7771 \end{bmatrix}} = \underset{\mathbf{V}}{\begin{bmatrix} .10 & .80 \\ .20 & .60 \\ .80 & .10 \\ .90 & .11 \end{bmatrix}}$$

interrelated variables in addition to the main factor with all the variables for this factor represented. If this small extra factor is rotated into the Varimax solution, the variance for the small additional factor will be inflated, pulling variance off the larger factor so that these variables appear to be considerably less related to the larger factor than they could be. The smaller factor, meanwhile, is inflated to a larger overall share of the variance than is appropriate.

It is very important in using Varimax, therefore, to avoid including extra factors in the rotations. It is often necessary to inspect several solutions, varying the number of factors rotated, before the most satisfactory solution is settled upon. It also may be necessary to rotate the Varimax solution further orthogonally or obliquely, or both, before an entirely satisfactory result has been obtained. Where such hand rotations are undertaken, starting from a Varimax solution represents a great time-saver compared with starting the hand rotations from the original unrotated matrix. The Varimax solution is most apt to prove satisfactory when the factor analytic investigation has been carefully designed so that the probable solution will be likely to approximate the demands of the Varimax criterion.

Kaiser recommends the use of the "normal" Varimax in preference to the "raw" Varimax method just described. With the normal Varimax, all data-variable vectors are scaled to unit length by dividing each variable's loadings by the square root of its communality. Then, the raw Varimax procedure can be applied in carrying out the rotations. At the conclusion, the loadings for each variable are rescaled back down to proper size by multiplying each loading for a given variable by the square root of the communality for that variable. After this operation, the sum of squares of the loadings for the variable will again equal the communality instead of 1.0.

The "normal" Varimax method gives equal weight to all variables in determining the rotations. This seems reasonable in a well-designed study where all variables have substantial communalities. Where many of the communalities are quite low, however, the investigator is encouraged by this author to consider the possibility that the "raw" Varimax solution may be superior to the normal Varimax solution. To inflate a very low communality to 1.0 and then treat that variable as if it were of equal importance to those with originally high communalities seems questionable as a general practice.

[7.5] THE TANDEM CRITERIA METHOD

The author has long felt that available information about the correlations among the data variables could be utilized profitably in factor rotations. Experimentation with this idea led to the development of two analytic rotational criteria that are most effective when used in tandem. The first of these, Criterion I, is based on the following principle.

Principle I. If two variables are correlated, they should appear on the same factor.

A mathematical interpretation of this principle is developed in the following section to provide the basis for Criterion I rotations.

[7.5.1]
ROTATION BY CRITERION I OF THE TANDEM CRITERIA

Principle I may be approximated mathematically by the following:

$$F = \sum_{k=1}^{m} \sum_{i=1}^{n} \sum_{j=1}^{n} r_{ij}'^2 a_{ik}^2 a_{jk}^2 \tag{7.11}$$

where

$r'_{ij} = \sum_{k=1}^{n} a_{ik} a_{jk}$, the reproduced correlation between variable i and variable j,

a_{ik} is the factor loading of variable i on factor k,

a_{jk} is the factor loading of variable j on factor k.

The function F in Eq. (7.11) is to be maximized. This will occur when the variables that are correlated with each other are put on the same factor. As with the Varimax method, maximization of the function in Eq. (7.11) is accomplished by rotating the factors two at a time, maximizing the function in each rotation separately. After each factor is rotated with each of the other factors, and a cycle is thus completed, the function value is computed. Additional cycles are completed until the value of the function fails to increase. For two factors, the value of the function may be expressed as follows:

$$f = \sum_{i-1}^{n} \sum_{j-1}^{n} r_{ij}'^2 (X_i^2 X_j^2 + Y_i^2 Y_j^2) \tag{7.12}$$

where X and Y are the rotated factor loadings as defined in Eqs. (7.3) and (7.4) for the Varimax method.

The remainder of the development requires calculus. The reader may skip to Eq. (7.20) if he desires to pass around this part of the proof. The function in Eq. (7.12) is differentiated with respect to φ and set equal to zero, with Eqs. (7.6) used to give the values for $dX/d\varphi$ and $dY/d\varphi$:

$$\frac{df}{d\varphi} = \sum_{i=1}^{n} \sum_{j=1}^{n} r_{ij}'^2 \times [X_i^2(2X_j Y_j) + X_j^2(2X_i Y_i) + Y_i^2(2Y_j)(-X_j) + Y_j^2(2Y_i)(-X_i)]$$

(7.13)

$$\frac{df}{d\varphi} = \sum_{i=1}^{n} \sum_{j=1}^{n} r_{ij}'^2 [2X_j Y_j(X_i^2 - Y_i^2) + 2X_i Y_i(X_j^2 - Y_j^2)]$$

Breaking (7.13) into parts and substituting Eqs. (7.4) for X and Y, we have

(7.14)

$$2X_i Y_i = -2(-x_i^2 \sin\varphi\cos\varphi + x_i y_i \cos^2\varphi - x_i y_i \sin^2\varphi + y_i^2 \sin\varphi\cos\varphi)$$

(7.15)

$$2X_i Y_i = -(x_i^2 - y_i^2)\sin 2\varphi + 2x_i y_i \cos 2\varphi$$

By analogy to Eq. (7.15),

(7.16) $$2X_j Y_j = -(x_j^2 - y_j^2)\sin 2\varphi + 2x_j y_j \cos 2\varphi$$

Continuing with other parts of (7.13), we get

$$\begin{aligned} X_i^2 - Y_i^2 &= x_i^2 \cos^2\varphi + y_i^2 \sin^2\varphi + 2x_i y_i \cos\varphi \sin\varphi \\ &\quad -(x_i^2 \sin^2\varphi + y_i^2 \cos^2\varphi - 2x_i y_i \sin\varphi \cos\varphi) \\ &= (x_i^2 - y_i^2)\cos^2\varphi - (x_i^2 y_i^2)\sin^2\varphi + 4x_i y_i \sin\varphi \cos\varphi \end{aligned}$$

(7.17) $$X_i^2 - Y_i^2 = (x_i^2 - y_i^2)\cos 2\varphi + 2x_i y_i \sin 2\varphi$$

By analogy with (7.17):

(7.18) $$X_j^2 - Y_j^2 = (x_j^2 - y_j^2)\cos 2\varphi + 2x_j y_j \sin 2\varphi$$

Substituting (7.15)–(7.18) in Eq. (7.13) and simplifying further leads to

(7.19) $$\begin{aligned} \frac{df}{d\varphi} &= -\sum_{i=1}^{n}\sum_{j=1}^{n} r_{ij}'^2 [(x_i^2 - y_i^2)(x_j^2 - y_j^2) - 4x_i y_i x_j y_j] \sin 4\varphi \\ &\quad + \sum_{i=1}^{n}\sum_{j=1}^{n} r_{ij}'^2 [2x_i y_i (x_j^2 - y_j^2) + 2x_j y_j (x_i^2 - y_i^2)] \cos 4\varphi \end{aligned}$$

Setting Eq. (7.19) equal to zero and solving for $\tan 4\varphi$, which equals $\sin 4\varphi / \cos 4\varphi$, gives

(7.20) $$\tan 4\varphi = \frac{2\sum_{i=1}^{n}\sum_{j=1}^{n} r_{ij}'^2 [x_i y_i (x_j^2 - y_j^2) + x_j y_j (x_i^2 - y_i^2)]}{\sum_{i=1}^{n}\sum_{j=1}^{n} r_{ij}'^2 [(x_i^2 - y_i^2)(x_j^2 - y_j^2) - 4x_i y_i x_j y_j]}$$

Any angle of rotation φ that maximizes the function in Eq. (7.12) must satisfy Eq. (7.20). To ensure, however, that a maximum is reached, rather than a minimum or a point of inflection, the second derivative evaluated at φ must be negative. When Eq. (7.20) is expressed as $\tan 4\varphi = \text{num}/\text{denom}$, Eq. (7.19) may be expressed as

(7.21) $$\frac{df}{d\varphi} = -\text{denom} \sin 4\varphi + \text{num} \cos 4\varphi$$

Differentiating (7.21) with respect to φ gives

(7.22) $$\begin{aligned} \frac{d^2 f}{d\varphi^2} &= -[\text{denom}(4\cos 4\varphi + \text{num}(-\sin 4\varphi)](4) \\ &= -4(\text{denom} \cos 4\varphi + \text{num} \sin 4\varphi) \end{aligned}$$

The quadrant in which 4φ is placed must be such that the expression in (7.22), the second derivative, remains negative. These relationships are shown in Table 7.2.

TABLE 7.2

Sign combinations	Second derivative	Sign	4φ
num +, denom +	$-4[(+)(+)+(+)(+)]$	−	0–90°
num +, denom −	$-4[(-)(-)+(+)(+)]$	−	90–180°
num −, denom −	$-4[(-)(-)+(-)(-)]$	−	180–270°
num −, denom +	$-4[(+)(+)+(-)(-)]$	−	270–360°

The angle of rotation that will maximize the function in Eq. (7.12) can be found, therefore, from the following:

(7.23)

		num −	num +
denom	+	$-\varphi$	φ
	−	$-(45°-\varphi)$	$(45°-\varphi)$

The angle φ itself is found from Eq. (7.20), by ignoring the signs on numerator and denominator, taking 4φ in the first quadrant, and dividing the angle 4φ by 4 to get φ. With this value of φ, the transformation matrix that will perform the Criterion I rotation should be chosen from (7.10) by the same rules used to select the transformation matrix $\mathbf{\Lambda}$ for the Varimax rotation.

As an illustration, the four-variable centroid matrix in Table 2.2 will be rotated by Criterion I. The computations are shown in Table 7.3. The rotated values in matrix $\mathbf{V}$ are very close to those in Table 2.2 although not so close as the Varimax values.

Criterion I rotations are not expected to and generally do not give a close approximation to simple structure, however. Criterion I rotations spread the variance out to make the rotated factors less unequal in size than the unrotated factors, as do other methods of rotation. To remain on the same factor, variables must be correlated with each other. This forces a spreading of the variance from larger unrotated factors to smaller unrotated factors because the large unrotated factors typically have substantial loadings for variables that in many cases are relatively orthogonal to each other. The demand that correlated variables should be placed on the same factor puts an effective damper, however, on the process of shifting variance from larger to smaller factors. The effect of this is that all the available unrotated factors that could possibly be worth retaining can be included in the Criterion I rotations without

TABLE 7.3

Criterion I Rotation of Two Centroid Factors from the Four-Variable Problem

x	y	$(x^2 - y^2)$	xy		1	2	3	4
.578	.562	.0182	.3248	1	(.65)	.50	.16	.17
.531	.344	.1636	.1827	2	.50	(.40)	.22	.24
.687	−.422	.2939	−.2899	3	.16	.22	(.65)	.73
.765	−.484	.3510	−.3702	4	.17	.24	.73	(.82)

Matrix of r' values

Numerator values:

For $i = 1, j = 1$	$1 \times .65[(.3248)(.0182)+(.3248)(.0182)] =$	.0077
For $i,j = 1,2$ and $2,1$	$2 \times .50[(.3248)(.1636)+(.1827)(.0182)] =$	.0565
For $i,j = 1,3$ and $3,1$	$2 \times .16[(.3248)).2939)+(-.2899)(.0182)] =$	.0289
For $i,j = 1,4$ and $4,1$	$2 \times .17[(.3248)(.3510)+(-.3702)(.0182)] =$	.0365
For $i = 2, j = 2$	$1 \times .40[(.1827)(.1636)+(.1827)(.1636)] =$	.0239
For $i,j = 2,3$ and $3,2$	$2 \times .22[(.1827)(.2939)+(-.2899)(.1636)] =$	.0028
For $i,j = 2,4$ and $4,2$	$2 \times .24[(.1827)(.3510)+(-.3702)(.1636)] =$	.0017
For $i = 3, j = 3$	$1 \times .65[(-.2899)(.2939)+(-.2899)(.2939)] =$	−.1107
For $i,j = 3,4$ and $4,3$	$2 \times .73[(-.2899)(.3510)+(-.3702)(.2939)] =$	−.3075
For $i = 4, j = 4$	$1 \times .82[(-.3702)(.3510)+(-.3702)(.3510)] =$	−.2131
	Sum	−.4733

Numerator (num) for Eq. (7.20) equals $2 \times (-.4733)$ or $-.9466$.

Denominator values:

For $i = 1, j = 1$	$1 \times .65[(.3248)(.0182)-4(.1055)] =$	−.2705
For $i,j = 1,2$ and $2,1$	$2 \times .50[(.3248)(.1636)-4(.0593)] =$	−.1841
For $i,j = 1,3$ and $3,1$	$2 \times .16[(.3248)(.2939)-4(-.0942)] =$	.1511
For $i,j = 1,4$ and $4,1$	$2 \times .17[(.3248)(.3510)-4(-.1202)] =$	.2022
For $i = 2, j = 2$	$1 \times .40[(.1827)(.1636)-4(.0334)] =$	−.0415
For $i,j = 2,3$ and $3,2$	$2 \times .22[(.1827)(.2939)-4(-.0530)] =$	.1169
For $i,j = 2,4$ and $4,2$	$2 \times .24[(.1827)(.3510)-4(-.0676)] =$	.1606
For $i = 3, j = 3$	$1 \times .65[(-.2899)(.2939)-4(.0840)] =$	−.2738
For $i,j = 3,4$ and $4,3$	$2 \times .73[(-.2899)(.3510)-4(.1073)] =$	−.7753
For $i = 4, j = 4$	$1 \times .82[(-.3702)(.3510)-4(.1370)] =$	−.5559
	Denominator (denom) and sum	−1.4703

$$\operatorname{Tan} 4\varphi = -.9466/-1.4703 = .6438, \qquad 4\varphi = 32°46', \qquad \varphi = 8°11\tfrac{1}{2}'$$

Rounding φ to 8°, $(45° - \varphi) = 37°$. The angle of rotation is $(45° - \varphi)$ because the denominator is negative. The cosine of 37° is .7986; the sine, .6018. Taking into account that the numerator is negative, the transformation matrix and the calculations for the rotation are as follows:

$$\underset{\mathbf{A}}{\begin{bmatrix} .578 & .562 \\ .531 & .344 \\ .687 & -.422 \\ .765 & -.484 \end{bmatrix}} \times \underset{\mathbf{\Lambda}}{\begin{bmatrix} .7986 & .6018 \\ -.6018 & .7986 \end{bmatrix}} = \underset{\mathbf{V}}{\begin{bmatrix} .1234 & .7967 \\ .2170 & .5943 \\ .8026 & .0764 \\ .9022 & .0739 \end{bmatrix}}$$

fear of dispersing the variance too much, thereby distorting the solution. It will be recalled that this is a problem with the Varimax method.

When all the unrotated factors of any size are rotated by Criterion I, there are often some Criterion I rotated factors that are still too small to be considered important, for example, those with no loadings at all as high as .3 or more. Some of the other Criterion I factors may be marginal, that is, with several loadings in the range from .3 to .45 but no higher loadings. Other factors usually will be larger, with several loadings above .45. By inspection of the Criterion I factors, it is ordinarily possible to narrow down the choice of the correct number of factors to a small range of possibilities. This is the major advantage of using Criterion I. By squeezing the variance down to as few factors as possible, taking into account the correlations among the variables, unnecessary factors may be discarded. Criterion I rotations, then, are not carried out to give the final factors but rather as an intermediate step to permit a determination of how many factors are really needed in the solution.

An exception to this manner of using Criterion I rotations occurs in the case where a general factor is suspected or the hypothesis of a general factor is being tested. Criterion I rotations will reveal the presence of a general factor, if it exists, since variables that are all correlated with each other will be retained as much as possible on the same factor rather than scattered around to different factors as with some other methods based on simple structure concepts. Where a general factor is expected, therefore, Criterion I rotations may well be utilized as the final solution.

Where there is no general factor hypothesis, however, the retained Criterion I factors will be rotated further by the use of a second criterion based on a somewhat different principle. Rotation of the retained Criterion I factors by this additional criterion typically requires more than one run. For example, with Criterion I rotations, it may have been established tentatively that there are either 10, 11, or 12 factors worth retaining. Trial rotations by the second criterion, called "Criterion II," may be carried out with 10, 11, and 12 factors separately and these solutions examined afterward to determine which of them, if any, is satisfactory. If a relatively small Criterion I factor has its moderate loadings increased in size so that the factor takes on greater importance as a result of the Criterion II rotations, it is ordinarily considered to be a factor worthy of retention. If a minor Criterion I factor fails to build up in this way or is built up only through the splitting of a major factor into two similar halves, one half overwhelming this minor factor, the Criterion II rotations are rerun without this Criterion I factor. To be retained, a minor Criterion I factor must be enhanced or magnified by the Criterion II rotations rather than fundamentally changed in character. If any major Criterion I factor is split into two parts by the Criterion II rotations, another run must be made with fewer factors and the Criterion I factor that was too weak to survive in the Criterion

II rotations eliminated. Criterion II rotations are based on the following principle.

Principle II. If two variables are not correlated, they should not appear on the same factor.

This principle can be approximated mathematically to provide the basis for Criterion II rotations.

[7.5.2]

ROTATION BY CRITERION II OF THE TANDEM CRITERIA

A mathematical approximation to the idea involved in Principle II is the following:

$$F = \sum_{k=1}^{m} \sum_{i=1}^{n} \sum_{j=1}^{n} (1-r_{ij}'^2) a_{ik}^2 a_{jk}^2 \tag{7.24}$$

where all terms are defined as in Eq. (7.11). By minimizing the function in (7.24), Principle II is approximately satisfied.

The mathematical development for Criterion II parallels that for Criterion I in Eqs. (7.11)–(7.20), except that wherever $r_{ij}'^2$ appears for Criterion I, $(1-r_{ij}'^2)$ must be substituted to obtain the corresponding results for Criterion II. Thus by analogy with Eq. (7.20), the angle φ that gives the desired Criterion II rotation is derived from

(7.25)

$$\tan 4\varphi = \frac{2\sum_{i=1}^{n}\sum_{j=1}^{n}(1-r_{ij}'^2)[x_i y_i(x_j^2 - y_j^2) + x_j y_j(x_i^2 - y_i^2)]}{\sum_{i=1}^{n}\sum_{j=1}^{n}(1-r_{ij}'^2)[(x_i^2 - y_i^2)(x_j^2 - y_j^2) - 4x_i y_i x_j y_j]}$$

To ensure that a minimum has been reached for the function, however, the second derivative evaluated at φ must be positive. By reasoning analogous to that used to find the appropriate angle of rotation for Criterion I, the desired angle of rotation for Criterion II is given by the following:

(7.26)

		num −	num +
denom	+	$(45° - \varphi)$	$-(45° - \varphi)$
denom	−	φ	$-\varphi$

The angle φ itself is found from Eq. (7.25), ignoring the signs on numerator and denominator, taking 4φ in the first quadrant, and dividing the angle 4φ

by 4 to get φ. The transformation matrix that will perform the Criterion II rotation is then chosen from the following possibilities:

$$\begin{bmatrix} \cos(45°-\varphi) & \sin(45°-\varphi) \\ -\sin(45°-\varphi) & \cos(45°-\varphi) \end{bmatrix}, \quad \begin{bmatrix} \cos(45°-\varphi) & -\sin(45°-\varphi) \\ \sin(45°-\varphi) & \cos(45°-\varphi) \end{bmatrix}$$

$$(\text{num}+, \text{denom}+) \qquad\qquad (\text{num}-, \text{denom}+)$$

$$\begin{bmatrix} \cos\varphi & \sin\varphi \\ -\sin\varphi & \cos\varphi \end{bmatrix}, \quad \begin{bmatrix} \cos\varphi & -\sin\varphi \\ \sin\varphi & \cos\varphi \end{bmatrix}$$

$$(\text{num}+, \text{denom}-) \qquad (\text{num}-, \text{denom}-)$$

Varimax rotations could be applied in many cases and results obtained similar to those for Criterion II rotations of Criterion I factors, provided that all superfluous minor factors have been eliminated and the same number of factors is rotated for both methods. If this control is not exercised, Criterion II will disperse the variance much more than will Varimax, even splitting large factors into two more or less equal halves. For this reason, Criterion II cannot be applied unless the number of factors has been determined within a relatively small margin of error. This is normally accomplished by the prior application of Criterion I, as already explained, so that only retained Criterion I factors are rotated by Criterion II. An example of a complete solution using Criterion II rotations following Criterion I rotations is given in Chapter 11.

The preceding solutions for the Tandem Criteria are the "raw" Criterion I and Criterion II solutions. As with Varimax, "normal" Criterion I and Criterion II solutions may be obtained by scaling all factor loadings to give communalities of 1.0 before starting the computations.[1] After the rotations are completed, the loadings are scaled back down to give the original communalities. The author feels that the use of "normal"solutions with Tandem Criteria, as with Varimax, is of questionable superiority if there are many data variables with low communalities. A computer program for carrying out Tandem Criteria rotations is described in Chapter 12.

As with all other rotational criteria, special examples can be conceived where the Tandem Criteria will not provide a suitable solution, quite apart from their limitation to orthogonal solutions. If it were possible to find two distinct clusters of variables that correlate highly within clusters but not at all across clusters, and all of which have substantial loadings on the same factor, Tandem Criteria would have difficulty finding this solution. Such a situation would require that both clusters have substantial loadings of opposite sign on

[1] The author is indebted to Henry Kaiser for suggesting the use of the "normal" procedures with Tandem Criteria and also for suggesting the use of r', the reproduced correlation, instead of r, the raw-data correlation, in the equations.

some other factor besides the one on which both clusters have substantial positive loadings. Although such a situation is conceivable, it is not apt to be encountered very often, and certainly it would be rare to have this kind of phenomenon involved in more than a fraction of the variables in a matrix of substantial size. The user of the Tandem Criteria must be alert to the possibility, however, that his data may present special characteristics that render this method unsuitable.

[7.6] ONE STRATEGY FOR FACTOR ANALYTIC WORK

In the typical factor analysis performed by the average investigator, there may not be available a great deal of information about the underlying factor structure beyond the expectation that several dimensions will be needed. Under these circumstances, the author has found the following procedures to be serviceable:

1. Extract factors by the minimum residual method until convergence on vectors of opposite sign occurs, not using communality estimates.
2. Inspect these factors and drop any obviously residual factors, for example, those with no loadings of .2 or more.
3. Rotate the retained minimum residual factors by Criterion I of the Tandem Criteria, throwing away all clearly residual rotated factors, for example, those with no loading as high as .3 or more.
4. Rotate the retained Criterion I factors by Criterion II. If any minor factor of the retained Criterion I factors fails to magnify properly or to retain its basic character in the Criterion II rotations, it is dropped, and the other Criterion I factors are rerotated by Criterion II leaving out this marginal Criterion I factor.
5. Continue this process of dropping Criterion I factors that do not hold up until no more factors can be eliminated in this way.
6. The final Criterion II solution can be rotated further as necessary to satisfy other criteria the investigator may have in mind. For example, this Criterion II solution can be used as a point of departure for rotating further to oblique simple structure, either by hand methods or by analytic methods. This solution may also be altered as needed to achieve maximum consistency with expectations based on known facts or theories.

CHAPTER EIGHT

Planning the Standard Design Factor Analysis

One of the most common errors committed by inexperienced research workers is to collect a mass of data before deciding how the data are to be analyzed. Despite the fact that proper planning is just as important in factor analytic research as it is in other areas, factor analysis is often called upon to rescue unplanned research endeavors from disaster. For every person who asks how to plan his project so that factor analysis can be used effectively, there seem to be ten who wait until after the data are already collected to ask for help in doing "something" with what they have. Needless to say, when factor analysis is used as an afterthought on data from a poorly planned investigation, the results are not apt to be gratifying nor representative of what factor analysis can do when used properly. Investigators who wish to get the most from this powerful analytical tool would do well to give careful attention to the principles of proper design of factor analytic research. Principles of design and other considerations involved in a well-planned factor analytic investigation are the subject of this chapter.

[8.1]
SOME PRINCIPLES OF DESIGN

Ideally, the investigator should be free to define the area he wishes to study, choose the variables he wishes to work with, obtain data from an appropriate sample of individuals he selects, and then carry out the analysis. Where the investigator has this kind of latitude, he is in the best position to design a good study. If he has no choice of data or subjects, his freedom may be so greatly reduced that a well-designed study might be impossible to achieve. It will be assumed here that the investigator is free to proceed in the most appropriate fashion. To the extent that he is not, he can only do his best to approximate the conditions for a properly planned study as closely as possible.

The first step in planning a factor analysis is to define the domain to be studied and then to develop a hypothesized factor structure for this domain (Guilford, 1950). That is, the investigator uses all the knowledge and deductive power at his disposal to formulate a factor model. This model will state explicitly the nature of the expected factors together with alternative hypotheses concerning their possible characteristics. Next, data variables that are expected to be related to the hypothesized factors in a particular way should be identified. For example, if a particular factor is expected to emerge from the analysis, for example, Extraversion–Introversion, what are the data variables that should have high factor loadings on that factor and what variables should have low or negligible factor loadings on that factor?

It is important to go through this type of analysis to ensure that any expected factor will be overdetermined if it does appear in the solution. That is, there will be several data variables with high loadings on the factor as well as a much larger number of data variables that will be essentially unrelated to the factor. If a factor is not overdetermined, it may appear only weakly and with a small number of defining data variables. This situation makes the factor difficult to position and hard to identify. Of course, the investigator may be wrong in his hypotheses about the factors and what variables will represent them. To the extent that he is in error, the factor solution will not verify his hypotheses. If he fails to think this through ahead of time, however, it is unlikely that he will make adequate provision to see that all important factors are properly represented. The tendency with unplanned investigations is to have one or a few factors unnecessarily overdetermined while the others are insufficiently represented in the analysis. Finding appropriate data variables to serve these functions in a factor analysis is not always easy. The research worker frequently finds that he must design or create new variables with the desired characteristics to serve as marker variables for the factors which he hypothesizes will emerge. This is because the best marker variables are relatively pure-factor measures while the bulk of the data variables ordinarily available to the investigator are factorially complex, measuring several factors rather than just one.

The need to overdetermine factors means that each factor should have five or more data variables as good markers to represent it if at all possible. Three good relatively factor-pure markers would constitute an absolute minimum number of variables to define a factor in the analysis. The more variables there are to define a factor, the more clearly will it be established in the analysis, and the easier it will be to locate it in the rotations. The total number of data variables included in a factor analysis, then, should be at least five or six times as great as the number of factors expected to emerge.

Factor positions are determined by the variables that are *not* related to the factor as well as by those that *are*. In fact, with simple structure, the variables not related to the factor, *rather than* those that are, determine the factor position. Simple structure can be most helpful, therefore, as a guide to rotations when there are many variables in the analysis that are essentially unrelated to the factor in question. This can be achieved best by including six or more hypothesized factors that are relatively independent of one another. Under these conditions, the great majority of the data variables will be in the hyperplane for any one factor, thereby providing a clear hyperplane to assist in locating the factor if it does emerge as predicted. A study designed in this way stands a good chance of yielding an excellent simple structure solution if the investigator's analysis of the situation has been reasonably good. At least, if the domain is at all close to his perception of it, simple structure will probably be a fairly good rotational model for the analysis.

In ascertaining whether a given hypothesized factor is likely to be overdetermined in the analysis, only relatively pure-factor data variables can be counted on. A pure-factor data variable has a major loading in only one factor. It is to contrasted with a complex data variable that has major loadings in more than one factor. There is nothing wrong with having some complex data variables in the analysis; indeed, one of the purposes of the analysis may be to gain a better understanding of the factorial composition of one or more such variables. These complex data variables are not very useful, however, in determining the proper positioning of factors. They are apt to have moderate loadings on several factors; hence they do not contribute to defining hyperplanes nor are they helpful in identifying the nature of the factors on which they have their major loadings. The Varimax method, for example, tries to avoid precisely this type of loading since it is not readily interpreted. Where such complex data variables are included in an analysis, then, they should not be counted on to define factors. The several "marker" variables for each factor should be as factor pure as possible. This is a very difficult requirement to satisfy, of course, if the investigator is constrained to choose his data variables from a set that does not include any factor-pure measures. Unfortunately, this is all too often the case where factor analysis is being applied as an afterthought.

If the hypothesized factor is not represented by several relatively factor-pure measures, and several complex data variables involving this factor appear in the matrix, there is serious risk of improper positioning of the factor. For example, if several complex variables are included that share more than one common factor, a factor may be positioned in such a way that these complex data variables appear to be pure-factor measures. The factor itself, in such cases, becomes a complex factor rather than a unitary, that is, univocal, factor, and the appropriate factor position will be missed altogether because there were not enough pure-factor measures of that factor in the matrix. The investigator may be quite unaware that the factor he has located is in fact not univocal but complex in nature.

A good example of this phenomenon would be an analysis in which the eight abnormal scale scores of the MMPI were included in the same matrix along with some total scores in other psychological tests. The MMPI scale scores are highly complex factorially (Comrey & Levonian, 1958; Comrey, 1959) and also overlap each other considerably in factorial content. The scales *Hs*, *D*, and *Hy* could be expected to define one factor. It might look as if these scales were pure-factor measures of this obtained factor. In reality, however, the factor defined by these variables would not be univocal but complex, based on the combination of several factors these complex measures all share. If some data variables that measure both factors A and B are included in the matrix, there must be other data variables that measure A but not B and B but not A if separate factors A and B are to be isolated in the analysis. The number of variables that measure both A and B should be considerably smaller than the number of variables that measure these factors alone to avoid the possibility that a factor will be placed through the cluster of complex variables that measure both A and B together.

Where the investigator does not have the option of including several data variables that should be relatively pure-factor measures of a hypothesized factor, it may be necessary for him to rotate the factors in a rather unorthodox fashion to make allowances for the peculiarities of his sample of variables. If he knows, or strongly suspects, for instance, that there is only one data variable that is likely to be a pure-factor measure of a given hypothesized factor, the investigator may be forced to line up a factor right through that data vector regardless of what other criteria of rotation might suggest. A proper design of the factor study, of course, would obviate the need for such arbitrary manipulations.

As a scientific tool, factor analysis has its greatest application in the process of determining what the important variables are in a given scientific domain. Determining what those important variables are is a necessary preliminary step before it is possible to begin the next phase of scientific work, which is devoted to determining the nature of the functional relationships between the important variables in the field of study. The social sciences are still very much in the phase of trying to determine what variables are important and how to measure them. Although factor analysis can be a helpful aid in this endeavor, it is apt to be most useful where applied to a series of interrelated investigations rather than in just a one-shot study. The process of refining factor constructs takes place slowly. Each investigation provides some clues and hypotheses which must be checked out in a subsequent study. The conceptual boundaries of a factor construct are only dimly perceived at first in many cases. Variables that relate to it initially are modified in an attempt to increase their factor purity. An effort is made to develop other variables that also will relate to the construct. Variables that fail to do so are dropped in subsequent investigations. By means of such hypothetico-deductive methods applied through a series of empirical studies, the factor construct becomes more and more clearly understood. As this is going on, the methods of assessing the factor construct are being improved. After the factor construct has been well described and good measures of it have been established, experiments of a nonfactor analytic kind may be undertaken to pin down still more tightly what the construct represents. These studies will also seek to determine how it is related to practical criteria of interest as well as to other variables of scientific interest in the domain of investigation. Those who would use factor analytic methods to increase scientific understanding, therefore, are urged to plan programmatic sequences of investigations, and to make changes and test hypotheses from one study to the next, rather than settling for a single analysis, however impressive in scope.

[8.1.1] *THE FACTOR HIERARCHY*

Constructs represented by individual factors in the results of a factor analysis can range from the highly specific and narrow to the very broad and complex, depending upon the nature of the data variables being analyzed. Factors that are very low in the hierarchy of factors are those that are based on relatively simple phenomena of limited general interest. Perhaps the most common way in which such low-level factors are produced is through the inclusion in the same analysis of very similar variables. For example, if diastolic and systolic blood pressure are included in a factor analysis of several physiological response measures, it is likely that a doublet factor will emerge from the analysis in which only these two variables have high loadings. The factor obtained in this case does not shed much light on basic underlying physiological mechanisms; it merely advises us that diastolic and systolic blood pressure overlap tremendously in what they measure, enough perhaps to be considered almost alternate forms of the same variable.

Other common examples occur where questionnaire items are being factor analyzed. If two or more of the questionnaire items bear a marked similarity to each other, much more so than with the other items, it is almost inevitable that a factor will emerge defined principally if not exclusively by these similar items. Again, the factor in question does not herald the emergence of some dramatic new construct of great taxonomic interest. It merely means that the choice of items for analysis resulted in the production of a relatively uninteresting, unimportant, low-level factor. Use of the Varimax criterion in factor rotation tends to make these low-level factors stand out in bold relief because it will tend to push the loadings for these variables as high as possible to maximize the variance of the squared factor loadings.

An example of this kind of phenomenon can be seen in Fig. 8.1. Suppose that variables 1–6 all represent items that measure to some degree a factor construct of sufficient generality to place it above the lowest level in the factor hierarchy. Suppose in addition that items 5 and 6 are rather similar items. The following personality-type questionnaire items, used with a graded frequency scale of response, might represent this situation:

1. I am tense and nervous.
2. I lack confidence.
3. I am unable to cope with life's problems.
4. My moods fluctuate easily.
5. I feel blue.
6. I suffer from depression.

Variables 5 and 6 correlate with each other very highly because they are really only alternate ways of asking the same question. Questions 1–4 are related to each other, and also to items 5 and 6, since they all are related to the personality factor Emotional Stability versus Neuroticism, but no two of these items are asking the same question in a different way except items 5 and 6. These last two items correlate at a higher level because they not only share the common variance due to the Emotional Stability versus Neuroticism factor which they both measure, but in addition they share a good measure of specific variance associated with this kind of question.

In Fig. 8.1, with the factors positioned as I and II, all items are represented by vectors that project onto factor I about equally and with substantial loadings. Variables 5 and 6 have substantial loadings on factor II, but the other variables do not. In this position, factor I represents the Neuroticism versus Emotional Stability factor which is related about equally to all the variables. Factor II represents the leftover common variance in variables 5 and 6 which cannot be accounted for by the Neuroticism versus Emotional Stability factor.

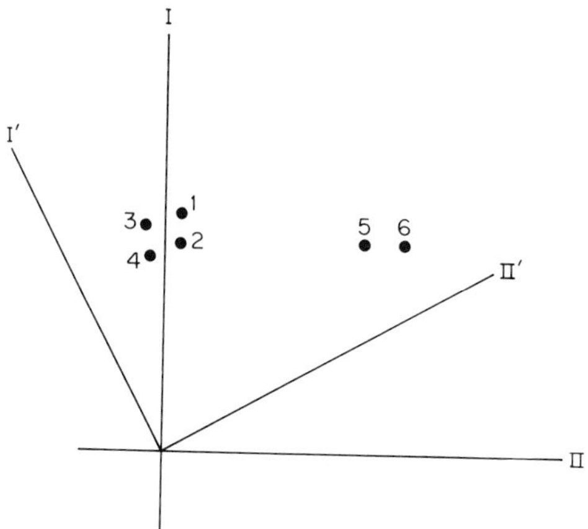

FIG. 8.1. Factors of different levels.

It is, therefore, a minor, low-level factor of limited importance in the hierarchy of factors. Factor I, on the other hand, is at a different level, higher in the hierarchy, and of much greater general scientific interest.

By poor design of this factor analytic experiment, through the use of two overly similar variables in items 5 and 6, factors of markedly different levels in the hierarchy of factors have been introduced into the same analysis. The effects of this design error can be handled by appropriate rotation of the factor axes to sort out the factors properly into one major and one minor factor, like I and II in Fig. 8.1. Unfortunately, however, typical rotational methods are not apt to give a solution like factors I and II but rather are more likely to place the factors as in positions I′ and II′. These positions for the factors treat both factors more or less as equals, one with major loadings for variables 1–4 and smaller loadings for variables 5 and 6. The other factor will have major loadings for variables 5 and 6 with small loadings for variables 1–4. An oblique solution would eliminate the small loadings on both factors, leaving 1, 2, 3, and 4 on one factor and 5 and 6 on the other factor with a substantial positive correlation between the two factors. In either the orthogonal or the oblique solution of this kind, the major Emotional Stability versus Neuroticism factor would not be properly represented in the solution because of the disturbing effect of the extra-large correlation between variables 5 and 6. Neither of these solutions would show all items about equally related to the same factor as they should.

The problem created by the overlapping specific variance of variables 5 and 6 shown in Fig. 8.1 could have been avoided if one of these two variables had been eliminated from the analysis or if both had been combined to form a single variable. This could be accomplished by merely adding the two scores for variables 5 and 6 together to yield a single score for each person on these two items.

The presence of low-level factors in the hierarchy of factors through the introduction of variables with overlapping specific variance is often not anticipated by the investigator, particularly in exploratory studies. Scrutiny of the emerging factors in an analysis will often reveal that one or more factors are based either totally or partially on this kind of phenomenon. Under such circumstances, it is often wise to run the analysis again after eliminating or combining the offending variables to eliminate factors that are too low in the hierarchy of factors.

Another way of forcing the factors to a higher level in the hierarchy of factors is through underextraction, that is, not taking out as many factors as the data seem to call for. Since the extracted factors are ordered in a steeply graded series with respect to size, by dropping off the later factors, smaller low-level factor precursors that will produce factors in the rotations like factor II in Fig. 8.1 tend to be eliminated, leaving only the larger factors like factor I. Eysenck (1960) has used this technique in his analyses of personality variables since he extracts only a few factors from his matrices of correlations between personality variables. The difficulty with this method of avoiding the production of low-level factors is that it requires a precise determination of the number of factors that must be extracted to be sure that only the major factors, and no low-level factors, will emerge from the analysis. Such information is normally not available to the investigator. If he is mistaken, the factor solution may be seriously distorted. In most cases, therefore, it is better to extract all the permissible factors and then gradually eliminate those of minor importance, reanalyzing if necessary after elimination of some variables from the analysis.

It should not be inferred from the foregoing discussion that a factor construct should be as high as possible in the hierarchy of factors. As a rule, factor constructs are apt to be most useful for scientific purposes if they are at an intermediate level in the hierarchy, although high-level factors can be useful for some purposes. In the domain of human abilities, General Intelligence represents a very broad factor construct, high up in the hierarchy of factors. It is based on the substantial intercorrelations that exist between such intermediate-level factor constructs as Verbal Ability, Numerical Ability, Memory, Reasoning Ability, and Perceptual Speed. Classification of people on the basis of intermediate-level factor traits permits better utilization of manpower than assessment on a single overall intelligence measure. In the latter case, all jobs requiring intelligence would be competing for the same people whereas with measures of intermediate-level ability traits, more high-level people would be available for assignment since the same individuals are not high on every trait. Correlating these intermediate-level constructs with other variables of scientific interest also provides better understanding in the field of inquiry than it is possible to achieve with the exclusive use of the fewer, broad, high-level factor constructs. Factor traits at the lowest level in the hierarchy of factors, however, are too numerous and too narrow in what they measure to be used effectively for general classification purposes and for scientific description.

Traits like General Intelligence are legitimate factor constructs of broad character located high up in the hierarchy of factors. All broadly based factors that emerge from factor analyses do not necessarily fall in this "legitimate" category, however. It is possible to create meaningless factors high in the hierarchy of factors through faulty design of the factor analytic investigation. This occurs commonly where investigators include several highly complex data variables that overlap each other in several factors without having those factors represented in the same analyses by several relatively factor-pure measures for each factor. This phenomenon has already been illustrated earlier in this chapter using MMPI scale scores as variables. To be most useful for scientific purposes, a higher order broad factor construct should be based on the actual overlap of substantially correlated factors at a lower level in the hierarchy of factors. It should not represent merely a linear combination of essentially unrelated lower level constructs. It is unfortunately true that many of the constructs popular in the social sciences today fall in this latter category because they have been developed by armchair thinking rather than through careful refinement by a series of factor analytic investigations. The philosophically inclined scientist who is content to believe that he has a unitary construct without checking it out with data is more often than not apt to be mistaken in his beliefs.

[8.2] DATA VARIABLES

As explained earlier in this chapter, the particular variables selected for a study have a very critical bearing on the kinds of results obtained. It is not only the general composition of the variables in some abstract sense that is important, however. The measurement characteristics of the data variables and also the subjects from which the data are obtained can have a great influence on the results obtained.

[8.2.1] SCALES OF MEASUREMENT

The ideal situation is to have data variables that are continuous and normally distributed with normal bivariate surfaces (that is, bell-shaped joint frequency distributions) for every pair of variables. The regressions for all pairs of variables should be linear. When these ideal circumstances occur, the usual Pearson product–moment correlation coefficient gives a reliable index of the degree of correlation between data variables, and the linear factor model represents a good approximation to the data. There appears to be some confusion, however, regarding the use of the product–moment correlation coefficient on scores from scales for which interval-scale properties have not been demonstrated. If there is a normal bivariate distribution in the population for the two variables involved, the sampling distribution for the correlation coefficient does not depend on the characteristics of the measurement scales for the two variables. If the distributions for the data variables are reasonably normal in form, the investigator need have no concern about applying a "parametric" factor analytic model without demonstrating that his scores are measured on an equal-unit scale. Many people mistakenly believe that because means and standard deviations are computed in making various statistical tests, an equal-unit measurement scale is required. The mathematical derivations involved in statistical tests, however, typically are only concerned with the nature of the distributions involved rather than the kinds of measurement scales that led to those distributions (Burke, 1953).

There are many other aspects of measurement scales used for data variables that do represent a source of concern, however. For example, the correlations, and hence the factor analytic results derived from them, are affected by a reduction of the continuous measurement scale to one which has fewer than a dozen categories. In the extreme, the variable is dichotomous, for example, "yes–no," "pass–fail," "married–single," and so on. Wherever possible, data variables with a reduced number of categories should be avoided. Correlations in these cases are less reliable and subject to distortions, sometimes to a very significant degree. The potential severity of these distortions increases as the number of measurement categories in the data variable decreases. It is not always possible to avoid using such variables since in many cases there is no way of getting more categories, for example, if sex membership is being related to other variables. In many cases, however, by planning ahead, the investigator can arrange to use variables measured on scales with 12 or more steps rather than just two or three.

Data variables included in the same matrix should not be dependent on each other in some artificial way. For example, consider an ability test that gives a verbal score, a quantitative score, and a total score. The total score is just the sum of the other two and hence is linearly dependent on them. Inclusion of all three scores in a factor analysis results in the production of a spurious factor based on the forced overlap between the total score and the other two scores. The total score should be omitted from such an analysis. Another example is the inclusion of "right" and "wrong" scores. The number right is either wholly or partially dependent on the number wrong, depending on how the wrong score is calculated. If the wrong score equals the total number of items minus the number right, there is a total linear dependence of the two scores. If the wrong score is the number of items attempted and missed, the dependence may be only partial but still unacceptable. Only one or the other of these two scores should be included in the analysis to prevent the appearance of a spurious factor.

So-called "ipsative" scores (Cattell, 1944b) also introduce this kind of unwanted dependence between scores. Suppose the respondent taking an interest inventory is given a choice between two activities: A—solving an arithmetic reasoning problem, and B—designing a flower garden. If he chooses response A, he gets a point on interest scale X. If he chooses B, he gets a point on interest scale Y. The score on each interest scale is the total number of times the respondent chooses the appropriate alternative for that scale. In such cases, the subject is limited as to the amount of interest he may express. He has to choose where he will place it. He cannot be "interested" in everything since he may not say, "I am interested in both these things." He also cannot say, "I am not interested in either." All subjects are treated as though they had the same amount of interest to distribute. Quite apart from being incorrect, this premise leads to interest scores that have an artificial dependence between them that make them unsuitable for factor analytic work. If a subject's responses give him a high score in one interest area, he must compensate for this by having lower scores in some other interest area. The use of such scores in factor analytic work introduces spurious common factor variance. Where there is no alternative but to use such scores, a careful evaluation of the results is necessary to ascertain to what extent the findings have been determined by the nature of the ipsative scores involved.

Spurious overlap between data variables can also be introduced through the commonly encountered practice of scoring items of a test on more than one variable. On the MMPI, for example, admission of a particular symptom may give the respondent a point on both the Depression scale and the Hysteria scale. There are so many items on the MMPI that are scored on more than one scale that substantial correlations between total scores are guaranteed. Use of these total scores in the same factor analysis, therefore, forces the introduction of overlapping specific variance into the common factor space, distorting the results accordingly.

Some data variables may have a sufficient number of scoring categories and may not be artificially dependent on other scores, but they may have badly skewed distributions. Some Rorschach test scores, for example, may have the bulk of the responses tallied for scores close to zero with a scattering of the remaining scores over a wide range. With such distributions it is prudent to make scatter plots to be sure that the regressions are linear. Where the regressions are not linear especially, it is advisable to transform the scores to eliminate such irregularities in the data. Sometimes, converting the scores to normalized standard scale scores, for example, with mean equal to 50 and standard deviation equal to 10, will be sufficient to eliminate the nonlinearity. In other cases, it may be necessary to combine categories, even reducing the variable to a dichotomy. When this is done, the variable should be dichotomized as near the median as possible to make both categories have about the same percentage of cases. If there is a genuine curvilinear relationship involved, however, it may be necessary to handle the analysis for the variable in question in an entirely different way. Some work has been done on nonlinear factor analytic models (McDonald, 1967), but as yet they have not become operational to any great extent.

Some type of scale transformation is particularly essential in cases where a few scores in the sample depart radically from the others. An example would be to have a few high-IQ subjects thrown in with a sample of low-IQ subjects. The high-IQ subjects would be much higher than the bulk of the subjects on all ability-test variables. This would result in correlations near 1.0 for the paired ability variables. Conversion of all scores to normalized standard scores would greatly reduce this distortion since only the rank positions of the subjects would be preserved by this transformation. These subjects would be high on all the tests, contributing to the elevation of the correlations, but they would not have a disproportionate effect in comparison with other subjects.

[8.2.2] SAMPLING

The particular sample of data objects, or subjects, upon which the investigator decides to take his measurements of the data variables has a great bearing on the final results of the factor analysis. The most obvious consideration in this respect is the size of the sample upon which the correlations are computed. As the number of observations increases, the reliability of the obtained correlations goes up, although with diminishing returns. Samples of size 50 give very inadequate reliability of correlation coefficients, whereas samples of size 1000 are more than adequate for most factor analytic purposes. The adequacy of sample size might be evaluated very roughly on the following scale: 50—very poor; 100—poor; 200—fair; 300—good; 500—very good; and 1000—excellent.

Guertin and Bailey (1970) have shown that with smaller samples the random errors of the less reliable correlation coefficients increase the absolute size of the correlations in the matrix. This results in greater communalities and a larger amount of common-factor variance, although the increase is due to spurious common-factor variance. This additional variance thrown into the analysis tends to produce distortions the seriousness of which is a function of the absolute amount of spurious variance added. The conclusion is that the investigator should use samples that are large wherever possible, preferably 500 or more. There is probably little to be gained in most situations by going over 1000 cases. If some other kind of correlation coefficient other than the Pearson product–moment correlation coefficient is used, larger samples are needed to achieve the same level of stability of the correlation coefficients. The tetrachoric coefficient, for example, may require twice as many cases to yield comparable stability. In those cases where the investigator has no alternative but to carry out a factor analysis with samples that are too small, he must be especially conservative in his interpretation of the results.

The composition of the sample also can have a dramatic effect on the correlations obtained and consequently on the ultimate factor analytic results. The range of scores in the sample on the data variables compared to the range of scores on these variables for the entire universe constitutes an important source of modifications in factor analytic results as a function of sampling procedures. To illustrate the importance of range of data-variable scores on factor analytic results, suppose that psychological tests on Verbal Ability, Numerical Ability, Arithmetic Reasoning, Memory, and Perceptual Speed were administered to a large, random sample from the general population together with many other variables unrelated to intelligence. A factor analysis of the intercorrelations of these variables would be likely to produce a very prominent factor of General Intelligence with substantial loadings for the above-named tests.

Next, take a new sample of the same size that is made up entirely of individuals who have an IQ of exactly 100. These individuals would be chosen at random from the entire population of individuals who have IQs equal to 100. In this case, the range on intelligence for the second sample has been reduced to zero. A factor analysis of the intercorrelations of the tests in the second sample would fail to produce a General Intelligence factor because no significant variations due to differences in intelligence among the subjects are left in the test scores. The correlations among these ability tests in such a sample would be likely to distribute approximately in a random fashion about zero.

If the restriction in range on intelligence were only partial instead of complete, such as would be the case if all the subjects were graduates of prestigious American colleges and universities, the intelligence factor would still appear in the factor analytic results, but it would account for less variance than it would in a random sample from the general population. In general, the greater the degree of restriction of range for a given factor in the sample studied, the smaller will be the loadings of marker variables on that factor in the resulting analysis. If the loadings become too small, of course, the factor is not likely to be identified at all.

If a given factor is to appear in a factor analysis, then data variables that represent that factor must appear in the analysis, and the sample of data objects, or subjects, selected must vary with respect to that factor. Sometimes, variance on a given factor is reduced through restriction in range on a variable that is correlated with the factor rather than on the factor itself. This was more or less true for the previously discussed example involving graduates of prestigious colleges and universities. The same type of selection effect might be achieved by obtaining subjects who volunteered after seeing an advertisement in a particular magazine. Readers of this magazine might on the average be considerably above average in intelligence and volunteers might be generally higher in intelligence than nonvolunteers.

For exploratory work, it is much more important to ensure that there is plenty of variance on the factor in the sample than it is to have a representative sample from the population at large. It is particularly important to be certain, in other words, that no restriction in range on the factor has occurred through some peculiarity of the method of sample selection. In the later stages of taxonomic research when the constructs are in the process of final refinement, representative samples from specified populations become more essential so that the sizes of factor loadings and correlations between factors can be pinpointed rather accurately.

Care must be exercised in combining subjects from different sources into the same sample for factor analytic purposes since the factor structure might be very different in the separate groups. Where this is the case, it is better to carry out the analyses in the groups separately. For example, a test that measures reasoning capacity in 6-year-old children might become a test of perceptual speed in young adults. It would be undesirable to combine 6 year olds and young adults into one sample for analysis purposes when dealing with tests that substantially vary factorially in what they measure in the two groups. Generally speaking, if two heterogeneous groups are to be combined for factor analysis, it is desirable to ascertain that the basic facture structure is essentially the same in the two groups.

Even with a specified sample of subjects, results of a factor analytic study can be influenced by the conditions under which the data are collected. The motivational circumstances surrounding the data collection, for example, are very important. The author once had occasion to collect data on two different ability tests that turned out to correlate with each other much more highly than expected. The tests were administered to navy recruits who were informed that the test results would not go on their service record. It was later learned that they also had their liberty curtailed for involuntary participation in the testing. The subjects were also permitted to leave as soon as they had completed the two tests. Some of the subjects tried conscientiously to do their best on both tests. Others went through the motions of marking the answer sheets, without attempting to do well, and left as soon as they could do so without being conspicuous. The subjects that tried had the high scores on both tests, while those who did not had the low scores on both tests. In effect, both tests became measures of the subjects' motivation rather than what was intended and hence correlated with each other as two alternate forms of the same test would correlate.

Substantial differences in factor results for the same subjects can occur if results are to be used for selection purposes as opposed to guidance purposes. An applicant for a job who takes a personality test, for example, as part of the evaluation process is often motivated to give the best possible impression and hence may answer the questions in a way calculated to have that effect. The same test administered to the same person when he comes to a guidance clinic to get help in solving his problems may be answered very differently. Tests administered to job applicants or other individuals who are concerned about the effect of the test results on their status tend to be heavily influenced by "social desirability" response sets. The respondents try to determine what kind of image is most favorable and then answer the questions in accordance with their perception. This same attitude is carried through all test performances, often introducing a considerable overlap among the variables that might not necessarily be there if the subjects had a different motivational set.

Any conditions that reduce the respondents' desire to participate intelligently, accurately, and willingly, therefore, are likely to have a deleterious effect on the data in one way or another. Thus it is very important, for the investigator to make sure that the subjects wish to participate and that they will be honest and conscientious in performing the assigned tasks. Forced participation, poor testing conditions, and inappropriate test materials are just a few of the ways in which data can be downgraded. In psychological testing, for example, even item format can significantly affect a subject's motivation. If he is asked to respond to questions among which it is difficult to decide the available response possibilities, even a motivated subject is apt to become disgusted in due time and quit trying to do a good job. He may even shift to responding capriciously. It cannot be emphasized strongly enough that the investigator must devote a great deal of care to securing the best possible data he can. No matter how skillfully the factor analytic work is done, it can never correct the deficiencies of poor data. To quote a time-worn expression that fits, "You can't make a silk purse out of a sow's ear."

The nature of the experimental treatment itself can affect the factor composition of the data variables under investigation quite apart from any motivational changes in the subjects. Fleishman and Hempel (1954) demonstrated that the factor composition of a task at an early stage in the process of skill acquisition may be considerably different from what it will be later on. If subjects practice on a given set of tasks for many trials, the factor structure at the beginning of the experiment may be considerably different from what it is at the end.

[8.3] CORRELATION COEFFICIENTS

The question of what correlation coefficient to use for factor analytic work is one that comes up frequently, especially where the analysis was not planned before the data were collected. In these cases, there are typically some variables which do not have continuous, approximately normal distributions. The best approach is to design the study in such a way that the data variables will have such distributions so that the usual Pearson product–moment correlation coefficient is appropriate. Where this is not possible, it may be necessary to consider other alternatives.

[8.3.1] *THE PEARSON PRODUCT–MOMENT CORRELATION COEFFICIENT*

The Pearson product–moment correlation coefficient may be computed by the following formula:

$$(8.1) \qquad r = \frac{N\sum XY - \sum X \sum Y}{\sqrt{N\sum X^2 - (\sum X)^2}\sqrt{N\sum Y^2 - (\sum Y)^2}}$$

The summations in formula (8.1) are from 1 to N, the number of cases. This is the most stable of the correlation coefficients. When it is used, however, the variables should be distributed in a manner that does not depart too far from normality, and the regression for every pair of variables should be rectilinear. Ideally, the investigator would check every regression plot to see that it is linear, but in practice this precaution is not often taken. Formula (8.1) is suitable for computer or desk-calculator applications. If the correlation must be computed without such aids, more convenient formulas are available (Guilford, 1965).

[8.3.2] *CORRELATION COEFFICIENTS FOR DICHOTOMOUS DATA*

Several different correlation coefficients have been developed to deal with the situation where one or both of the variables being correlated are dichotomous. Some of these are now given.

1. *Phi* (φ). This coefficient is used where both variables represent absolute dichotomies, such as "male versus female."

2. *Point Biserial.* This coefficient is used where one variable is an absolute dichotomy, but the other variable is continuous and approximately normal.

3. *Biserial.* This coefficient may be used where one of the variables is known to be normally distributed but has been artificially dichotomized into two categories, while the other variable is continuous and approximately normal.

4. *Tetrachoric.* This coefficient requires that both variables be normally distributed but artificially dichotomized.

If dichotomous variables are scored 1 for the higher category and 0 for the lower category, computation of the Pearson correlation coefficient with these scores will give the same results as computing the φ coefficient and the point biserial correlation coefficient. If an entire matrix of correlation coefficients is being obtained by computer, it is easier to use the product–moment formula to calculate all coefficients rather than to use φ or point biserial formula for those coefficients where these methods would be appropriate. The ease of applying the same program for all coefficients would more than make up for the slightly greater computation time. The only reason for using the special formulas for these coefficients (Guilford, 1965) would be to save labor in hand computations. The author seldom uses anything but the Pearson product–moment formula (8.1) for computing the matrix of correlations for a factor analysis, regardless of the number of categories in the variables correlated.

The only reason for computing a biserial instead of a product–moment correlation coefficient would be to save labor unless for some reason the information about the full range of scores were unavailable. The investigator should plan to avoid this. If he does, there is no need to use the biserial correlation coefficient except where the investigator does not have access to computing machinery.

The tetrachoric correlation coefficient was used in precomputer research efforts by Thurstone and others as a labor-saving device. Continuous variable regression plots were reduced to fourfold tables of frequencies which could be used to determine correlation coefficients from computing diagrams (Chesire, Saffir, & Thurstone, 1933). These tetrachoric coefficients represented shortcut approximations to the product–moment correlation coefficients. This method would be used today only if the investigator could not gain access to a computer, since such coefficients are less reliable than product–moment correlation coefficients.

One of the disturbing effects of decreasing the number of categories of measurement in variables to be correlated is a tendency to reduce the size of the maximum correlation obtainable below an absolute value of 1.0. The fewer the categories, the more pronounced this effect is likely to become. With dichotomous data, the maximum obtainable correlation may be quite small if the proportions of data objects in the categories diverge markedly for the two variables. Consider the following data for two personality test questions: I. I am happy most of the time, and II. I feel excitedly happy, on top of the world. A fourfold table of response percentages for these two questions might be the following:

		I: no	I: yes	
II	yes	$b = .03$	$a = .07$	$p_2 = .10$
	no	$c = .37$	$d = .53$	$q_2 = .90$
		$q_1 = .40$	$p_1 = .60$	Sum $= 1.00$

In this example, p_1 is the proportion saying "yes" to question I and p_2 is the proportion saying "yes" to question II. The proportions saying "no" are given by q_1 and q_2, respectively. The proportion saying "yes" to both questions is given by the value $a = .07$. Other cells give the proportions for other combinations of "yes" and "no" responses. The φ coefficient may be computed by the following formula:

$$\varphi = \frac{ac - bd}{\sqrt{p_1 q_1 p_2 q_2}} \tag{8.2}$$

For the sample above, the computation of φ by formula (8.2) gives

$$\varphi = \frac{(.07)(.37)-(.03)(.53)}{\sqrt{(.60)(.40)(.10)(.90)}} = \frac{.0259-.0159}{\sqrt{(.24)(.09)}} = \frac{.01}{.147} = .068$$

The value .068 is very small for a correlation coefficient. Compare it, however, with the value of $\varphi_{\max}$, the largest value of φ that could possibly be obtained with the same marginal totals, that is, values of p and q for the two items. This kind of hypothetical result is shown as follows:

		I: no	I: yes	
II	yes	$b = .00$	$a = .10$	$p_2 = .10$
	no	$c = .40$	$d = .50$	$q_2 = .90$
		$q_1 = .40$	$p_1 = .60$	Sum $= 1.00$

$$\varphi_{\max} = \frac{(.10)(.40)-(.00)(.50)}{\sqrt{(.60)(.40)(.10)(.90)}} = \frac{.04}{.147} = .272$$

There is no way to arrange this table to obtain a higher φ coefficient without changing the marginal totals; hence this table gives φ_{max}. The actually obtained φ coefficient for the data in the previous table seems more impressive when compared with a maximum possible φ of .272 than it does when compared with 1.0, which is the maximum that the product–moment coefficient can reach with continuous data. Many investigators have embraced the practice of dividing φ by φ_{max} as a means of correcting the φ coefficient for the fact that it often cannot achieve a value of 1.0 under any circumstances, given the marginal totals in the data. If this were done here, φ/φ_{max} would be .068/.272 or .25, which though not large is more imposing than .068, the uncorrected φ coefficient.

One of the reasons that φ/φ_{max} has been used for computing correlations on dichotomous variables for factor analytic work has been the widespread belief that factor solutions based on uncorrected φ coefficients will be dominated by "difficulty" factors. These represent factors presumed to arise because restrictions on the size of the maximum φ coefficient would classify variables into different categories according to the amount of restriction rather than according to content *per se*.

In carrying out a series of factor analyses of items on the MMPI scales, the author first attempted to use φ/φ_{max} as the means of correlating these dichotomous personality-test items. Many of the MMPI items have rather extreme splits, with only a very small percentage of subjects in one category and with the great majority in the other. Under these conditions, the maximum φ is very small and the corrected φ/φ_{max} coefficients can become very large. In some factor analyses of these coefficients, communalities in excess of 1.0 were being accumulated very quickly in the factor extraction process. Such results are totally unreasonable, of course, since individual test items are unreliable at best and will seldom have a proportion of true variance that approaches 1.0. These results were disturbing since φ/φ_{max} coefficients produced impossible results and uncorrected φ coefficients were generally considered to be inappropriate for factor analytic work. The next step was to try the tetrachoric coefficient even though it requires artificially dichotomized normal distributions, an assumption clearly not met with test items. Using the $\cos \pi$ approximation (Guilford, 1965) to the tetrachoric correlation coefficient, these correlation matrices were recomputed and the factor analyses done again. The results still gave badly overinflated correlation coefficients and communalities, if anything worse than the φ/φ_{max}.

After these experiences, the author decided to try the uncorrected φ coefficient even though it was contrary to accepted practice to do so. There was little else to do since the other available methods failed to yield acceptable results. The factors that emerged from the analyses of φ coefficients did not exhibit unreasonably large item communalities, and they also were meaningful from the standpoint of content. The major factors, therefore, did not appear to represent "difficulty" factors as feared but instead seemed to be based on substantive content of the items. If difficulty factors were present, they appeared to be minor in nature rather than factors of any great significance in the analyses. These phenomena are illustrated in published analyses of the same MMPI items using φ, φ/φ_{max}, and the tetrachoric coefficient (Comrey & Levonian, 1958). Interestingly enough, the major factors obtained were readily matched across these analyses, although the general size level of loadings was considerably higher for the analyses based on φ/φ_{max} and the tetrachoric coefficient than for that based on the uncorrected φ coefficient. The smaller factors showed no correspondence across the three analyses.

The conclusion seems warranted that if dichotomous variables, for example, such as responses to test items, are to be correlated and factor analyzed, the method of choice is the φ coefficient. Where the proportion of individuals in the two categories departs from .5, item variance is diminished, and if the proportions in a given category are different in the two items being correlated, ceiling effects can be expected on the size of the coefficient obtainable. As the difference between the proportions increases between the two items, these effects become more and more pronounced. With a 99–1 split in one variable and a 50–50 split in the other, the maximum φ is only about .10, which means that only about 1 percent of the variance in the 50–50 variable can be predicted from knowledge about the results in the 99–1 variable. This is easily grasped intuitively by considering the amount of information available to the investigator if he knows the scores on the 99–1 variable. He can only say something "different" about one person out of 100. For the other 99, he must make the same prediction since they all had the same score on that item. If he tries to predict scores on the 50–50 item, the best he can hope for is to improve very slightly on a chance prediction.

Dichotomous variables with extreme splits, therefore, give very little information and provide poor prediction of other variables. The φ coefficient gives small values for the correlations in these cases and rightly so because they provide little basis for intelligent classificatory decisions. Such variables must necessarily show up with small loadings in any factor analysis because they do not contribute much common-factor variance in most cases. As with the product–moment coefficient, however, a few extreme cases thrown into an otherwise homogeneous group can give some very high correlations with the φ coefficient. For example, three or four mental cases thrown in with about 96 or 97 normals could produce a number of 97–3 splits on personality items asking about psychological symptoms. If the same three or four persons are the ones admitting to the symptoms in each case, the φ coefficients will be very high. If only one subject out of 100 said "yes" to each of two items, and the other 99 said "no" to both items, the φ coefficient would be 1.0.

The moral to this story is that the investigator should try to avoid dichotomous data variables wherever possible. He should try very hard to obtain continuously measured data variables, that is, with at least 12 categories of response, and he should seek to obtain good distributions, approximately normal if possible. Under these circumstances, most of the aberrations in the correlational results mentioned can be avoided. Where dichotomous variables must be used, every effort should be made to dichotomize as close to the median as possible. The correlation should be computed by the φ coefficient or, equivalently, by the Pearson product–moment coefficient with scores of 1 and 0 for the dichotomous variables.

[8.4] COMMON ERRORS IN THE USE OF FACTOR ANALYSIS

Some of the more common errors in the use of factor analysis are enumerated in this section to provide a convenient check list which the investigator may find helpful to review when planning a factor analytic study. Many of these errors have been described by Guilford (1952). Most of them have been mentioned already so this list will be an abbreviated summary.

1. Collecting the data before planning the factor analysis.
2. Use of data variables with bad distributions and inappropriate regression forms:

   a. badly skewed distributions, for example, with ability tests that are too easy or too hard for the subjects tested;
   b. truncated distributions;
   c. bimodal distributions;
   d. distributions with a few extreme cases;
   e. extreme splits in dichotomized variables;
   f. nonlinear regressions.

3. Use of data variables that are not experimentally independent of one another:

   a. scoring the same item response on more than one variable;
   b. in a forced-choice item, scoring one response alternative on one variable and the other on a second variable;
   c. having one data variable as a linear combination of others, for example, verbal, quantitative, and total score.

4. Failure to overdetermine the factors. The number of variables should be several times as large as the number of factors. There should be at least five good marker variables for each factor anticipated.

5. Use of too many complex data variables. The best variables for defining factors are relatively factor pure. Only a few multiple-factor data variables should be used. If complex variables measuring both factors A and B are included, there must be some variables that measure A and not B and others that measure B but not A.

6. Including highly similar variables in the analysis that produce factors at a very low level in the hierarchy of factors. Two similar items in an analysis of personality-test items or alternate forms of the same test represent examples of this error.

7. Failure to provide good marker variables for a factor that may be present in other factor-complex data variables that are included. Without the markers, the factor will be hard to locate, although variance for that factor will be present in the analysis and must appear on some factor.

8. Poor sampling procedures:

a. taking a sample of cases that is too small to obtain stable correlations;

b. combining two distinct groups with different factor structure into the same sample for factor analytic purposes;

c. loss of a factor through biased sampling that restricts the range of variability on that factor.

9. Not including enough factors in the analysis. To have good hyperplanes for using simple structure as a guide in rotations, it is necessary to have a large number of points in the hyperplanes. Without six or more factors in the analysis that are relatively independent of one another, it is difficult to have good hyperplanes.

10. Use of inappropriate correlation coefficients such as φ/φ_{max} or use of a coefficient such as the tetrachoric in a situation which violates the assumptions underlying its use.

11. Use of inappropriate communality estimates, for example, 1.0 in the diagonals when the objectives of the investigation are concerned only with common-factor variance.

12. Extracting too few factors, forcing a factor solution for m factors into a space of fewer dimensions with consequent distortion of the factor solution.

13. Poor rotation procedures:

a. failure to rotate at all;

b. using an orthogonal solution when an oblique solution is necessary to give a good picture of the results;

c. permitting an unwarranted degree of obliquity between factors in the pursuit of simple structure;

d. use of rotational criteria which have not been determined to be appropriate for the kind of data involved;

e. rotation of extra small factors by an analytic rotation method that spreads the variance too much under such circumstances;

f. failure to plan the study so that there will be a suitable rotational criterion that can be employed.

14. Interpreting the first extracted factor as a general factor.

15. Leaping to conclusions about the nature of a factor on the basis of insufficient evidence, for example, low loadings and lack of outside confirmatory information. The interpretations of factors must be verified on the basis of evidence outside the factor analysis itself. Follow-up factor analyses and construct validation studies are an important part of this verification process.

CHAPTER NINE

Alternate Designs in Factor Analysis

Up to this point, attention has been devoted exclusively to the traditional, most widely used factor analytic design, called "*R*-technique" by Cattell (1952) to distinguish it from other factor analytic designs such as *P*- and *Q*-techniques. The purpose of this chapter is to explore briefly some of these alternatives to the traditional *R*-technique factor analytic design.

[9.1] *Q*-TECHNIQUE

The most commonly considered alternative to the *R*-technique factor analysis is *Q*-technique, which is sometimes referred to as "inverse" factor analysis. An important difference between the *Q*- and *R*-techniques is illustrated in Fig. 9.1 which shows a typical data matrix. There are N data objects, or subjects, as rows of the data matrix $\mathbf{Z}$ in Fig. 9.1. Each column of $\mathbf{Z}$ consists of standard scores for one data object or person for all the n data variables. In the *R*-technique factor analysis, a correlation is computed by taking a pair of rows of matrix $\mathbf{Z}$ and determining the average cross-product term:

$$r_{ij} = \frac{1}{N}\sum_{k=1}^{N} z_{ik} z_{jk} \tag{9.1}$$

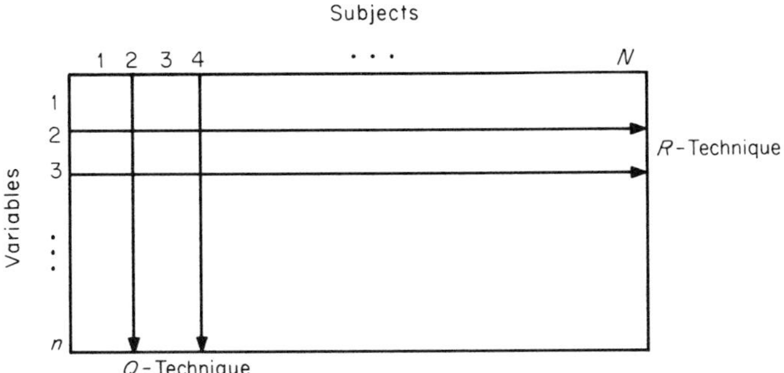

FIG. 9.1. Data matrix **Z** for *Q*- and *R*-techniques.

For *R*-technique factor analysis, N, the number of cases, is normally much larger than n, the number of data variables. The number of cases N must be large to obtain stable correlation coefficients.

In *Q*-technique factor analysis, the starting correlation matrix is computed in a different way. Instead of correlating two data variables over the sample of data objects (or subjects), two data objects (or subjects) are correlated over the sample of data variables. In Fig. 9.1 this is shown by two perpendicular lines indicating that in order to give a correlation coefficient, two columns of the data matrix are related rather than two rows as in *R*-technique. For the *Q*-technique factor analysis, then, the correlation is computed as follows:

$$r_{ij} = \frac{1}{n} \sum_{k=1}^{n} z_{ki} z_{kj} \tag{9.2}$$

The limits on the summation in Eq. (9.2) are from 1 to n, the number of data variables, whereas in Eq. (9.1) the limits are from 1 to N, the number of data objects. In Eq. (9.1) the column subscript varies from 1 to N, whereas in Eq. (9.2) it is the row subscript that varies, and from 1 to n rather than 1 to N. When the data objects are people, the matrix of correlations that is factor analyzed in *Q*-technique contains correlations between persons rather than between variables as in *R*-technique.

Both formulas (9.1) and (9.2) are average cross products of standard scores. If the raw scores are correlated instead of standard scores, the raw-score correlation formula itself has the effect of reducing the scores to standard scores before the correlation is computed. In *R*-technique analysis, then, differences in means and standard deviations between the two variables correlated have no effect on the correlation coefficient since both variables are automatically scaled to zero means and unit variances. When two individuals are correlated over the data-variable scores in *Q*-technique analysis, shown by two columns of the data matrix in Fig. 9.1, there is also a rescaling of the means and standard deviations of the two individuals to 0 and 1, respectively. Differences in mean scores and variabilities of the individuals do not affect the correlation between two persons. If the scores are in the same rank order with about equal spacing between the scores for both individuals, the correlation between them will be very high even if one person's scores are considerably higher on the average and more spread out than the other person's scores.

An important difference between Q- and R-technique factor analysis is evident right at this point. With R-technique, the elements in a given row of the data matrix (see Fig. 9.1) are all obtained on the same measurement scale, for example, with the same test or other data-gathering instrument. With Q-technique, however, this is not the case. The elements in a column of the data matrix (see Fig. 9.1) are obtained ordinarily from different measurement scales since each row of the data matrix is obtained with a different measuring device.

Consider the hypothetical miniature data matrix shown in Table 9.1. Measurements of five physical size variables are shown for five fictitious subjects. Height is measured in inches, weight in pounds, shoe size on an arbitrary scale, and waist and neck size in inches. If the first two individuals (columns 1 and 2 of Table 9.1) are correlated, the value obtained is about .99. The correlations for each pair considered will be close to this figure.

Table 9.2 shows the scores for these same individuals after they have been standardized by rows. That is, each height measurement, for example, is converted to a standard score by the formula $z = (X - M)/\sigma$. In the case of person 1, this would be $(72 - 70.2)/2.32 = .78$ for the height variable. The weight scores would be standardized by the formula $(X - 168.0)/30.59$. Each row of Table 9.1 is converted to standard scores by subtracting the mean of the row from each raw score and then dividing by the standard deviation to obtain the corresponding score in Table 9.2.

TABLE 9.1

Unstandardized Miniature Data Matrix

Variables	Subjects						
	1	2	3	4	5	M	σ
1. Height	72	74	70	69	66	70.2	2.32
2. Weight	220	180	150	160	130	168.0	30.59
3. Shoe Size	12	14	10	9	7	10.4	2.42
4. Waist Size	36	34	33	34	28	33.0	2.68
5. Neck Size	17	15	15	15	14	15.2	.98

TABLE 9.2

Standardized Miniature Data Matrix

Variables	Subjects						
	1	2	3	4	5	M	σ
1. Height	.78	1.64	−.09	−.52	−1.81	0	1
2. Weight	1.70	.39	−.59	−.26	−1.24	0	1
3. Shoe Size	.66	1.49	−.17	−.58	−1.40	0	1
4. Waist Size	1.12	.37	.00	.37	−1.87	0	1
5. Neck Size	1.84	−.20	−.20	−.20	−1.22	0	1

If the first two columns of scores in Table 9.2, for the first two subjects, are correlated, the value obtained is .59, considerably below the .99 value obtained by correlating the unscaled raw scores. The correlations for other pairs of subjects will also be much reduced when computed with the scaled scores instead of the raw scores.

The use of raw scores for correlating persons in Q-technique factor analysis spuriously inflates the correlations, capitalizing on the fact that the scales of measurement differ from row to row in the data matrix. An example was chosen here that exaggerates this phenomenon, resulting in extremely high correlations between subjects. Even under ordinary conditions, however, there is often an intolerable inflation of the correlations between subjects unless the data matrix is standardized by rows before computing the correlations between persons. Of course, the data may be scaled to some mean and standard deviation other than 0 and 1, respectively, if desired. It is only essentially that they be scaled to the same mean and standard deviation. For example, T-scores, with a mean of 50 and a standard deviation of 10, represent a popular choice.

One frequently encountered situation in Q-technique studies where it is easy to overlook the need for standardizing the data matrix by rows is the case where the data variables are test-item scores. In such cases, the possible scores may be only 0 and 1, or values over a very small range. The differences between means and standard deviations of the various rows of the data matrix are not so great under these circumstances; so the investigator may not recognize that the correlations between persons have been spuriously elevated. This is because the correlations are more moderate in size than for the example in Tables 9.1 and 9.2. Even in the case where means and standard deviations of the rows do not fluctuate greatly, however, it is necessary to scale the rows of the data matrix to equal means and standard deviations before computing the correlation matrix for Q-technique factor analysis.

The fact that computing a correlation coefficient between the scores for two individuals over a collection of data variables automatically scales the two sets of scores to the same mean and standard deviation for the two individuals has important implications for the kinds of factors derived in Q-technique analysis. Two individuals will be highly correlated if they have the same pattern of highs and lows in their scores, regardless of the absolute level of their scores. Thus, two individuals will be highly correlated and will tend to fall on the same Q-technique factor if their profiles look alike even if one profile is much higher than the other. If the investigator is looking for types of individuals who have not only similarly shaped profiles but in addition profiles that are at the same level, he will need to look beyond the Q-technique factor results *per se*. For example, profiles of individuals with high loadings on the same factor may be compared visually to see how much they differ in absolute level. In this way, individuals may be grouped who not only have similar profiles but actually have similar scores on the data variables sampled.

In R-technique the data variables are correlated with each other, and the factors show the relationships of the data variables with each other. In Q-technique factor analysis the subjects, or data objects, are correlated with each other, and the factors show the relationships of the subjects or data objects to each other. To put it in a somewhat oversimplified way, R-technique factors represent clusters of similar variables, whereas Q-technique factors represent clusters of similar persons or data objects.

Stability of the correlation coefficient depends on taking a large representative sample of cases from the population. In R-technique, this ordinarily means collecting data for a large number of people on a limited number of variables. With Q-technique, on the other hand, this ordinarily means taking a very large number of observations for a limited number of people or other data objects. In both cases the representativeness of the sample is important as well as the number of paired observations over which the correlation is computed. In R-technique, even if a large sample is obtained but the sampling method is biased, that is, certain kinds of individuals are systematically excluded from the sample, the sample correlations may depart substantially from the population values. In the case of Q-technique, the correlation between two individuals is also subject to distortion because of a biased sampling of data variables over which to compute the correlation.

Correlating the item responses on a personality test for two people, for example, even if item scores are standardized, may give a very different result than if the correlation is computed over a collection of standardized ability-test scores. Theoretically, if we wish to know *the* correlation between two individuals, the correlation should be computed over a representative sample of all possible data variables.

It is manifestly impossible, however, and perhaps not desirable, to correlate people over all possible data variables or even a representative sample from that universe of variables. Therefore in practice *Q*-technique analyses must be based on a limited and usually nonrepresentative subset of the universe of all possible data variables. The clusters of similar individuals that emerge as factors in such analyses must be considered as "types" only with respect to the variables sampled. The more limited is the sampling of data variables, the more limited is the sense in which the individuals may be classified as belonging to the same type.

The practice of building up the number of observations for *Q*-technique analysis by using items of a single test as variables clearly does little to make the sample of observations broadly representative. On the contrary, it is picking out a very minute corner of the universe to study. This may be worthwhile where the characteristics of a particular test are being studied, but it is not a desirable procedure where the intent is to locate types of individuals in a more general sense, for example, psychopathological types, normal personality types, and intellectual ability types. When general objectives of this kind are involved in *Q*-technique analysis, it is essential to obtain a broader sampling of data variables on which each subject is measured, thereby coming closer to the ideal of a representative sampling of data variables from a broad and well-defined population of variables.

[9.1.1] *THE ROTATION PROBLEM IN Q-TECHNIQUE ANALYSIS*

One of the greatest difficulties with *Q*-technique analysis is the tendency for traditional rotational criteria to be inappropriate for such studies. Typically, a sample of readily available subjects is used, hopefully with an adequate battery of observations for each subject in order to give good correlations between subjects. There is usually no particular reason to suppose that such a random collection of subjects would cluster so as to give a good simple structure, for example, in rotations of *Q*-technique factors. To obtain a good simple structure solution, it would be necessary to have several clusters of subjects such that those within a cluster correlate substantially with each other but have very low correlations with subjects from other clusters.

If subjects and variables are selected for *Q*-technique analysis without concern for the problems of rotating the factor results, it is likely that subjects will not form very distinct clusters. The correlations between subjects are apt to be rather low generally after the data matrix has been properly standardized by rows so that all variables have the same mean and standard deviation. The factor structure in such cases is likely to be rather undifferentiated, giving the investigator only an arbitrary choice among several more or less equally uncompelling alternatives.

Q-technique analysis, therefore, is not apt to provide a very fruitful rescue vehicle for saving unplanned investigations from disaster. Its proper use requires as much, if not more, planning than the standard R-technique analysis. The investigator should make sure that he takes the following steps:

1. Define the domain of variables over which the individuals are to be correlated.
2. Select a set of variables on which the individuals are to be measured such that the domain defined in step 1 is adequately represented. The set of variables must contain enough elements to provide stable correlations between individuals. Several hundred elements would be desirable.
3. Select several individuals to represent each pure "type" that is hypothesized to exist. At least a half-dozen, and more if possible, should be included for each type.

If the hypothesis that a given type exists is correct, and if the individuals selected to represent it are examples of this type, with respect to the set of variables over which the correlations between individuals are computed, a Q-technique factor should emerge for which these individuals all have substantial loadings. These individuals selected to represent a given type should correlate more with each other than they do with individuals selected to represent other types.

Individuals who do not represent pure types have the same status in Q-technique factor analysis that complex data variables have in R-technique factor analysis. They can be included, but they do not provide much help in determining where the factors should be located. Such individuals will have loadings of moderate size on several Q-technique factors rather than high loadings on only one factor and negligible loadings on all other factors. The pure-type individual, on the other hand, does for Q-technique factor analysis what the pure-factor data variable does for R-technique factor analysis. Such pure-type individuals, when included with others that belong to the same type, define the factors in a sharp way, permitting the common rotational criteria to locate relatively clear-cut and compelling factor positions.

It is clear that merely taking a sampling of "available" subjects for Q-technique factor analysis is not likely to provide several pure-type individuals to represent each type factor while introducing only a relatively small number of individuals who are not pure types. Since pure types are the exception rather than the rule, a random sampling of individuals is most likely to contain a heavy preponderance of individuals who are not pure types. It would be surprising, therefore, if such sampling methods were to yield enough pure-type individuals to provide a good definition for more than a small proportion of the *Q*-technique factors. It is quite possible that none of them would be adequately represented. In any event, the large number of impure types, corresponding to complex data variables in *R*-technique analysis, would greatly complicate the task of achieving a good rotational solution.

These considerations emphasize the fact that subjects must be carefully selected to give meaningful *Q*-technique factor analytic results, just as data variables must be carefully selected to yield good *R*-technique factor analytic results. As with *R*-technique analysis, the investigator's hypotheses about the factors and individuals defining them in *Q*-technique analysis may prove to be incorrect, or partially so. Only through a series of carefully planned analyses, each taking advantage of what was learned in the previous one, will the investigator gradually establish what the type factors are in a given domain. This technique of analysis is very powerful and has tremendous potential for scientific explication in those areas where many data are available for relatively few data objects, such as in clinical research, but it should be emphasized that these benefits are not apt to accrue without careful planning of the research investigations in which it is to be used.

One might ask, "What good is a type after it has been located?" The location of well-defined types and knowledge of type membership can be very useful in the description, prediction, and control of human behavior. In clinical psychology, for example, "cookbooks" have been developed for use in dealing with types. A set of rules is established that permits unequivocal determination of type membership. The similarities of behavior, symptoms, etiology of pathological symptoms, typical responses to different forms of treatment, prognosis, and so on, are cataloged for each type. Once an individual is classified as belonging to a given type, a great deal is immediately known about him, how he should be treated, and what the outcome is likely to be. An example of this approach is to be found in the MMPI cookbook developed by Gilberstadt and Duker (1965) although the types described by these authors were not established on the basis of factor analytic studies.

[9.2] *P*-TECHNIQUE

Much less commonly used than *R*- and *Q*-technique factor analyses but potentially very valuable for clinical study of the individual case is *P*-technique factor analysis. One of the pioneer investigations using this method was carried out by Cattell, Cattell, and Rhymer (1947). The data matrix for *P*-technique analysis is shown in Fig. 9.2. The correlations are computed between pairs of variables over *N* occasions.

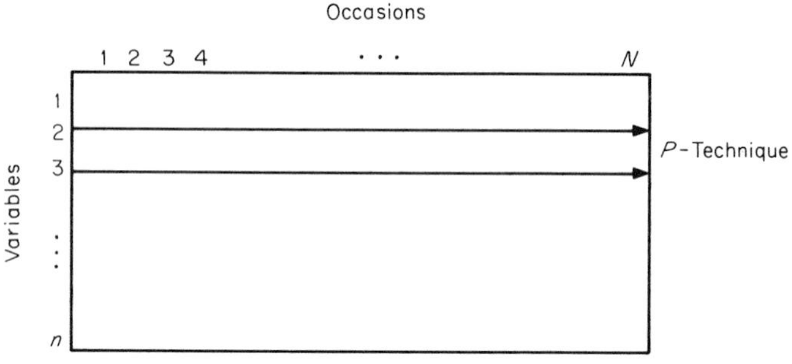

FIG. 9.2. *P*-technique data matrix.

In *P*-technique analysis, then, the subject of study is a single person, or data object. A measurement on each data variable is taken on the first occasion, for example, on day 1. On the second occasion, for example, day 2, the measurements are taken again to give the scores in column 2 of the data matrix in Fig. 9.2. The process is repeated on the third occasion, and so on, ending up with the Nth occasion. The occasions could be repeated trials with very little time between trials or with large time intervals between trials. There is no need to standardize the scores for the matrix of correlations in *P*-technique analysis since the correlations are computed between data variables as in *R*-technique. The correlation coefficient formula automatically adjusts for the differences in mean and standard deviation between variables. Since the measurements over occasions are repeated on the same measuring instruments, there is no variation in the measurement scale from column to column that would artificially inflate the correlations. In *Q*-technique analysis, standardization of scores by rows was necessary to avoid row variations in scale. It is unnecessary in *P*-technique analysis.

The factors derived in *P*-technique, as in *R*-technique, represent clusters of substantially correlated variables. The meaning of the correlation between variables is different in the two techniques, however. If tests of Verbal Knowledge and Verbal Fluency, for exauple, are highly correlated and appear on the same factor in *R*-technique analysis, it means that if a subject has a high score on one of these two tests, he is also apt to have a high score on the other, and vice versa. If he has a low score in one, he is likely to have a low score in the other, and vice versa. In *P*-technique, on the other hand, a high correlation between variables means that as one variable goes up or down, the other variable moves with it over the series of occasions. For example, Heart Rate and Palmar Sweating would be two variables that tend to move up and down together over a series of occasions manipulated to expose the individual to varying amounts of stress. Heart Rate and Palmar Sweating would both be high under high-stress conditions. They would both be low in the resting state. These two variables might be selected as potential pure-factor measures of an hypothesized "Emotional Arousal" factor expected in a *P*-technique analysis of physiological data variables.

P-technique analysis is suitable only for certain kinds of variables. There must be reliably measurable variations in scores on the variable over the series of occasions studied. This is not apt to be the case for some commonly studied variables. Ability-test scores, for example, may vary over time for a given individual, but the difference in scores for a particular test on two occasions is usually attributed to errors of measurement, practice effect, and so on, rather than primarily to real variations in the amount of ability a person has from time to time. A variable like Heart Rate, on the other hand, is ideal for *P*-technique analysis because it can be accurately measured and it fluctuates over wide limits as a function of events in the individual's external and internal environment, coming periodically to a steady state.

Another very important requirement for variables in *P*-technique analysis is that the process of taking the measurements *per se* should not affect the present or subsequent measurement values obtained to any appreciable degree. This is clearly not the case with those kinds of tests where the person's performance steadily improves with each succeeding trial as he becomes more and more familiar with the task and expert at doing it. It is not necessarily impossible to carry out *P*-technique studies with such variables, but the danger of artifacts in the results is great. If the subject improves on all tasks as the trials progress, for example, this will force a fairly substantial positive correlation between all pairs of variables, tending to give one large, dominant factor which could be labeled "Practice Effect." Variables like Heart Rate do not suffer from this difficulty since the value obtained is not affected in any systematic fashion by the measurement if the individual is properly trained and appropriate measuring apparatus is employed.

P-technique analysis, therefore, has the greatest potential in clinical research involving variables that exhibit substantial real fluctuations in the individual over time. It has relatively little potential for the study of those characteristics that are stable in the individual and for which variations in scores over time can be attributed largely to practice effects and errors of measurement.

[9.3] LESSER KNOWN DESIGNS

R-, *Q*-, and *P*-techniques are the best known of the basic factor analytic designs, but they by no means exhaust the possibilities, as Cattell (1952) has pointed out. With subjects, variables, and occasions representing three dimensions, it is possible to obtain three two-dimensional data matrices, each of which gives two ways of computing correlations. This gives six possible factor analytic designs. Figure 9.1. gives one of these data matrices from which *R*- and *Q*-technique designs are obtained. Figure 9.2 gives the *P*-technique design. It would also be possible, however, to compute the correlations in Fig. 9.2 by taking pairs of columns instead of pairs of rows. This would give correlations of pairs of occasions over the series of data-variable scores. Correlating columns from the data matrix in Fig. 9.2 yields a correlation matrix that can be factored to give what Cattell (1952) calls an "*O*-technique" factor analysis.

Because of the variations in means and standard deviations of the row variables in the data matrix in Fig. 9.2, the rows would have to be standardized, giving means and standard deviations that are equal from row to row, before an *O*-technique factor analysis can be carried out. This same step was required for a proper *Q*-technique analysis. The factors derived from an *O*-technique analysis show which occasions are clustered. Such an analysis could reveal the presence, for example, of one or more periodic states in a given individual. A particular physiological condition, characterized by a given pattern of scores on the variables, might emerge as a factor, with a period of approximately 10, 20, or even 30 days from one appearance to the next.

Correlating occasions over a sample of individuals for a given variable produces a matrix of correlations each of which is essentially a measure of test–retest reliability. This type of design is derived from a third possible data matrix, different from those shown in Figs. 9.1 and 9.2. Thus, many persons would be measured many times on the same data variable. Where the correlations are computed between occasions over a sample of individuals, the type of factor analysis obtained from the resulting correlation matrix is called "*T*-technique" by Cattell (1952). The factors derived would represent clusters of occasions upon which the individuals performed in a similar manner.

The same data matrix that provides the basis for *T*-technique can also give rise to a second design, referred to as "*S*-technique." In *S*-technique, two individuals are correlated over the series of occasions on which measurements are taken for the variable under investigation. *S*-technique, as Cattell points out, can be used to cluster people who respond in the same way over a series of varying social situations where all the measurements are taken with respect to a single variable, for example, the amount of overt aggression exhibited. Since *O*-, *S*-, and *T*-technique designs are used so little, they are not discussed further here. The interested reader is referred to Cattell (1952) for a more extended discussion of these methods.

CHAPTER TEN

Interpretation and Application of Factor Analytic Results

[10.1]

INTRODUCTION

The usual procedures followed in factor interpretation are deceptively simple. Those data variables with high factor loadings are considered to be "like" the factor in some sense and those with zero or near-zero loadings are treated as being "not like" the factor, whatever it is. Those variables that are "like" the factor, that is, have high loadings on the factor, are examined to find out what they have in common that could be the basis for the factor that has emerged. High loadings in both the positive and negative direction are considered.

If a data variable were to correlate perfectly with a factor, it would ordinarily be considered identical with the factor in what it measures. Since data variables are not perfectly reliable, they cannot correlate perfectly with a factor, of course, but with a data-variable reliability of .81, a factor loading of .90 would indicate a total overlap in true variance between the data variable and the factor (see Guilford, 1965, p. 442 index of reliability). In general, a factor loading for a data variable equal to the square root of its reliability coefficient is suggestive of a completely pure-factor test or data variable. All its true variance is placed on that one factor, making it essentially identical with the factor except for its error variance.

In practice, of course, the correlations of data variables with factors generally fall well below the square roots of the data-variable reliabilities, indicating that the data variables have other true variance that is not shared with this factor. The additional true variance may be due exclusively to specific variance, but more likely it will be due to a combination of specific variance and some common variance associated with other common factors in the analysis.

Correlations of data variables with the factors, it should be remembered, are elements of the factor structure matrix rather than the factor pattern matrix (see Chapter 6) where the solutions are oblique. Only with orthogonal factor solutions are the correlations of the data variables with the factors equal to the factor loadings With oblique solutions, the factor loadings, or pattern coefficients, might equal or exceed the square roots of the data-variable reliabilities in some cases This would not indicate necessarily that the variable is factor pure, however, since the structure coefficient, the correlation of the data variable with the factor, might be considerably lower.

Ordinarily, then, the correlations of the data variables with a factor they define are well below the level that would permit an easy inference of synonymity between the factor and one or more of the data variables. There must be a painstaking analysis of each highly loaded data variable into its constituent components so that the elements common to all data variables that define the factor may be identified and labeled. The process of trying to identify the nature of a factor is facilitated by several conditions.

1. The higher the factor loadings, the greater is the degree of overlapping true variance between the data variable and the factor and the more the factor is like the data variable in question. The square of the correlation of the variable with the factor is a good indication of the extent of the overlap. Dividing this figure by the reliability coefficient for the data variable gives the proportion of the data variable's true variance that it shares with the factor. It is also sometimes helpful to divide the correlation of a data variable with the factor by the square root of the reliability coefficient to get an idea about what the correlation between the two would be if the data variable were perfectly reliable.

2. The more factor pure a variable is that defines a factor, the easier it is to make inferences regarding the nature of the factor. If a complex data variable has a substantial loading on a given factor, it is difficult to determine from this information alone which of its many constituent parts is responsible. Data variables of great factor complexity offer little more help in the factor interpretation stage than they did in the factor rotation stage.

3. The greater the number of variables with a substantial loading on the factor, other things being equal, the easier it is to isolate what the factor probably represents. It is still possible to obtain an erroneous impression of the factor's character despite the availability of many highly loaded variables

to define it, however, if the sample of variables is biased in some sense important to the factor definition. If the data variables included fail to represent some important aspect of the factor, for example, the picture of what the factor is like will be incomplete. The situation is somewhat analogous to the task of describing a person. Several important correct and revealing characteristics of an individual could be presented, but they might not tell the whole story. A man's employment history and background might look exceedingly impressive, but the inclusion of one additional item, for example, the information that he has spent the last two years in jail following conviction for embezzlement, could make a significant difference to a prospective employer.

Even though a factor is prominently displayed in a given study, therefore, it cannot be assumed that all major facets of the factor have been represented by the variables defining the factor. Once a major factor has been isolated in a given domain, an important focus of further research is the search for other data variables related to the factor in a substantial way that may help to give a more complete picture of what the factor is like.

A question that frequently arises is that of how high the correlation between a data variable and a factor must be before it can be regarded as "significant" for interpretive purposes. There can be no answer to this question in any precise statistical sense since there is not available at the present time any statistical test that can establish the significance level of a rotated factor loading. The loading of a given variable on a factor can be altered easily by rotating the factor a little closer to or a little farther away from the particular data-variable vector in question. A crude index of the usability of a given data variable for interpretive purposes is the square of the correlation between the factor and the data variable. It has already been mentioned that if this value is high enough, it indicates a total overlap of the data variable's true variance with the factor. This value is also helpful at the low end to decide when the data variable is too unlike the factor to be considered in the factor interpretation.

A fairly commonly used cutoff level for orthogonal factor loadings is .30; that is, no variable with a factor loading below .30 is listed among those data variables defining the factor. A squared value $(.30)^2$ gives .09, which indicates that a data variable correlating with the factor less than .30 has less than 10 percent of its variance in common with the factor. The other 90 plus percent lies elsewhere, in specific and common factors plus error. Some investigators have been known to list loadings less than .30 in interpreting factors, but in the orthogonal case where such loadings are interpretable as correlations, such values seem excessively low to this author. The danger, of course, is that aspects of the variable from the 90 percent variance *not* associated with the factor will be erroneously attributed to the factor itself. This danger is particularly great where the investigator has hypothesized that the data variable in question will relate to this factor, is anxious to believe that it does, and then accepts a loading of between .2 and .3 as a confirmation of his hypothesis.

Whereas loadings of .30 and above have commonly been listed among those high enough to provide some interpretive value, such loadings certainly cannot be relied upon to provide a very good basis for factor interpretation. Table 10.1 can be used to give a rough idea of the value of variable–factor correlations (orthogonal factor loadings or structure coefficients) for factor interpretation purposes. Adjustments must be made with oblique factor solutions to consider both the factor loadings (pattern coefficients) and the correlations of the data variables with the factors (structure coefficients). If the data variable has its principle oblique factor loading on a particular factor, then the correlation of the variable with that factor from the structure matrix may be used in Table 10.1 to judge the potential usefulness of that variable for factor interpretation.

TABLE 10.1

Scale of Variable–Factor Correlations

Orthogonal factor loading	Percentage of variance	Rating
.71	50	Excellent
.63	40	Very good
.55	30	Good
.45	20	Fair
.32	10	Poor

Common sense dictates that factor interpretation based on a few variables with only "poor" to "fair" ratings in Table 10.1 must be made very cautiously and with every expectation that substantial revisions may be necessary in the future. If several data variables are available with loadings in the "very good" to "excellent" category, the investigator can afford to be a somewhat more definite in what he says about the factor. Even here, however, the investigator must avoid dogmatic statements since subsequent work may establish that some aspects of the factor have not been well represented, if at all, in the group of variables he included in his investigations. Later work may force significant modifications of the factor interpretation.

Factor interpretation ideally should go beyond merely determining which data variables are like the factor and which are not and the resultant identification and naming of what appears to be the common element or elements represented by the factor. The best kind of factor interpretation will also shed some light on several other areas of concern. Where does this factor lie in the hierarchy of factors? Is it a relatively specific-level factor that is too limited in scope to be of much scientific value or is it very broad and complex, perhaps too much so? How does the factor relate to previously developed taxonomic systems of interest in the domain under investigation? How firmly established is the factor identification? Will other studies be needed to identify more completely just what the factor is? What potential value does the factor have for purposes of general scientific description and theory building? To what extent are variations in factor scores assignable to hereditary and environmental influences, respectively?

These questions are not easy to answer and in fact few factor analytic studies are published in which there is much of an attempt to do so. All too often, the report of a factor analysis presents a list of factors with the data variables that are highly loaded on the factors together with factor names but not a great deal else in the way of factor interpretation. The investigator should at least be aware of the need for more extensive interpretation and should attempt to provide it as his research endeavor unfolds over a series of studies.

[10.1.1] ARE FACTORS "REAL"?

A persistent and recurring controversy in factor interpretation centers about the meaning that can be attached to a factor. On the one extreme there is the position expressed by Kelley (1940, p. 120): "There is no search for timeless, spaceless, populationless truth in factor analysis; rather, it represents a simple, straightforward problem of description in several dimensions of a definite group functioning in definite manners, and he who presumes to read more remote verities into the factorial outcome is certainly doomed to disappointment."

An able spokesman for the position on the other side of this controversy, Cattell feels that the psychologist or social scientist cannot be content with factors that are merely mathematical conveniences. He states that

> in the case of factor analysis, the scientist wants each factor to correspond to some unitary influence with which he is familiar on *other* and general scientific grounds—some influence which he has reason to believe is a functional unity in nature. Consequently, he argues that there is *one* position in the rotation which corresponds to the real factors and that all other positions encountered are mathematical transformations of this real position—false claimants which we have not yet succeeded in eliminating [p. 66].[1]

[1] From R. B. Cattell, *Factor analysis*. New York: Harper, 1952.

The controversy, then, centers about the question of whether factors are to be considered as merely arbitrary descriptive variables with no more intrinsic meaning than any other set of descriptors or whether they are to be considered as "real" entities in nature, lying there waiting to be discovered by the enterprising research worker. Cattell, of course, does not take the position that the factors derived from any given factor analysis will necessarily represent examples of the real functional unities that he seeks. The location of such factors represents a desirable goal of factor analysis, in his view, whereas those who take the viewpoint expressed by Kelley deny that such a goal is attainable. Cattell obviously believes that the goal is attainable and furthermore claims that in any given study, rotation of factor axes to oblique simple structure is the most likely way to achieve success in this search.

Among active factor analysts doing empirical taxonomic research today, there appears to be more leaning toward Cattell's position than toward Kelley's in the sense that these workers like to believe they are coming up with constructs that are not just arbitrary descriptive variables. Each taxonomic investigator hopes to find constructs that are better in some sense than other possible descriptors, although he may not be able to prove it nor even to specify precisely why he thinks his are better.

A science consists of descriptions of the functional relationships between *the* variables in that field. To determine what the functional relationships are, it is clearly necessary first to know what *the* variables to be related are. It makes sense to believe that all possible sets of variables are not equally good for developing a particular science. Some sets may be more parsimonious than others. Some sets may be more easily integrated into unifying theoretical superstructures. Some sets of variables may have greater heuristic value than others. Variables used as the basis for development of a science, other things being equal, would seem to be superior to alternate possibilities when they are more unitary rather than complex, in closer alignment with causal forces that can be identified in the external or internal environment of man, easier to understand, of greater practical utility, easier to measure, and so on.

It seems clear to this author that the goal of factor analysis should be to isolate constructs that *do* have greater intrinsic value for the purposes of building a science. It is going too far to say that there exists one and only one set of "correct" constructs, or "real" factors, in a given domain. On the other hand, many social science fields, for example, are burdened down with a veritable avalanche of possible variables from which to choose in trying to develop a viable science. These are clearly not all equally good. An appropriate and reasonable goal of factor analysis, in this author's opinion, is to isolate constructs that can be shown to be superior to others in certain defined ways for the purpose of building a science.

As in most human endeavors, however, the goals of factor analysis are not always achieved in practice. The plethora of "factors" ground out by busy electronic computers is dwarfed only by the productions of armchair variable builders. If it were possible to ensure that every factor analysis would turn up at least one "good" construct, progress in the social sciences would be much faster than it is in fact. It is unfortunately true that most factor analyses do not reveal anything particularly useful in the way of scientific constructs that have not been previously isolated. Part of this is due to the fact that it is so easy to make serious mistakes in planning a factor analytic study. Another reason is that well-defined factor constructs usually emerge only out of a well-designed and integrated series of investigations. Finally, useful constructs are not necessarily revealed as such upon inspection. It is necessary to establish their claim to such status by appeal to evidence outside the factor analytic investigations themselves. This requires a great deal of painstaking work and hence is frequently left for "other" investigators to carry out.

The task of proper interpretation of factor analytic results, therefore, is a complex one. It can be approached very simply as one of describing the putative nature of the common elements among the data variables that define the factor and giving the factor a name. On the other hand, it can be approached as a part of a long-range task of developing the best possible set of factor constructs. In this latter case, factor interpretation must be concerned with the matters discussed above and such questions as: (a) Is this factor essentially an artifact or not? (b) What is the potential value of this factor? (c) Do the variables that define the factor reveal all its major aspects? (d) Can other data variables be found that will measure it better? (e) Will it hold up in another investigation using other variables and subjects? To answer such questions makes factor interpretation a great deal more difficult than mere factor naming. It also often involves delaying answers until additional investigations have been carried out. Factor interpretation at its best becomes an activity to be developed over a series of related studies rather than something to be based solely on a single investigation.

[10.2] FACTOR ANALYSIS AND MULTIPLE REGRESSION ANALYSIS

Although factor analysis and multiple regression analysis have some common objectives, there is a considerable difference between them with respect to what they can be expected to accomplish. In multiple regression analysis, a data matrix is available which is used to determine the intercorrelations among predictor variables and also correlations of the predictors with one or more criterion variables, such as measures of success in training, success on the job, or success in some other kind of life adjustment. The object of the multiple regression analysis is to pick a subset of the predictor variables that will best predict the criterion variable and to determine their relative weights for making that prediction.

Regression analysis has, therefore, an immediate practical goal, namely, to predict a given criterion of adjustment. The results of such studies often have considerable utility, for example, in the selection and classification of personnel. If the primary objective of a given research study is the selection of the best predictors from a specified given set of data variables, multiple regression analysis would be preferred to factor analysis as the principal method of treating the data. Factor analysis alone would not provide the most useful approach to dealing with the immediate practical problem.

Factor analysis can be a very useful supplementary method of analysis, however, in more extensive research efforts designed to improve the prediction of a given criterion of adjustment. Consider an example of a large firm that employs several hundred individuals in a particular occupational specialty. The firm finds that a substantial number of the individuals hired for this type of job fail within one year and either quit or are discharged. The firm decides to develop a testing program to select future employees in an attempt to reduce the number of failures on the job.

A simple-minded multiple regression approach to this problem could begin with the selection of a battery of available tests that might have some validity for predicting success in this occupational activity. These tests could be administered to all currently employed individuals in this kind of job. Test scores would be correlated with a numerical index of success in performing this work after dividing the sample into two randomly selected halves. Multiple regression analysis performed in each group separately would reveal those variables that contribute significantly to predicting the criterion of success on the job. Variables that prove to be significant contributors in both samples could be utilized to form a test battery for selection purposes. The total sample could be used to find optimum regression weights for these tests. This regression equation could be used to predict success for new applicants who would be hired regardless of their test scores. By follow-up evaluation of these new employees, the effectiveness of the test battery for selection purposes could be assessed.

The effectiveness of this conventional approach depends on the battery of tests selected for trial and on the quality of the criterion. Factor analysis could be employed to improve both in a long-range research program designed to upgrade the effectiveness of personnel selection. Only by determining what the criterion consists of is it possible to know in any precise way what kinds of data variables should be used to predict it. Analysis of the criterion not only helps to suggest what predictor variables should be tried, but it also can reveal possible areas of criterion improvement. This point is discussed in somewhat greater detail later in the chapter.

Available published test instruments may prove to be less than optimum as predictors for the particular criterion being predicted. Their factor composition, for example, may be complex and may contain too much variance not found in the criterion. It is also unlikely that such tests would be constructed in such a way that the proportions of variance assigned to various factors in the test would coincide with those found in the criterion itself. Factor analyses of predictor variables together with criterion variables can help to answer questions of this kind. Where it is necessary to develop new tests that are relatively pure measures of the factors identified in the criterion itself, factor analysis can be helpful in the test-development program. The use of pure-factor predictors representing the various identifiable factors in the criterion permits the choice of optimum weights for those factors in predicting the criterion. Careful research and test development of this kind, therefore, can often improve the level of prediction for criteria of adjustment attainable with unaltered existing predictors and criteria.

Although factor analysis can be of great practical value in this way as a means of improving prediction of important criteria, multiple regression analysis still represents the major statistical tool in this work. It is in the area of scientific understanding and taxonomic research that factor analysis plays the greater role with multiple regression analysis retreating to a secondary position of importance. Predicting specific criteria by means of test batteries selected through multiple regression analysis represents an activity of limited general scientific interest because, among other things, the number of potential criteria to predict is very large. Even for the same general kind of occupational activity, a test battery that would be good for predicting success in one firm might not be satisfactory in another. Scientific goals call for results that have greater generality, results that are not tied to a specific criterion of practical adjustment in a specific cultural, industrial, or educational setting.

The investigator with broader scientific objectives in mind, but who nevertheless may be hopeful of achieving results with ultimate potential practical utility, is more concerned with using factor analysis to establish those variables that can provide the basis for a taxonomy describing an entire area of scientific concern. If factor analysis can be utilized to produce an adequate taxonomy for describing personality, for example, and to develop instruments to measure the data variables that make up the taxonomy, these same data variables can be utilized for predicting many different kinds of practical criteria of adjustment that contain important personality components. The multiple regression equation weights assigned to the relatively pure-factor personality measures will vary from one criterion to another, but the same pool of personality measures can be tried out for each criterion-prediction project. Many of these measures would be discarded for purposes of predicting a specific criterion, of course, since their weights would not be sufficiently different from zero.

The ultimate goal of taxonomic research using factor analysis is to provide a complete catalog of variables, and methods of measuring them, to cover the entire scientific domain under investigation. In psychology, for example, this would ideally involve a set of variables adequate to describe the individual completely for all practical purposes, except for errors of measurement. Such a set of variables would not only constitute the basis of a science of human behavior but would also serve practical objectives of prediction as well. Any practical criterion would be predicted by using a subset of these variables, each optimally weighted to predict this particular criterion. Taxonomic research of this kind has been conducted most intensively to date in the area of human abilities. The domain of human personality is also receiving a great deal of attention, although the progress in this field lags behind that in the mental abilities field. Work in other areas besides these two has been much less extensive. The general availability of computers, technological advances in factor analysis, and increased interest in factor analysis by investigators from many fields, however, offer considerable hope for an accelerated rate of progress from factor analytic applications in the future.

[10.3] FACTOR SCORES

Once a factor has been isolated, investigators are eager to relate this variable to other variables of interest in a given scientific area. If a single pure-factor test of high reliability is available for a given factor, scores on this test can be used as factor scores for the factor, although they would contain error to the extent that the test is not perfectly reliable. It is more usual, however, to have several variables that are related to the factor but no one of which has anything approaching 80 percent of its variance concentrated in this one factor. In such cases, it is necessary to attempt to estimate factor scores utilizing the scores of those variables related to the factor in a more modest way. Some of the methods of estimating factor scores are considered briefly in the following sections.

[10.3.1] METHOD 1

One of the simplest ways of estimating factor scores is first to single out all those data variables that have factor loadings on the factor above a certain selected cutoff value, for example, .50. The raw scores for these data variables may be added up to provide a rough estimate of the factor score on this factor for a given individual. The same procedure is followed for each other individual. The raw score for any variable with a negative loading of $-.50$ or less would be subtracted rather than added because the data variable is negatively related to the factor. This method has the disadvantage that only an arbitrary cutoff point is used to determine which variables have high enough loadings to be used as estimates of the factor score. The higher the cutoff is set, the fewer will be the data variables used as factor score estimators. The lower the cutoff, the less related will some of the variables be to the factor and hence the more impure will be the factor scores. This method also has the disadvantage of giving disproportionate weight to those variables with greater raw-score variabilities. It is a rather crude method, therefore, but under some circumstances it may be quite adequate, such as for rough exploratory work and where the variables do not differ greatly among themselves in variability. The factor scores obtained in this fashion are not likely to be uncorrelated among the various factors even if the factor solution is an orthogonal one.

[10.3.2] *METHOD 2*

A somewhat more sophisticated approach than the previous one is to scale the raw scores for all variables to the same mean and standard deviation before adding scores for those variables with loadings above the cutoff. This ensures that every variable will receive the same weight as every other variable in determining the factor scores. This refinement is usually worth the effort involved unless the variables are reasonably similar in the size of their standard deviations in the raw-score form. Since standard deviations of raw scores can and do vary over a wide range, failure to standardize scores in factor score estimation can result in markedly uneven weights for the different factor score components.

[10.3.3] *METHOD 3*

The two methods just described do not base the weights assigned to factor score component variables on their loadings on the factor. That is, a variable with a high loading on the factor does not necessarily have a higher weight in computing the factor score than a variable with a lower loading. A further refinement in factor score estimation, therefore, is to weight the various scores after they have been scaled to the same mean and standard deviation.

The weight for each score is the factor loading of that variable on the factor or perhaps an integral value approximately proportional to the factor loading. This method can be applied just to those variables with factor loadings above a specified cutoff value, or it can be applied to all the variables. Variables with small loadings, of course, would have little effect on the total factor scores.

The advantage of this method is that it allows those variables with the highest loadings on the factor to have the greatest effect in estimating the factor scores. A disadvantage is the greater computational effort required to obtain the factor scores. A more subtle disadvantage is the possibility that differences in factor loadings among those variables with loadings above the cutoff point are due more to vagaries of variable selection and rotation than to any real differences in their value for estimating the factor scores. To the extent that this is true, this method would not represent an improvement over Method 2.

[10.3.4] *METHOD 4*

If n factors are extracted from an $n \times n$ correlation matrix, with 1's used as communalities, residuals will vanish and factor scores may be calculated as follows:

$$\mathbf{Z} = \mathbf{PF} \tag{10.1}$$

where

$\mathbf{Z}$ is an $n \times N$ matrix of scaled scores on the data variables,
$\mathbf{P}$ is an $n \times n$ matrix of factor loadings,
$\mathbf{F}$ is an $n \times N$ matrix of scaled factor scores.

Then

$$\mathbf{P}^{-1}\mathbf{Z} = \mathbf{P}^{-1}\mathbf{PF}$$

and

$$\mathbf{F} = \mathbf{P}^{-1}\mathbf{Z} \tag{10.2}$$

This solution for the matrix of factor scores $\mathbf{F}$ requires that $\mathbf{P}$, the matrix of factor loadings, have an inverse. This will not be true unless as many factors are extracted as there are data variables. Computing the inverse of an $n \times n$ matrix can become very time-consuming as n increases, making this a laborious method for computing factor scores. Other objections to the method, however, are even more serious. Since a primary objective of factor analysis is to account for the overlap among many data variables through the use of a much smaller number of factor constructs, the idea of extracting as many factors as there are data variables has little appeal to the empirically oriented scientist. It might well have appeal in certain kinds of problems for the theoretician.

Where unities are placed in the diagonals of the correlation matrix and a principal component solution is obtained, it is possible to derive factor scores for both rotated and unrotated factors without computing an inverse and without even extracting n factors. These procedures and other methods for obtaining factor scores have been described rather completely by Harman (1967, pp. 345–374). Some of these methods permit the calculation of factor scores that are uncorrelated with each other, a property that is important for some research purposes.

[*10.3.5*]
METHOD 5

Multiple regression methods can also be employed to estimate factor scores using the following basic equation:

$$z_{fi} = \beta_1 z_{1i} + \beta_2 z_{2i} + \beta_3 z_{3i} + \cdots + \beta_n z_{ni} \tag{10.3}$$

where

z_{fi} is a standard score in factor f for person i,
z_{1i} is a standard score in variable 1 for person i,
z_{2i} is a standard score in variable 2 for person i,
β_i is the standard regression coefficient for variable i.

The standard scores on the n variables used to predict the factor scores are known. These variables could consist of all the data variables in the factor analysis, in which case many of the β_i weights would be very low because their loadings on the factor would be low, or the variables included could be a subset of these, restricted to only those with loadings above a selected cutoff point. The development here, however, will presume that all variables are being used.

Equation (10.3) is like the standard multiple regression equation where n predictors are being used to predict a single criterion variable. To obtain the β_i weights for this equation, it is sufficient to know the correlations among the predictors and the correlations of the predictors with the criterion, that is, the validity coefficients. In the application to the problem of estimating factor scores, the factor scores become the predicted criterion scores, the variables in the factor analysis are the predictors, and the orthogonal factor loadings or oblique structure coefficients are the validity coefficients. The unknown β_i

weights are obtained through the solution of the following normal equations (Guilford, 1965, p. 409), derived using the principle of least squares:

$$
\begin{aligned}
\beta_1 + \beta_2 r_{12} + \beta_3 r_{13} + \cdots + \beta_n r_{1n} &= r_{1f} \\
\beta_1 r_{21} + \beta_2 + \beta_3 r_{23} + \cdots + \beta_n r_{2n} &= r_{2f} \\
\beta_1 r_{31} + \beta_2 r_{32} + \beta_3 + \cdots + \beta_n r_{3n} &= r_{3f} \\
\beta_1 r_{n1} + \beta_2 r_{n2} + \beta_3 r_{n3} + \cdots + \beta_n &= r_{nf}
\end{aligned} \tag{10.4}
$$

Equation (10.4) may be expressed in matrix form as follows:

$$\mathbf{R}\beta = r_f \tag{10.5}$$

where **R** is the matrix of known correlations among variables 1 through n in Eq. (10.4); β is a column matrix containing the unknown β_i weights; and r_f is a column matrix of correlations between the variables and the factor, that is, orthogonal factor loadings of oblique structure coefficients. Provided the matrix **R** has an inverse, Eq. (10.5) and hence Eqs. (10.4) may be solved as follows:

$$\beta = \mathbf{R}^{-1} r_f \tag{10.6}$$

Thus, the column of β_i weights to be used in Eq. (10.3) for predicting the factor scores from the data-variable scores is obtained by multiplying the inverse of the matrix of correlations among the data variables by the column matrix of correlations of the data variables with the factor.

An advantage of using all variables in the regression equation (10.3) is that the inverse $\mathbf{R}^{-1}$ may be obtained once and used in Eq. (10.6) for all factors merely by changing the r_f column, depending on which factor is being considered. If only some of the variables are used to obtain a given set of factor scores, the $\mathbf{R}^{-1}$ matrix for predicting that factor must be derived from an **R** matrix containing only those variables that are being used.

The factor scores obtained in this way are least-squares estimates, given the correlations that constitute the data. These factor scores will not be independent of one another from factor to factor, even for an orthogonal factor solution. This may be a handicap for some theoretical investigations, but it is not for most practical purposes. Obtaining these least-squares factor scores does require a considerable amount of computation. The precision obtained is a positive feature, but it must be remembered that somewhat different rotations might alter the factor loadings sufficiently to nullify any real gain from this additional refinement in factor score computation. Capitalization on chance errors in multiple regression analysis is also a fact of life that must be remembered in considering whether least-squares factor scores are worth the additional cost over simpler methods.

[10.3.6]
OVERVIEW

Where factor scores are being derived from psychological test items, Method 1 is often the most convenient. It is desirable to compute test scores by adding item scores without weighting them since this makes test scoring easier. Where factor components have divergent variabilities that are substantial and not in accord with their relative degree of association with the factor, it may be necessary to weight the components differentially in obtaining factor scores. Total scores over three separate subgroups of items might be combined to give an overall factor score by the following formula:

$$F = 4\text{I} - \text{II} + 2\text{III} \qquad (10.7)$$

Rounded-off integral weights are used in Eq. (10.7) with variable II showing a negative weight, indicating that it was negatively loaded on the factor for which the factor score F is being computed. If the factor score were to be computed by adding item scores, instead of by adding total scores as in Eq. (10.7), the item scores for variable II would have to be reversed prior to being added so that the resulting reversed total of the variable II items would be positively instead of negatively related to the factor. There rarely would be any need to use weights more finely specified than rounded-off integral values for practical applications. The unstandardized factor scores obtained by Eq. (10.7) could be scaled to a mean of 50 and a standard deviation of 10, if desired, to place them on a more commonly used scale.

Where test or questionnaire items are used in determining factor scores, the kind of operation indicated by Eq. (10.7) can often be avoided by appropriate selection of items to be included in the factor score. By adding an item here, deleting an item there, and so on, the component variabilities can often be adjusted close enough to equality that a simple sum of item scores will suffice to give a practically acceptable factor score. Factor scores on the author's personality test (described in the next chapter) are obtained in this way by adding item scores where each factor is assessed by five subscales and each subscale is measured by four items. Since the subscales are not substantially different in variability and all have positive loadings on the factors, the factor scores are obtained merely by adding up the item scores over all the items representing subscales highly related to the factors. Scores for items that are negatively stated with respect to a given factor are reversed before adding them into the factor score. This is done by subtracting the item score from one point more than the maximum item score. On a seven-point item response scale, for example, if a person answers "7" to a negatively worded item, after reversal the item score becomes "1."

The factor scores obtained in this manner by adding item scores are not independent of each other from factor to factor but then neither are the factors they represent. Such procedures may be unacceptable in certain theoretical research endeavors but for most practical research objectives they are apt to be satisfactory.

[10.4] REPORTING A FACTOR ANALYTIC STUDY

A proper job of reporting a factor analytic study requires that the investigator provide the reader with enough information to permit the results to be checked independently. All too often, the investigator reports only the end results of his work, forcing the reader to take his findings on faith. This practice is to be discouraged. If the investigator reports all of his original data, his work may be checked from the very beginning. Rarely, however, is it expected that an investigator will go this far. It is expected, however, that the investigator should at least make available the correlation matrix that was computed from those original data and later subjected to factor analytic treatment. If this correlation matrix is made available, the reader may, if he so desires, repeat all the factor analytic work or apply to the data other methods of analysis that he may deem more suitable.

The reader may be aware that few professional journal editors these days are willing to publish correlation matrices with articles reporting factor analytic results because of the high cost involved and the fact that few readers will ever make use of this information. The investigator must, therefore, in most cases choose some alternate method of making the correlation matrix available. Some authors publish in monograph form where all the relevant tables may be included. Others, through footnotes in published articles, offer to furnish copies of the necessary tables upon written request to the author.

Perhaps the most economical and best recommended method of making such tables available is to take advantage of auxiliary publication services that permit readers to obtain the documents in question even though they are not published with the article, and without having to contact the author. Such facilities provide a more dependable and permanent record of research findings. Suppose, for example, that an author wishes to make available the correlation matrix for a factor analysis which is reported in an article he is submitting for publication. At an appropriate point in the article, usually in the results section, reference to a footnote such as the following would be made:

> [1] The correlation matrix upon which the factor analysis in this study was based has been deposited with the National Auxialiary Publications Service. Order document number ______ from ASIS National Auxiliary Publications Service, c/o CCM Information Sciences, Inc., 22 West 34th Street, New York, New York 10001, remitting in advance ______ for microfiche or ______ for photocopies.

The correlation matrix would be prepared for photocopying, even single spaced, and would be submitted along with the article manuscript to the editor, but it would be labeled for deposit with NAPS. If the article is accepted for publication, the editor will send the correlation matrix to NAPS and will obtain the document number and charges. These figures will be inserted by him in the blank space indicated in the sample footnote. The footnote itself would be published in the article. The interested reader will then be able to order the correlation matrix directly from NAPS if he so desires.

Although another investigator could repeat the factor analytic work from the correlation matrix alone, good reporting calls for submitting other matrices as well. For an orthogonal factor solution, for example, the following matrices should be submitted to NAPS or otherwise made available: correlation matrix, unrotated factor matrix, and rotated factor matrix. If available, the transformation matrix that carries the unrotated matrix into the rotated matrix should also be given. In reporting the results of an oblique factor analysis, the matrices that should be made available are the correlation matrix, unrotated factor matrix, matrix of correlations of the data variables with the reference vectors, transformation matrix that transforms the unrotated factor loadings to the reference vector projections, and matrix of correlations among the reference vectors. Unless an investigator has been forced to do the calculations without an electronic computer, he should also furnish the factor pattern, factor structure, and the matrix of correlations among the factors. Some research workers, for example, Eysenck (1960; Eysenck & Eysenck, 1969), feel that a second-order analysis of the matrix of correlations among the factors should also be provided. Where factor analytic methods are used that do not provide certain of these matrices, there would be no necessity to calculate them just for reporting purposes.

The reader of an article, book, or monograph may not be sufficiently interested to order the available documents and check the author's work for himself. He may be interested enough, however, to attempt an evaluation of the adequacy of the procedures used in carrying out the reported factor analysis. It is incumbent upon the author, therefore, to provide the reader with sufficient details to enable him to perform the evaluation. It is important, for example, for the author to report the following:

1. *The Sample.* How many cases of what kind were used and how they were selected.

2. *The Data Variables.* The data variables should be described in sufficient detail to enable the reader to have an understanding of what is being measured. It may be necessary to submit material on this to the auxiliary publication source.

3. *The Correlations.* What method of correlation was used? Was any

scaling carried out before computing the correlations? To what extent were the assumptions underlying the correlation method met by the data?

4. *Factor Extraction.* What method of factor extraction was used? What criterion was used to determine when to stop factoring?

5. *Communalities.* If communalities were used, how were they decided on? Was there any iteration of the communalities?

6. *Rotation.* What criteria were used to carry out the rotations? What evidence is available that would suggest that these criteria are appropriate?

When the reader is provided with adequate information in the preceding areas, he is in a reasonably good position to evaluate the results of the factor analytic study and their interpretation. If any question develops in his mind, the availability of the matrices through auxiliary publication sources will permit him to explore the matter to his satisfaction. If the author does not provide these resources, his readers can only consider the results and interpretations as unsupported hypotheses. Properly documented reports generally command greater attention and respect in the scientific community. Furthermore, the life expectancy of such studies is greater because even if newer, more powerful methods of data analysis become available, they can be applied to the data at a later date, if those data are made available through such auxiliary publication services as NAPS.

Although an investigator cannot usually expect to publish whole matrices with his article describing a factor analytic investigation, he can usually expect to publish a description and interpretation of each major factor. This might take the form of a major section heading entitled "Factor Results" with several subheadings, each devoted to a single major factor. The title of a subheading, for example, *Factor I. Extraversion–Introversion*, could be given followed by a description and interpretation of the factor, beginning with a table of the major loadings on the factor, such as that shown in Table 10.2. Those loadings of .30 or more, or some other figure such as .25, .35, and so on, would be included in the table for each factor. The reader should be told what cutoff

TABLE 10.2

Major Loadings on the Factor

Variable	Description	Loading
26	Lack of Reserve	.68
27	Lack of Seclusiveness	.60
28	No Loss for Words	.84
29	Lack of Shyness	.79
30	No Stage Fright	.49

figure is being used for including variables in the table. It is also common to list the variables in descending order of the size of their loadings on the factor rather than in the order of the variable numbers as shown in Table 10.2. In some cases, elaboration on the meaning of the factors and their relationship to theory and previous results will be reserved for the "Discussion" section of the article.

Careful description of procedures and results is necessary for adequate scientific reporting in any area. It is especially true, however, with factor analytic studies since a factor analysis merely provides the results of one investigator's way of interpreting the data. If the reporting job is adequate, the reader may evaluate how satisfactory that interpretation is for his purposes, and if he is not satisfied with the interpretation, the data are available from which he may develop his own factor interpretation. The investigator who takes the trouble to do this reporting properly will increase the present and potential usefulness of his work to the scientific community.

[10.5] APPLICATIONS OF FACTOR ANALYSIS

Many hundreds of articles, monographs, and books have been published in which factor analytic studies have been reported. The list of areas in which it has been applied is steadily growing. The following represents a very incomplete but suggestive list of areas in which factor analytic methods have been used.

Aptitudes and Abilities
- Intellectual abilites
- Clerical abilities
- Motor skills
- Language skills
- Social skills
- Academic achievement

Personality Characteristics
- Motivation
- Character
- Temperament
- Values
- Interests
- Emotions
- Mental symptoms

Attitudes and Opinions
- Political attitudes
- Social attitudes
- Judgments
- Aesthetic preferences
- Morale
- Ratings
- Humor

Physical Characteristics
- Dimensions of physique
- Biological variables
- Physiological responses
- Perceptual phenomena
- Physical symptoms

Selected Other Applications

Errors	Therapy methods
Accidents	Group characteristics
Job performance	Structure of language
Job characteristics	Freudian theory
Cultural differences	Theories of color vision
Political variables	Criteria of adjustment
Economic variables	Effects of training
Geological variables	Food preferences
Item analysis	Multidimensional scaling

Although the applications of factor analysis are increasing in number and extending into more and more new areas, it is still possible to classify a high percentage of such applications into one or more of the categories described in the following sections.

[*10.5.1*] *TAXONOMIC RESEARCH*

Perhaps the most common reason for undertaking a factor analytic study is to obtain some information concerning what are "the" factors in a comparatively unstructured area of investigation. The degree to which quantification has been achieved is sometimes used as an index of the state of development of a science. Before quantification can advance very far, however, there must be available information concerning what to quantify. Pioneer investigators in a new area frequently make use of factor analysis hoping that it will help them to navigate accurately in uncharted waters. Out of their studies, they hope to develop better ideas of what should be measured in the domain of investigation. It is a major step in the development of any science to reach concensus on just what constructs are to provide the structural foundation for further work in the field. Most of the social sciences and many other scientific fields are still very much in the stage of trying to settle on what variables are going to be most helpful for their further development. Although factor analysis does not provide a foolproof avenue to easy answers in this endeavor, it can be a valuable aid in dealing with the problems encountered at this stage of scientific development.

[*10.5.2*] *ANALYZING AND IMPROVING MEASURING INSTRUMENTS*

When a new construct is proposed for a given scientific field, attempts to develop methods of measuring that construct almost inevitably follow.

Before it is possible to determine precisely the nature of the functional relationships in a given scientific field, it is necessary to have adequate methods of measuring the variables that are to be related. One of the most common uses of factor analysis is to analyze the characteristics of a given measuring instrument or method to assess how well it is fulfilling its mission. This sometimes takes the form of a one-shot evaluation in which the investigator is checking on the adequacy of a particular measurement method of construct, often one proposed or developed by another investigator. Another related usage of factor analysis is that by the developer of a measurement method who is trying to improve the excellence of his instrument. Through factor analysis, he tries to locate areas of possible improvement, returns to the drawing board, makes refinements, and then tries again with another analysis. Such improvement attempts may involve factor analysis of individual items, subscales, or total variable scores with each other or with outside measures. The goal is to develop a highly reliable instrument that provides as pure a measure as possible of the construct in question.

[10.5.3] *ANALYZING AND IMPROVING CRITERIA*

A considerable amount of important applied work in psychology centers around the task of predicting criteria of adjustment, such as school achievement, job success, marital adjustment, and so on. As described earlier in the chapter, a crude frontal assault on such problems might involve the commonly used procedure of correlating available test measures and other convenient potential predictors with available criterion measures. A reasonable degree of success may be achieved by this approach if the available criterion measure is a good one and if the selected predictors account for a substantial proportion of the true variance in the criterion scores.

The use of factor analysis permits a more sophisticated approach to these important applied problems. The first objective is to determine the factor composition of the existing criterion measure or measures. This information is obtained from a series of factor analyses in which the criterion measure(s) is (are) included as variables in the analysis. In some instances, factor analyses may be carried out using only criterion measures as variables in an analysis where many possible criterion measures are available. Through such analyses, it is often possible to make substantial improvements in the criterion itself by identifying sources of unwanted variance. For example, it may be determined that ratings of "success" on the job have been too heavily influenced by personal friendship between raters and ratees or by other irrelevant sources of variance.

These factor analyses of criterion measures also provide information that can facilitate decisions about what separate criterion measures should be combined into one overall variable and how many separate criterion variables should be utilized. Where one single overall criterion variable is to be derived, these factor analysis results can provide important information useful in determining what components should be included in the overall composite criterion and what their relative weights should be to provide the best possible definition of "success."

Knowledge about the factors that are represented in the criterion permits a more scientific choice of the variables that are to be tried out as potential predictors. In the absence of knowledge about the factor composition of the criterion, such decisions must be based to at least some extent upon pure guess work.

[10.5.4] IDENTIFYING TYPES

The search for types, whether among people or other kinds of data objects, goes on and on because it is convenient to think in such terms. This "type" of person will do such-and-such and must be treated in this or that fashion. If people could be classified into a small number of categories or "types" for any given decision-making purpose, life would become considerably easier for the decision maker, be he an educator, employer, psychotherapist, or whatever. *Q*-technique analysis is specifically aimed at the objective of locating types and for this purpose represents a powerful tool which as yet has not been fully exploited. It is not the only kind of factor analysis useful for this purpose, however, because for many *R*-technique analyses, the experimenter's unspoken long-term goals are more concerned with locating types than with taxonomic considerations *per se.*

Information about types can provide aid in developing a taxonomy, and taxonomic information is useful in locating types if they exist. Unfortunately, much research aimed at locating types presumes that they exist and tends to force such distinctions on the data whether or not the data exhibit such discontinuities in fact. There is a discontinuity between male and female that permits human beings to be divided into two clear-cut types along sex lines. To divide people into "bright" and "dumb" types, however, is inappropriate because there is no discontinuity on the scale of human intelligence. Factor analysis can be a valuable aid to locating types where they exist, but it cannot create them.

[10.5.5] THEORY TESTING

Many factor analytic studies are designed to test theory-deduced hypotheses concerning the number of parameters needed to account for a given body of data. Is a two-factor theory adequate to account for certain data in color vision? Is there a single factor underlying political and social attitude variables? How many dimensions are involved in olfactory judgments? If theory X is true, variables a, b, c, and d should emerge with high loadings on the same factor; do they in fact come out this way? Factor analysis can be used to answer these and many other kinds of theory-oriented questions. Such theory applications of factor analysis are not apt to occur often in sciences where there are few variables and measurement is highly refined. In those developing sciences plagued by a plethora of imperfectly measured variables, however, factor analysis can be expected to play a more and more important role in the construction and testing of theories.

CHAPTER ELEVEN

Development of the Comrey Personality Scales: An Example of the Use of Factor Analysis

One of the common uses of factor analysis mentioned in the last chapter occurs in the development of psychological tests. This chapter presents an illustration of the use of factor analysis for this purpose, describing the development of the Comrey Personality Scales (Comrey, 1970a, b), an inventory of factored personality traits.

The Comrey Personality Scales, hereafter abbreviated as the CPS, were designed to measure the variables that make up a taxonomy of personality traits developed over a period of years by the author with the aid of many of his students. The factored traits included in this taxonomy are the following: T—Trust versus Defensiveness; O—Orderliness versus Lack of Compulsion; C—Social Conformity versus Rebelliousness; A—Activity versus Lack of Energy; S—Emotional Stability versus Neuroticism; E—Extraversion versus Introversion; M—Masculinity versus Femininity; and P—Empathy versus Egocentrism. The factored personality scales for measuring these traits evolved in a series of steps as the taxonomy itself developed, with improvements in both occurring at each step. The attempt to develop this taxonomy was inspired originally by discrepancies in the personality taxonomies proposed by Guilford, Cattell, and Eysenck, three of the best known writers in the field of personality measurement. The taxonomic research undertaken by the author had as its initial objective a resolution of the differences among these well-known writers. The intention was to independently seek out and identify the major factors of personality with the idea of comparing the resulting taxonomy of traits with those of the three authors mentioned.

As the research progressed, it became clear that the author's own taxonomy would not closely match that of any one of these previous writers. Instead, a taxonomic system began to emerge that, while sharing much with these previous systems, was nevertheless distinct from any of them. As the form of this taxonomy took shape, methods of measuring the constituent factors were developed and refined, ultimately leading to the published version of the CPS.

[11.1] STUDIES OF MMPI ITEMS

Since the objective of the planned research program called for the identification of the main factors needed to describe human personality, the decision was made to study the items of several well-known personality test instruments under the assumption that they would define at least some of the needed personality factors well enough to permit their identification. The MMPI was selected as the first instrument to be studied because it was the most widely used personality inventory and also because the usual scores derived from the MMPI appeared to be factorially complex and overlapping. Factor analyses of the items scored on these scales would serve the double function of (a) yielding clues to the important personality dimensions, and (b) clarifying the factor composition of these widely used scales.

It was decided, therefore, to factor analyze the items on each of the main scales of the MMPI (Comrey, 1957a,b,c, 1958a,b,c,d,e,f; Comrey & Marggraff, 1958). Because computers were new in psychology when this work was begun, it took three years to develop a completely computerized procedure for factor analyzing the matrices of correlation among the items on the MMPI scales. This work involved a considerable amount of experimentation with different kinds of correlation coefficients, methods of factor extraction, and methods of factor rotation. It was also necessary to devise machine-language programs to handle large matrices on an erratic early computer with a very small high-speed memory (The Bureau of Standards Western Automatic Computer at UCLA).

These investigations showed that the φ coefficient was the best available method for computing the interitem correlations, despite common belief at the time to the contrary (Comrey & Levonian, 1958). Tetrachoric and φ/φ_{max} coefficients, popular at that time, proved to be unsatisfactory. These investigations also showed that the Varimax method of Kaiser (1958), who had just developed this procedure, was the best available method for analytic rotation. A semianalytic method developed by Thurstone (1954) was also tried out, but it failed to produce acceptable results (Comrey, 1959).

These studies demonstrated that the scales of the MMPI are indeed factorially complex with many of the same factors showing up on more than one scale. Moreover, the connections between the scale names and the apparent content of the factors found in the scale items are rather tenuous for most of the scales. Beyond this, however, these factor studies of MMPI items provided the initial identification of three of the final CPS personality taxonomy factors. The Cynicism factor in the MMPI analyses ultimately led to the Trust versus Defensiveness factor. Agitation, Sex Concern, and Neuroticism factors from the MMPI analyses led eventually to the present Emotional Stability versus Neuroticism factor, and finally, the MMPI Shyness factor is now represented in the CPS as the Extraversion versus Introversion factor. The initial identification of these factors was poor, as were the methods of measuring them. The present-day factor measures and the factors themselves were developed through a long series of refinements over the subsequent 12 years. As newer factors were being sought and developed, work continued on the refining of factors already identified.

[11.2] THE FHID APPROACH

These studies with MMPI items were followed by investigations designed to locate important constructs in the personality domain through a consideration of other personality tests (Levonian et al., 1959; Comrey & Soufi, 1960, 1961). Ideas derived from these studies and from the published works of other investigators provided the basis for the first empirical investigation by the author aimed directly at developing a factor analytic taxonomy of personality traits (Comrey, 1961). This article also outlined a first approximation to the research strategy that was to guide subsequent research efforts culminating in the publication of the CPS.

A very important feature of this research strategy is the use of the Factored Homogeneous Item Dimension (FHID) as the basic unit of analysis in factor analytic studies designed to locate the main factors for a taxonomy of personality. A study of past factor analytic investigations, the author's as well as those of others, revealed that many obtained factors are virtual artifacts produced by the combination of variables that are too similar in character. If a random selection of personality-test items from a specified pool is factor analyzed, for example, it is probable that some of the items will be very similar to other items in the pool, creating subsets of items that are essentially alternate ways of asking the same question. Because of the great similarity of the items in such a homogeneous subset, they will correlate considerably higher among themselves than they will with items outside the subset. Because of this, these highly related items will be very apt to define their own separate factor in the solution.

A factor produced by the unwitting or even intentional insertion of highly similar items into an analysis is not the kind of broad factor construct that is needed for a taxonomy to describe the whole of human personality. In factor analytic studies designed to locate broader and more meaningful constructs, it is necessary to take steps to avoid the production of factors that are not at an appropriate level in the hierarchy of factors.

The development and use of the FHID as the basic unit in the factor analysis has helped to accomplish this objective in the studies leading to the production of the CPS. The FHID is a total score variable calculated by summing scores over several items that are required to satisfy the following two criteria: (a) The items must be originally developed and logically conceived as measures of the variable under consideration; (b) the items must be found to define the same factor in a factor analysis of items. By meeting both these requirements, the items forming a FHID are shown to have both conceptual and demonstrated statistical homogeneity. To control for acquiescence response bias, it is usually desirable to phrase half the items positively with respect to the FHID name, and half negatively. When this has been done, item scores for the negatively worded items must be reversed before adding them to the positively stated item scores to obtain the total FHID score. This score reversal is accomplished by subtracting the number of the response to the negatively worded item from one more than the number associated with the highest numbered response. If the items are scored on a five-choice response scale, the item score for a negatively worded item would be subtracted from six to reverse the item score.

The early studies carried out in this programmatic research revealed many factors that were essentially nothing more than FHIDs. Subsequent experience showed that it was relatively easy to develop a pool of conceptually homogeneous items that would define a single factor when included in a factor analysis of items. Not all items, of course, would emerge with loadings of sufficient magnitude to be acceptable. Items that failed to yield an adequately high loading were modified or replaced. Through a succession of analyses, a set of items could be developed that would meet the criteria for a FHID. Although it proved to be relatively easy to develop these item factors, or FHIDs, to represent numerous defined concepts, success was not always attained, by any means. A poorly defined concept usually leads to failure to develop a FHID. The items may split up into more than one item factor when included in a factor analysis of items or they may fail to define any factor at all. In such cases, the concept must be redefined for another attempt or dropped.

The relative ease with which such item factors are produced militates against accepting them as the important variables to be used as the basis for scientific development in a given field. In areas of measurement that do not depend on items, factors of a similarly arbitrary sort can be produced by using highly overlapping measures in the same analysis, for example, diastolic and systolic blood pressure included together in the same analysis of physiological variables. Such factors are considered to be at a very low level in the hierarchy of factors.

Factor analysis of FHIDs themselves, however, forces the production of factors at a higher level in the factor hierarcy. Such factors, therefore, are much more likely to represent the broader constructs needed for a taxonomy to provide the basis for development of a science. The only way to produce a low-level factor in an analysis of FHIDs is by introducing into the same analysis FHIDs that overlap too highly, that is, are alternate forms of each other. This would be analogous to including items that are different ways of asking the same question in a factor analysis of items, thereby producing a FHID. In analyses of FHIDs, this problem is avoided by making certain that each FHID included as a variable is conceptually distinct from every other FHID. If it is conceptually distinct from the other FHIDs, it cannot be merely an alternate form of some other variable. If it were, it could lead to the production of a low-level factor based on the great amount of overlap between the two variables. Factors produced by the common variance found among conceptually distinct FHIDs, then, are presumed to represent constructs at a level of generality above the FHID level.

This next level in the factor hierarchy, where factors that come from analyses of FHIDs lie, might be called the "primary" level, following Thurstone's use of the term in his famous studies of the Primary Mental Abilities. The factors at this primary level are conceived as providing the kind of constructs most likely to be useful for developing a science of personality, for example. Such factors are limited in number, whereas the number of factors at the FHID level is almost without limit. Primary-level factors are numerous enough, however, and broad enough to cover the personality domain. Although these primary-level factors are considerably broader in character than FHIDs, they are not so broad as to represent "type" factors. Type factors are found at a higher level in the factor hierarchy, being based on the correlations among primary-level factors. Type factors are too few, too broad, and too complex to be most useful for the purpose of building a general taxonomy for scientific purposes, although they may be very useful for certain applications. General Intelligence would be an example of a well-established type-level factor in the domain of human abilities. It is a high-level factor in the hierarchy of factors based on the correlations among the primary factors of mental ability.

Application of the FHID approach to the task of building a taxonomy of factored personality traits gradually led to the following general strategy.

1. Formulate verbal definitions of many concepts that might lead to the development of FHIDs.
2. Develop a pool of items for each of these defined concepts, half stated positively and half stated negatively with respect to the concept name.
3. Carry out factor analyses of items in which all the items for a given concept are included in the same analysis together with items designed to measure other concepts. Avoid including items in the same analysis for concepts that are expected to measure the same primary-level factor construct. If two item pools from the same primary-level factor are included in the same analysis, the factor analytic results may not produce only FHIDs but may produce the primary-level factor or some hybrid factor that represents a mixture of the two levels.
4. Revise the items and item pools and reanalyze until an acceptable FHID is available for each concept retained. Concepts that fail to produce acceptable FHIDs in the refinement process are dropped out.
5. Factor analyze the FHIDs to develop a first approximation impression of the number and nature of the primary-level factor constructs. On the basis of these results, formulate a verbal conception of each factor construct identified at the primary-factor level.
6. Develop new FHIDs and revise or drop old ones with the objective of providing several variables with high loadings on each factor. Diversity of content is desirable in the variables chosen to define a factor so that all important aspects of the factor construct will be represented. FHIDs that prove to be factorially complex, that is, have major loadings on more than one primary-level factor, are eliminated.
7. Carry out new analyses in which the primary-level factors expected are specified along with the FHIDs that are supposed to define them. Drop,

modify, or redefine factors and FHIDs that fail to replicate in successive studies with different subjects. Retain the factors, and the FHIDs defining them, that replicate successfully in a series of studies. Continue, however, to search for FHIDs that might do an even better job of defining these replicated factors.

8. Formulate hypotheses about new factor constructs that might constitute useful additions to the taxonomy of replicated factors. Try to expand the taxonomy, in other words, into as yet unrepresented but important areas of the domain. Develop FHIDs that should define such hypothesized factors, and carry out new analyses to determine if such factors do emerge and whether or not they can be replicated. Continue the attempt to expand the taxonomy until no new primary-level factors can be found.

[11.3] RESEARCH STUDIES ON THE CPS TAXONOMY

The research strategy using the FHID, as described in the previous section, was applied in a series of investigations undertaken during the 1960s to develop the personality taxonomy upon which the CPS are based. The early studies (Comrey & Soufi, 1960, 1961; Comrey, 1961, 1962b; Comrey & Schlesinger, 1962) concentrated on methodology, developing FHIDs, and some preliminary identification of primary factors. Trust versus Defensiveness, Orderliness versus Lack of Compulsion, Emotional Stability versus Neuroticism, and Extraversion versus Introversion were already identified by this time, although under other names, for example, Hostility, Compulsion, Neuroticism, and Shyness, respectively.

These factors continued to emerge in subsequent studies, undergoing successive refinements and improvements in the FHIDs designed to measure them, although the Shyness factor did fail to separate from Neuroticism in the very next study (Comrey, 1964). The first strong identification of the Empathy versus Egocentrism factor occurred in 1965, under the name "Empathy." The previously identified four factors were replicated in this same study (Comrey, 1965). These early studies also showed promise for the development of a Dependence factor. Attempts to refine and replicate this factor in essentially the same form through a series of investigations were not successful, however; so it was not included in the final CPS taxonomy. Some aspects of this factor, however, suggested the possibility of a new factor that was ultimately identified and replicated as the Social Conformity versus Rebelliousness factor.

In the next study (Comrey & Jamison, 1966), the four previously identified factors were well defined. Dependence was also a major factor in this analysis. The Dependence factor split into two parts in the next study, however (Jamison & Comrey, 1968), one part retaining the name Dependence, which was ultimately dropped for failure to replicate, and the other part, named "Socialization," which exhibited a character very close to the final Social Conformity versus Rebelliousness factor. The previously established factors were also replicated in this study.

The remaining two factors in the taxonomy, Activity versus Lack of Energy and Masculinity versus Femininity, were added to the taxonomy as a result of the next investigation (Comrey, Jamison, & King, 1968) in which FHIDs from the previous studies were analyzed together with quasi-FHIDs derived from an analysis of the Guilford–Zimmerman Temperament Survey. These "quasi-FHIDs" were groups of items found in the Guilford–Zimmerman survey that appeared to meet the criteria for FHIDs reasonably well. In this particular analysis, all the previously established CPS factors were replicated but in addition, the Guilford–Zimmerman General Activity and Masculinity factors emerged as distinct from the CPS factors. Additional studies were carried out in which these two factors were replicated using new items and FHIDs to define them so they were added to the taxonomy as Activity versus Lack of Energy and Masculinity versus Femininity. The Guilford–Zimmerman Emotional Stability factor proved to be similar to the Neuroticism factor (Emotional Stability versus Neuroticism) and the Guilford–Zimmerman Personal Relations factor was closely identified with the Hostility factor (Trust versus Defensiveness).

The next investigation (Comrey & Duffy, 1968) failed tc establish any new factors for the taxonomy, although it provided information about replication of the factors ultimately selected for the final taxonomy. This study did establish the essential similarity between Eysenck's Neuroticism and the CPS Neuroticism, previously replicated so often. Eysenck's Extraversion–Introversion was also found to be closely identified with the CPS Shyness factor. In view of the similarities among these Guilford–Zimmerman, Eysenck, and CPS factors, the CPS Neuroticism factor was renamed as "Emotional Stability versus Neuroticism" and the CPS Shyness factor was renamed as "Extraversion versus Introversion." Other name changes in the CPS factors were initiated to provide a bipolar name for each factor with the more culturally approved pole mentioned first.

In this same investigation, an unsuccessful attempt was made to locate FHIDs from the Cattell 16 PF test to analyze in the same matrix with the CPS FHIDs for the purpose of locating possible additional factors. Since the Cattell items could not be broken down into subgroups meeting the criteria for FHIDs, the Cattell total factor scores were analyzed with the CPS FHIDs. The Cattell factor scores overlapped with many of the CPS factors but did not match any of them on a one-to-one basis. Cattell's second-order Neuroticism and Extraversion–Introversion factors, however, appeared to be reasonably well matched with the correspondingly named CPS primary-level factors. More recently, Howarth and Browne (1971) have also had difficulty replicating Cattell's factors at the primary-factor level.

The last published study in this series (Duffy, Jamison, & Comrey, 1969) represented an unsuccessful attempt to replicate the Guilford–Zimmerman Thoughtfulness factor as a possible addition to the CPS taxonomy. This study also failed to replicate the Masculinity factor satisfactorily, although later unpublished studies did so. The results of the final empirical replication study, described fully in the CPS manual (Comrey, 1970b), are presented briefly in the next section.

[11.4] FACTOR ANALYSIS OF THE FHIDS IN THE CPS

After the final decision had been made about what factors had demonstrated sufficient independence and replicability to be included in the personality taxonomy, a new sample of subjects was drawn to serve as the basis for deriving normative data and for obtaining a final factor solution. The completed CPS were administered to 746 volunteer subjects, 362 males and 384 females. About one-third of these subjects were visitors to a university Open House day at UCLA, comprising friends, acquaintances, and family of students for the most part. The remaining two-thirds of the cases consisted of university students and their friends and UCLA employees. The sample, therefore, would represent predominantly the educated upper-middle class with a heavy concentration of subjects in the college-age group but some representation for all adult age groups.

[11.4.1] THE VARIABLES

Forty-four variables were included in the analysis, 40 FHIDs defining the eight CPS factors, two validity scale scores, V and R, plus the variables of Age and Sex. Each of the personality factors was represented by five FHIDs. Each FHID in turn contained four items, two positively stated with respect to the FHID name and two negatively stated. Each item was answered by using one of the following response scales:

Scale *X*: 7. Always 6. Very Frequently 5. Frequently
4. Occasionally 3. Rarely 2. Very Rarely 1. Never

Scale *Y*: 7. Definitely 6. Very Probably 5. Probably 4. Possibly
3. Probably Not 2. Very Probably Not 1. Definitely Not

The validity scales *V* and *R* were designed to provide information about the genuineness of the test protocol. The *V* scale presents items to which the respondent should give a particular extreme response, either a "1" for a positively stated item or a "7" for a negatively stated item. Since there are eight items on this scale, the expected total score is 8 due to the fact that an item score of 7 becomes 1 when the score is reversed for a negatively stated item. If the respondent has a score that departs too far from 8 on the *V* scale, for example if values approach 25 or more, the test protocol becomes suspect. Random marking should produce on the average a score of 32 on the *V* scale.

The *R* scale, on the other hand, is designed as a "response bias" scale, to indicate whether the subject is responding in a "socially desirable" way, that is, making himself look good. A very low score on the *R* scale would suggest that the subject might be deliberately giving a socially undesirable impression of himself for some reason. The 44 variables included in the analysis are listed below. Following each FHID name (variables 1 to 40) a letter is given in parentheses to show what factor this FHID is supposed to measure. A sample item is also given with its item number in the CPS test booklet. The *X* or *Y* after the item number indicates the preferred scale to be used by the respondent in choosing his answer to the test statement. If the item number is underlined, the item is negatively stated with respect to the FHID name. The 44 variables are the following.

1. Lack of Cynism (T).	1*X*. The average person is honest.
2. Lack of Defensiveness (T).	19*X*. You can get what is coming to you without having to be aggressive or competitive.
3. Belief in Human Worth (T).	37*Y*. Most people are valuable human beings.
4. Trust in Human Nature (T).	10*X*. Other people are selfishly concerned about themselves in what they do.
5. Lack of Paranoia (T).	28*X*. Some people will deliberately say or do things to hurt you.
6. Neatness (O).	2*Y*. I could live in a pigpen without letting it bother me.
7. Routine (O).	20*Y*. Living according to a schedule is something I like to avoid.

8. Order (O).	38*X*. My room is a mess.
9. Cautiousness (O).	11*X*. I am a cautious person.
10. Meticulousness (O).	29*X*. I will go to great lengths to correct mistakes in my work which other people wouldn't even notice.
11. Law Enforcement (C).	3*Y*. This society provides too much protection for criminals.
12. Acceptance of Social Order (C).	21*Y*. The laws governing the people of this country are sound and need only minor changes, if any.
13. Intolerance of Nonconformity (C).	39*Y*. Young people should be more willing than they are to do what their elders tell them to do.
14. Respect for Law (C).	12*X*. If the laws of society are unjust, they should be disobeyed.
15. Need for Approval (C).	30*X*. I ignore what my neighbors might think of me.
16. Exercise (A).	4*X*. If I think about exercising, I lie down until the idea goes away.
17. Energy (A).	22*X*. I seem to lack the drive necessary to get things done.
18. Need to Excel (A).	40*Y*. Being a big success in life requires more effort than I am willing to make.
19. Liking for Work (A).	13*X*. I love to work long hours.
20. Stamina (A).	31*X*. I can work a long time without feeling tired.
21. Lack of Inferiority Feelings (S).	6*X*. I feel inferior to the people I know.
22. Lack of Depression (S).	24*X*. I feel so down-in-the-dumps that nothing can cheer me up.
23. Lack of Agitation (S).	42*X*. My nerves seem to be on edge.
24. Lack of Pessimism (S).	15*X*. I expect things to turn out for the best.
25. Mood Stability (S).	33*X*. My mood remains rather constant, neither going up nor down.
26. Lack of Reserve (E).	7*X*. I am a very talkative person.
27. Lack of Seclusiveness (E).	25*X*. At a party I like to meet as many people as I can.
28. No Loss for Words (E).	43*X*. It is easy for me to talk with people.
29. Lack of Shyness (E).	16*X*. I find it difficult to talk with a person I have just met.
30. No Stage Fright (E).	34*Y*. It would be hard for me to do anything in front of an audience.
31. No Fear of Bugs (M).	8*X*. Big bugs and other crawling creatures upset me.
32. No Crying (M).	26*X*. A sad movie makes me feel like crying.
33. No Romantic Love (M).	44*X*. I like movies which tell the story of two people in love.
34. Tolerance of Blood (M).	17*Y*. I could assist in a surgical operation without fainting if I had to.

35. Tolerance of Vulgarity (M).	35*X*. I can tolerate vulgarity.
36. Sympathy (P).	144*X*. I am rather insensitive to the difficulties that other people are having.
37. Helpfulness (P).	27*Y*. I enjoy helping people even if I don't know them very well.
38. Service (P).	45*Y*. I would like to devote my life to the service of others.
39. Generosity (P).	18*Y*. I would hate to make a loan to a poor family I didn't know very well.
40. Unselfishness (P).	36*X*. I take care of myself before I think about other people's needs.
41. Validity Scale (V).	5*Y*. If I were asked to lift a ten-ton weight, I could do it.
42. Response Bias Scale (R).	41*X*. My morals are above reproach.
43. Age (reported as a two-digit number).	
44. Sex (Male = 1, Female = 0).	

[11.4.2] THE ANALYSIS

The 44 variables described in the previous section were intercorrelated using the Pearson product–moment correlation coefficient. This 44 × 44 matrix of correlation coefficients is shown in Table 11.1.[1] The correlation matrix in Table 11.1 was factor analyzed by the minimum residual method (see Chapter 4), with 15 factors extracted to be certain that more than enough factors would be taken out. By the fifteenth factor, the loadings were of negligible importance and the residuals contained no appreciable true variance remaining to be extracted. No iteration of the minimum residual solution was carried out; so the factor loadings obtained were entirely a function of the off-diagonal entries in the correlation matrix. The first 12 of these minimum residual factors are shown in Table 11.2.

The 15 minimum residual factors were rotated by Criterion I of the Tandem Criteria for orthogonal analytic rotation (see Chapter 7). This procedure spreads the variance from the minimum residual solution out but only as far as the intercorrelations among the variables permit since the method seeks to place as much variance on one factor as possible given that the variables placed on the same factor must be correlated with each other. Most extraction methods can squeeze the variance down even more than Criterion I rotations do because they can place variables on the same factor that are not correlated with each other and typically do.

Only eight of the Criterion I factors were of appreciable importance, the remaining Criterion I factors being too small in the proportion of variance

[1]Tables 11.1—11.10 can be found at the end of this section beginning on page 260.

accounted for to be considered further. The largest 12 Criterion I factors are shown in Table 11.3, revealing clearly the negligible character of the last four factors. Only the first 12 minimum residual factors are shown in Table 11.2 since it was clearly unnecessary to extract more than 12 and since only 12 Criterion I factors were large enough even to be reported. Since only the first eight of the Criterion I factors had enough high loadings to be considered as possible major factors, only these eight factors were rerotated by Criterion II of the Tandem Criteria in the attempt to produce a solution in which the variance is distributed more evenly among the retained factors, such as is customarily obtained in a simple structure solution.

The eight rotated Criterion II factors are shown in Table 11.4. These results represent the final orthogonal rotated loadings for the 44 variables on the eight taxonomy factors. The sharp break in the number of important Criterion I factors at exactly eight when there were eight factors expected would not normally occur. These results were achieved after a long series of studies in which the factors and the variables defining them had been carefully refined to produce a very sharp factor structure.

Although the orthogonal solution in Table 11.4 presents a sharp factor structure, plots of the factors with each other suggest the need for oblique rotations if simple structure criteria are to be satisfied. To obtain a simple structure solution, rotations of factor axes to oblique positions were carried out in two stages. The first stage consisted of "hand" rotations based on inspection of visual 2×2 graphs of the factor loadings plotted on a Cartesian coordinate system (see Chapter 6). These rotations were completed with the use of a computer program that is described in Chapter 12. These rotations were planned to make the obvious gross oblique adjustments of the reference vector axes suggested by simple structure criteria without introducing any changes that clearly contradicted available psychological knowledge about the probable relationships among the factors.

Once the rough oblique positions for the reference vectors had been established, fine adjustments were carried out analytically using another computer program described in Chapter 12. This program searches for alternate new positions on either side of the current reference vector positions that have a greater number of points in the hyperplanes than at present. In this particular use of the program, the permissible search angles were kept small so that only modest adjustments in the reference vector positions could be achieved. Even with small adjustments, however, the hyperplane counts for the eight factors were increased appreciably, as follows: I. 33 to 35; II. 31 to 34; III. 25 to 29; IV. 34 to 35; V. 34 to 35; VI. 35 to 38; VII. 23 to 31; and VIII. 25 to 30.

The final oblique reference vector projections are shown in Table 11.5. The transformation matrix that carries the eight Criterion II factor loadings into the oblique reference vector projections is given in Table 11.6. Had hand rotations been used for the orthogonal rotations, the transformation matrix carrying the minimum residual factor loadings into the oblique reference vector projections would have been obtained because the final orthogonal transformation matrix could have been modified as the oblique rotations were carried out. The computer program that carries out the Tandem Criteria rotations, however, does not output the transformation matrix that would accomplish these rotations. The oblique rotation process, therefore, proceeded as if the Criterion II rotated matrix were the starting point in the entire rotational process. In reproducing the rotated solution using the transformation matrix shown in Table 11.6, therefore, the Criterion II solution must be taken as the starting point rather than the minimum residual solution. The matrix of correlations among the reference vectors, the **C** matrix, is shown in Table 11.7.

The final oblique solution consists of the factor pattern, shown in Table 11.8, the factor structure, shown in Table 11.9, and the matrix of correlations among the factors, shown in Table 11.10. These matrices were computed using the methods described in Chapter 6 with the aid of a computer program that is described in Chapter 12.

TABLE 11.1[a]

Correlations among FHIDs[b]

	1	2	3	4	5	6	7	8	9	10	11	12	13	14	15	16	17	18	19	20	21	22
1																						
2	37																					
3	44	32																				
4	54	43	44																			
5	45	42	50	46																		
6	-01	-19	-04	-04	-10																	
7	09	-12	-01	03	00	43																
8	14	-04	03	04	07	54	48															
9	04	-08	02	-03	00	23	33	22														
10	14	-10	03	03	02	28	28	42	13													
11	16	-27	00	02	-04	32	30	21	19	19												
12	33	-08	11	15	08	22	28	15	12	07	61											
13	13	-19	02	10	-03	42	41	25	23	21	64	50										
14	30	-07	18	16	09	29	34	26	30	22	58	52	56									
15	15	-12	12	09	-02	28	34	14	26	13	34	33	40	36								
16	06	06	08	-02	13	-02	-02	00	-21	00	07	02	-03	-01	-09							
17	11	02	12	07	16	10	10	22	-12	27	11	01	08	14	00	38						
18	03	-25	01	-03	-03	27	21	24	-05	37	25	14	23	20	18	16	43					
19	20	00	13	17	12	15	24	25	-02	43	14	09	20	16	05	24	42	36				
20	13	03	09	03	18	-03	00	11	-22	23	09	05	-01	03	-11	41	63	35	45			
21	08	07	12	05	21	-02	-02	12	-12	09	07	02	02	07	-16	19	44	27	12	34		
22	23	21	32	18	34	00	04	14	-04	07	14	13	04	17	-01	20	43	22	12	34	51	
23	07	14	19	12	26	-08	-03	06	-08	-08	07	08	-03	08	-14	21	27	07	07	31	46	51
24	26	19	33	21	30	-06	-04	09	-15	03	06	04	-01	09	-06	19	38	20	15	32	48	62
25	14	10	11	07	19	00	13	15	06	01	12	16	12	14	-01	14	22	08	10	26	36	42
26	-01	-01	09	09	06	-02	02	04	-09	02	04	-05	-01	04	01	04	36	16	04	15	28	29
27	06	02	24	10	14	02	-02	07	-08	11	08	02	08	16	08	12	38	26	11	16	29	31
28	09	11	19	13	23	-02	06	13	-11	06	06	-02	01	08	-05	19	44	14	14	30	47	41
29	11	12	14	14	20	-03	09	12	-10	05	03	-02	04	10	-05	15	38	12	14	24	40	31
30	01	05	02	06	09	-07	-02	-02	-17	03	-04	-01	-05	-04	-07	16	22	20	15	23	28	20
31	00	05	-05	-02	07	-27	-15	-15	-24	-08	-12	-08	-17	-24	-26	20	60	-03	07	20	10	-02
32	-11	-10	-14	-13	-02	-06	-04	-06	-12	-08	02	-04	-06	-11	-11	22	-03	10	-03	17	16	05
33	00	07	-07	-03	10	-09	-09	-04	-08	-02	-05	-06	-06	-10	-14	09	00	-06	06	17	10	00
34	01	-02	-01	00	05	-14	-06	-11	-20	00	-02	-04	-10	-12	-14	22	08	08	13	30	12	10
35	-19	00	-12	-17	-04	-31	-20	-15	-27	-18	-33	-27	-44	-42	-29	10	-08	-12	-18	06	00	-07
36	12	08	32	14	14	02	-03	05	09	20	-02	-07	-01	12	06	05	29	19	21	16	10	18
37	19	18	36	27	25	-09	-08	-02	-04	17	-02	-08	-02	14	-03	13	30	16	31	26	14	24
38	12	14	30	18	10	-02	-03	00	02	13	-12	-13	-04	08	04	-02	15	16	22	10	00	16
39	11	18	23	20	14	-08	-14	-06	-03	11	-17	-15	-04	04	-03	04	16	07	24	10	05	09
40	15	08	24	23	17	06	03	06	04	13	19	11	23	24	07	10	24	16	24	19	12	18
41	-04	-06	-13	-03	-03	-02	00	-01	-13	06	12	10	12	00	-04	10	-02	12	06	06	00	-06
42	27	09	17	25	23	18	21	26	20	29	28	23	35	39	15	04	26	20	33	18	17	17
43	19	04	13	12	16	07	18	16	10	06	11	12	17	18	05	-06	08	-11	14	06	06	03
44	-13	-14	-16	-18	-08	-08	-03	-09	-16	-06	-05	-06	-13	-19	-09	18	-03	11	-02	10	05	-04

[a] From Comrey, 1970b.

[b] Variables 1–40 are FHIDs: 41 = V; 42 = R; 43 = age; 44 = sex.

23	24	25	26	27	28	29	30	31	32	33	34	35	36	37	38	39	40	41	42	43	44
44																					
51	36																				
13	23	08																			
19	31	07	46																		
31	36	26	66	55																	
24	29	21	53	57	82																
21	15	14	36	27	46	43															
17	-01	14	-07	-09	-02	05	22														
23	-01	30	-07	-15	-01	-06	14	35													
08	-03	18	-14	-17	-04	-04	10	34	32												
14	06	11	04	-01	11	09	26	42	26	20											
05	-06	02	05	-12	01	00	08	31	28	14	26										
05	24	00	16	31	20	17	04	-14	-26	-25	-11	-22									
18	26	05	16	39	28	26	15	06	-14	-03	07	-20	60								
02	13	-05	13	25	12	08	08	-06	-13	-15	01	-15	50	57							
05	14	-01	12	26	17	16	10	-05	-18	-08	-07	-17	46	61	52						
13	19	12	10	23	16	15	12	-04	-09	-07	-04	-31	52	54	36	38					
05	-04	08	-02	-02	-03	-02	07	15	18	06	04	04	-13	-06	-06	01	00				
15	16	14	00	18	13	16	04	-10	-09	-03	-11	-43	32	37	21	34	37	13			
-03	09	17	-02	-03	14	18	00	-05	-10	14	-01	-06	-01	-02	-12	-08	05	-04	14		
07	-08	12	-04	-15	-11	-13	13	38	49	25	18	37	-20	-16	-11	-19	-09	11	-24	-15	

TABLE 11.2[a]

Minimum Residual Matrix

	1	2	3	4	5	6	7	8	9	10	11	12	h^2
1	.39	.13	−.11	.52	.07	.08	.18	−.20	.06	.03	.05	.09	.56
2	.19	−.22	−.31	.47	−.03	.16	.06	.02	−.01	.12	.06	−.12	.46
3	.45	−.02	−.30	.37	.00	.01	.00	−.11	.22	−.13	.06	.00	.51
4	.37	.03	−.26	.46	.02	.04	.21	−.10	.09	.15	.04	.01	.50
5	.41	−.14	−.13	.50	.00	.13	.07	.00	.06	.00	.05	−.04	.49
6	.14	.52	.22	−.16	−.01	.22	−.06	.05	.12	.07	.03	−.06	.46
7	.19	.48	.28	−.02	−.04	.25	.09	.13	.17	−.08	.02	−.02	.47
8	.29	.35	.23	−.06	−.02	.46	−.04	.14	.10	.07	.06	.04	.52
9	.01	.47	−.05	.04	−.07	.09	−.10	.27	.08	−.16	−.03	.04	.35
10	.31	.27	.10	−.20	.24	.32	.08	−.03	.02	.07	.04	.23	.45
11	.26	.52	.40	.04	−.01	−.35	.03	−.12	−.07	−.06	.04	.08	.65
12	.22	.45	.29	.27	−.03	−.28	.07	−.18	.00	.00	.02	.05	.52
13	.27	.63	.28	.03	.00	−.23	.10	.02	−.07	.05	.03	.02	.62
14	.40	.57	.13	.13	−.04	−.21	.02	.00	−.01	−.08	.07	.07	.58
15	.13	.52	.05	.01	−.06	−.09	.06	−.10	.23	−.08	.00	−.04	.37
16	.25	−.26	.24	−.01	.20	−.02	.02	−.15	−.04	−.05	.11	−.35	.39
17	.63	−.16	.20	−.25	.07	.17	−.08	−.20	−.14	−.08	.19	−.14	.69
18	.39	.11	.28	−.34	.21	.02	−.04	−.19	.18	.14	.08	.12	.52
19	.46	.08	.11	−.11	.39	.24	.19	−.07	−.11	−.02	.10	.04	.51
20	.48	−.28	.30	−.08	.30	.12	−.01	−.18	−.18	−.13	.20	−.07	.63
21	.49	−.30	.28	.00	−.14	.01	−.21	.08	−.08	.09	.09	.12	.52
22	.60	−.21	.14	.17	−.14	−.02	−.36	−.05	.10	−.01	.15	.09	.64
23	.40	−.31	.23	.21	−.04	−.12	−.30	.19	.00	.06	.14	−.02	.52
24	.56	−.24	.04	.16	−.11	.00	−.33	−.12	−.01	.01	.10	.11	.55
25	.35	−.14	.34	.26	−.02	−.04	−.21	.31	.02	−.06	.08	.01	.48
26	.39	−.21	.02	−.27	−.43	−.06	.16	−.04	.14	.02	.12	.00	.52
27	.55	−.11	−.10	−.27	−.28	−.13	.09	−.05	.08	.06	.09	.01	.52
28	.61	−.30	.09	−.16	−.48	.00	.25	.13	−.01	−.04	.21	−.03	.85
29	.55	−.25	.07	−.15	−.43	.02	.32	.16	−.07	−.02	.14	−.01	.73
30	.31	−.32	.17	−.14	−.08	−.08	.27	.08	.08	.08	.04	.04	.36
31	−.06	−.44	.24	.12	.30	−.10	.24	.13	.02	−.03	−.03	.06	.46
32	−.10	−.28	.48	.09	.23	−.14	−.03	.21	.20	.06	−.02	−.10	.50
33	−.07	−.22	.26	.20	.22	.05	.11	.20	−.12	−.05	−.04	.03	.29
34	.05	−.35	.26	.03	.20	−.05	.20	−.04	.05	−.13	.03	.11	.31
35	−.30	−.55	.17	−.01	−.01	.10	.07	−.03	.18	−.11	−.05	.00	.48
36	.48	.04	−.45	−.23	.17	−.06	−.12	.04	.06	−.16	.10	.02	.57
37	.58	−.12	−.43	−.12	.30	−.16	.06	.14	.02	−.08	.15	.05	.72
38	.36	−.03	−.45	−.18	.25	−.07	.00	.06	.25	−.01	.12	.06	.52
39	.38	−.08	−.49	−.16	.24	−.09	.04	.16	−.02	.14	.12	−.02	.54
40	.50	.10	−.22	−.06	.24	−.23	−.01	.13	−.03	−.06	.05	−.06	.44
41	−.01	−.02	.20	−.01	.15	−.14	.10	.03	−.06	.29	−.01	−.03	.18
42	.52	.32	−.07	.05	.20	−.02	.02	.19	−.20	.20	.05	.02	.54
43	.16	.14	.08	.21	−.10	.17	.14	.11	−.23	−.25	−.04	.09	.29
44	−.19	−.28	.39	−.02	.25	−.09	.03	.05	.31	−.02	−.06	−.09	.45

[a] From Comrey, 1970b.

TABLE 11.3[a]

Criterion I Loadings

	1	2	3	4	5	6	7	8	9	10	11	12	h^2
1	.10	.25	−.09	.67	.02	.00	.02	−.08	.01	.00	.01	.13	.56
2	.10	−.22	−.09	.60	−.06	.01	−.03	.03	−.12	.04	.03	−.13	.46
3	.20	.07	−.28	.56	−.09	−.05	−.05	.02	.22	−.06	−.02	−.01	.51
4	.10	.11	−.19	.64	−.03	−.03	.10	.02	−.05	.10	.00	.04	.50
5	.26	.00	−.11	.63	.06	.03	−.06	.01	−.02	−.05	−.02	−.06	.49
6	−.02	.52	.05	−.14	−.12	.35	−.02	−.02	.05	.09	.00	−.11	.45
7	.03	.54	.10	−.01	.01	.35	.07	.05	.08	−.09	.00	−.12	.47
8	.16	.41	.05	.03	−.07	.56	−.06	.01	−.04	−.02	.01	−.06	.52
9	−.18	.37	−.04	−.01	−.18	.14	−.09	.24	.03	−.21	.00	−.14	.35
10	.08	.34	−.20	−.04	.03	.45	.02	−.14	−.06	.00	.00	.23	.45
11	.10	.73	.08	−.09	.09	−.25	−.03	−.07	.01	.00	.01	.10	.65
12	.03	.62	.13	.17	.07	−.26	−.02	−.06	.03	.05	.01	.10	.52
13	.01	.77	−.01	−.05	.02	−.11	.05	.00	−.11	.04	.00	.01	.62
14	.10	.72	−.11	.11	−.07	−.13	−.03	.04	.01	−.07	.02	.03	.58
15	−.10	.52	−.02	.04	−.12	−.01	.11	.03	.24	.03	−.02	−.04	.37
16	.30	−.05	−.05	.02	.30	−.05	−.03	−.39	.04	.08	−.02	−.22	.39
17	.62	.12	−.20	−.05	.02	.17	−.06	−.45	.00	−.02	−.02	−.04	.69
18	.30	.32	−.18	−.21	.14	.23	−.02	−.20	.16	.25	−.01	.22	.52
19	.20	.25	−.32	.08	.20	.31	.06	−.35	−.13	−.06	.00	.10	.51
20	.48	.02	−.16	.01	.31	.11	−.12	−.49	−.04	−.09	.05	.03	.63
21	.66	.01	.02	.01	.10	.04	−.23	.04	−.10	.02	−.02	.08	.52
22	.65	.09	−.07	.22	−.02	−.01	−.36	.02	.14	.03	.01	.05	.64
23	.53	−.02	−.02	.13	.20	−.10	−.37	.13	−.05	.03	.04	−.11	.52
24	.59	.00	−.10	.24	−.07	−.04	−.34	−.05	.06	.02	−.03	.11	.55
25	.42	.14	.06	.12	.28	.01	−.32	.19	−.06	−.12	.02	−.15	.48
26	.59	−.02	−.04	−.08	−.16	−.02	.34	.08	.14	.05	.01	.01	.52
27	.57	.07	−.26	−.03	−.20	−.07	.24	.04	.08	.10	−.05	.04	.52
28	.82	.00	−.06	.04	−.08	.00	.35	.10	−.05	−.11	.06	−.09	.85
29	.72	.01	−.06	.05	−.07	.01	.40	.11	−.14	−.12	.01	−.07	.73
30	.45	−.07	−.07	−.03	.24	.01	.28	.07	−.02	.06	−.02	.05	.36
31	.06	−.26	.01	.04	.61	−.04	.06	.01	−.07	−.05	.00	.07	.46
32	.08	−.07	.17	−.12	.61	.00	−.15	.11	.07	.12	.00	−.14	.50
33	−.01	−.10	.12	.07	.44	.04	−.08	.02	−.19	−.15	−.01	−.02	.29
34	.16	−.14	.03	.02	.45	−.02	.09	−.10	.07	−.10	.02	.14	.31
35	.04	−.51	.27	−.06	.30	.06	.07	.01	.20	−.05	.00	.01	.48
36	.20	.04	−.68	.02	−.22	.02	−.05	−.05	.12	−.10	−.02	−.01	.58
37	.27	−.01	−.78	.14	.03	−.04	.03	−.01	−.01	−.07	.02	.02	.72
38	.09	−.04	−.66	.09	−.07	.08	.03	.07	.18	.06	.04	.04	.52
39	.12	−.10	−.68	.09	−.09	.01	.05	.04	−.13	.10	.04	−.03	.54
40	.19	.22	−.57	.07	.03	−.12	−.04	−.01	−.04	−.04	−.06	−.08	.44
41	.00	.08	.03	−.07	.24	−.05	.03	−.01	−.20	.26	.00	.03	.18
42	.16	.43	−.40	.15	−.02	.08	−.09	.00	−.35	.06	−.03	−.02	.54
43	.07	.20	.10	.19	−.02	.06	.04	−.01	−.18	−.40	−.05	.00	.29
44	−.03	−.14	.15	−.15	.55	.04	−.03	.02	.24	.13	−.03	−.07	.45

[a] From Comrey, 1970b.

TABLE 11.4[a]

Criterion II Loadings[b]

	T	O	C	A	S	E	M	P	h^2
1	.68	.09	.21	−.10	−.08	−.01	−.01	−.06	.54
2	.59	−.13	−.21	.04	−.12	.01	−.05	−.07	.43
3	.57	−.03	.06	−.04	−.19	.07	−.10	−.27	.45
4	.66	.01	.10	.00	−.04	.08	−.03	−.17	.49
5	.63	.00	−.02	−.07	−.25	.08	.04	−.09	.48
6	−.12	.57	.26	−.05	.02	−.04	−.14	.05	.43
7	.02	.59	.28	−.02	.02	.05	.00	.09	.44
8	.06	.69	.05	−.10	−.13	.05	−.08	.06	.52
9	−.02	.35	.22	.28	.00	−.13	−.15	−.08	.29
10	.00	.54	.05	−.25	.03	.01	−.03	−.17	.39
11	−.08	.16	.76	−.12	−.10	.02	.02	.06	.64
12	.17	.10	.67	−.05	−.08	−.04	.02	.12	.51
13	−.03	.31	.72	−.03	.00	.02	−.02	−.03	.61
14	.12	.26	.67	−.01	−.12	.04	−.12	−.13	.57
15	.06	.26	.44	.06	.12	.00	−.13	−.03	.31
16	.04	−.12	.00	−.51	−.13	.07	.19	.01	.33
17	.00	.15	.00	−.67	−.30	.31	−.11	−.13	.68
18	−.17	.34	.15	−.38	−.13	.14	.06	−.16	.38
19	.14	.34	.06	−.51	.02	.05	.10	−.26	.48
20	.04	.04	−.03	−.70	−.27	.12	.17	−.09	.62
21	.01	.05	−.03	−.20	−.59	.32	.07	.01	.50
22	.20	.03	.05	−.19	−.69	.22	−.07	−.08	.62
23	.09	−.08	.01	−.09	−.66	.14	.18	−.06	.51
24	.22	−.04	−.01	−.22	−.61	.18	−.13	−.10	.54
25	.09	.10	.10	−.03	−.57	.08	.27	.02	.44
26	−.02	−.01	−.01	−.08	−.12	.68	−.11	−.04	.50
27	.02	.00	.08	−.13	−.17	.60	−.18	−.26	.50
28	.11	.03	−.01	−.15	−.26	.84	−.02	−.06	.82
29	.12	.04	.00	−.12	−.17	.79	−.01	−.06	.69
30	.02	−.02	−.05	−.14	−.09	.49	.28	−.07	.36
31	.04	−.18	−.16	−.12	−.01	.02	.61	.01	.45
32	−.14	−.03	−.03	−.04	−.21	−.07	.60	.14	.46
33	.05	−.02	−.09	−.06	−.08	−.10	.43	.11	.23
34	.03	−.11	−.08	−.23	−.03	.10	.43	.04	.27
35	−.07	−.21	−.44	−.02	.02	.06	.34	.28	.44
36	.06	.04	−.02	−.16	−.10	.11	−.26	−.66	.55
37	.18	−.05	−.01	−.20	−.11	.19	.00	−.77	.71
38	.12	.06	−.10	−.05	.00	.10	−.07	−.66	.48
39	.12	−.04	−.11	−.06	.00	.13	−.10	−.67	.51
40	.09	.01	.23	−.15	−.13	.08	−.02	−.57	.43
41	−.07	.00	.11	−.06	.01	.00	.23	.02	.07
42	.18	.28	.31	−.12	−.16	.03	−.07	−.40	.41
43	.20	.15	.14	−.02	−.03	.06	−.03	.11	.10
44	−.16	−.04	−.11	−.08	−.03	−.08	.54	.14	.37

[a] From Comrey, 1970b.

[b] The signs of all loadings for factors A, S, and P must be reversed to be consistent with the direction of the factor names.

Table 11.5

Oblique Reference Vector Projections (V)[a]

	I	II	III	IV	V	VI	VII	VIII
1	.64	.01	.10	.09	.00	−.07	−.04	−.06
2	.54	−.06	−.27	−.04	.07	.00	−.06	−.04
3	.49	−.06	.01	−.01	.10	.01	−.09	.15
4	.62	−.03	.03	−.02	−.04	.05	−.01	.04
5	.57	.01	−.09	.03	.16	.02	.03	−.02
6	−.13	.43	.24	.07	−.02	−.05	−.10	−.05
7	.03	.46	.27	.02	−.03	.04	.06	−.10
8	.02	.62	.03	.09	.09	.01	−.03	−.08
9	−.04	.24	.26	−.26	.08	−.07	−.08	.07
10	−.03	.49	.05	.23	−.10	−.05	.01	.15
11	−.08	−.08	.72	.10	.08	−.03	.01	−.03
12	.17	−.12	.59	.05	.06	−.08	.00	−.13
13	−.03	.06	.69	.02	−.01	−.01	.01	.03
14	.08	.02	.63	−.02	.09	−.01	−.08	.10
15	.06	.09	.41	−.04	−.12	.00	−.09	.00
16	.03	−.08	−.05	.46	.01	−.06	.10	.01
17	−.09	.13	−.07	.59	.09	.09	−.15	.09
18	−.21	.29	.17	.32	.03	.03	.07	.18
19	.10	.32	.03	.46	−.15	−.09	.08	.23
20	−.01	.08	−.09	.62	.09	−.08	.07	.10
21	−.08	.06	−.03	.10	.47	.21	.07	.00
22	.07	.01	.02	.09	.57	.09	−.08	.04
23	.00	−.05	.04	−.02	.58	.06	.16	.07
24	.10	−.05	−.06	.14	.49	.05	−.15	.04
25	.03	.10	.15	−.07	.53	.04	.26	.01
26	−.05	−.02	−.02	.00	−.02	.62	−.01	.00
27	−.05	−.04	.08	.04	.02	.50	−.08	.20
28	.05	.03	−.02	.04	.07	.74	.08	.01
29	.08	.04	.00	.02	.00	.71	.10	.01
30	.02	.04	.00	.04	−.02	.45	.32	.09
31	.10	−.04	−.09	.06	−.01	.03	.55	.07
32	−.10	.06	.06	−.02	.23	−.05	.55	−.01
33	.09	.06	−.05	.03	.09	−.09	.37	−.05
34	.06	−.02	−.06	.17	−.03	.07	.38	.01
35	.00	−.02	−.41	.02	−.01	.10	.29	−.21
36	−.06	.02	.04	.08	.00	.03	−.19	.59
37	.07	−.03	.08	.08	−.02	.10	.06	.71
38	.03	.08	.00	−.03	−.06	.06	.01	.60
39	.03	.00	−.02	−.01	−.08	.08	−.02	.61
40	.00	−.06	.29	.06	.04	.00	.02	.53
41	−.04	−.01	.13	.04	−.02	.00	.21	.03
42	.09	.16	.33	.06	.09	−.04	−.02	.35
43	.20	.09	.08	.03	.00	.04	−.03	−.14
44	−.09	.07	−.03	.04	.05	−.06	.49	−.02

[a] To compute this matrix by the formula $\mathbf{V} = \mathbf{A}\mathbf{\Lambda}$, where $\mathbf{A}$ is the matrix of Criterion II factor loadings, requires a rearrangement of the columns of $\mathbf{A}$ from the present order, TOCASEMP, to the order ECPTMOSA.

TABLE 11.6

Transformation Matrix (Λ)

	1	2	3	4	5	6	7	8
1	−.003	−.001	.014	−.099	−.161	.959	.153	−.031
2	.009	−.318	.945	−.008	.001	−.033	−.007	.014
3	.112	−.017	−.184	.111	.045	.029	−.102	−.973
4	.976	−.005	−.148	.010	−.083	−.030	−.026	−.174
5	.101	.131	.155	−.105	.020	.062	.964	.148
6	−.019	.939	.039	−.004	−.003	.011	.100	−.012
7	.150	.000	−.038	.106	−.960	.079	.023	−.014
8	.028	−.005	.155	−.977	.210	.258	.161	.015

TABLE 11.7

Correlations among Reference Vectors (C)

	1	2	3	4	5	6	7	8
1	1.000	−.014	−.143	.000	−.212	−.004	.067	−.265
2	−.014	1.000	−.240	−.012	−.002	.026	.223	.020
3	−.143	−.240	1.000	−.203	.075	.028	.195	.243
4	.000	−.012	−.203	1.000	−.288	−.342	−.283	−.139
5	−.212	−.002	.075	−.288	1.000	−.172	.003	−.005
6	−.004	.026	.028	−.342	−.172	1.000	.249	−.041
7	.067	.223	.195	−.283	.003	.249	1.000	.243
8	−.265	.020	.243	−.139	−.005	−.041	.243	1.000

TABLE 11.8

Factor Pattern for the CPS (**P**)

	I	II	III	IV	V	VI	VII	VIII
1	.71	.01	.11	.11	.00	−.08	−.04	−.06
2	.60	−.07	−.30	−.06	.08	−.01	−.07	−.04
3	.53	−.06	.01	−.01	.11	.01	−.10	.17
4	.68	−.03	.03	−.03	−.05	.06	−.02	.05
5	.63	.01	−.10	.04	.18	.02	.04	−.02
6	−.14	.47	.27	.08	−.02	−.06	−.11	−.06
7	.03	.50	.30	.03	−.03	.05	.06	−.11
8	.02	.67	.04	.11	.10	.01	−.04	−.10
9	−.04	.25	.29	−.31	.08	−.08	−.09	.08
10	−.03	.53	.06	.28	−.11	−.06	.01	.17
11	−.09	−.09	.80	.12	.09	−.04	.01	−.04
12	.18	−.13	.66	.06	.07	−.09	−.01	−.14
13	−.04	.07	.77	.02	−.01	−.01	.01	.04
14	.08	.02	.70	−.02	.10	−.01	−.09	.11
15	.07	.10	.46	−.05	−.13	.00	−.10	.00
16	.03	−.09	−.06	.55	.01	−.07	.11	.01
17	−.10	.14	−.08	.71	.11	.10	−.17	.10
18	−.23	.31	.18	.38	.04	.03	.08	.20
19	.11	.35	.04	.55	−.18	−.10	.09	.26
20	−.01	.08	−.10	.74	.10	−.09	.08	.11
21	−.09	.07	−.03	.12	.54	.25	.08	.00
22	.08	.01	.02	.11	.65	.11	−.09	.04
23	.00	−.06	.05	−.02	.66	.07	.18	.08
24	.10	−.06	−.06	.17	.56	.06	−.17	.04
25	.03	.11	.16	−.08	.60	.05	.30	.01
26	−.06	−.02	−.02	.01	−.02	.72	−.01	.00
27	−.06	−.05	.09	.05	.02	.58	−.09	.23
28	.05	.03	−.02	.04	.08	.86	.10	.01
29	.08	.04	−.01	.02	.00	.82	.11	.01
30	.02	.04	.00	.05	−.02	.52	.38	.10
31	.11	−.04	−.10	.07	−.01	.03	.64	.08
32	−.11	.07	.07	−.03	.26	−.06	.63	−.02
33	.10	.06	−.06	.04	.10	−.11	.43	−.05
34	.07	−.02	−.06	.21	−.03	.08	.44	.02
35	.00	−.02	−.46	.02	−.01	.11	.34	−.24
36	−.07	.02	.04	.10	.00	.03	−.22	.65
37	.08	−.03	.09	.09	−.02	.11	.07	.79
38	.04	.09	.00	−.04	−.07	.06	.01	.67
39	.04	−.01	−.03	−.01	−.09	.10	−.02	.68
40	.00	−.06	.32	.08	.05	.00	.02	.60
41	−.05	−.01	.15	.05	−.02	.00	.24	.03
42	.10	.18	.36	.07	.10	−.05	−.03	.39
43	.22	.10	.09	.03	.00	.04	−.03	−.16
44	−.10	.08	−.04	.05	.06	−.06	.56	−.02

TABLE 11.9

Factor Structure for the CPS (S)

	I	II	III	IV	V	VI	VII	VIII
1	.71	.11	.23	.17	.19	.07	−.14	.14
2	.57	−.09	−.24	.00	.20	.07	−.09	.20
3	.63	.01	.06	.13	.28	.17	−.22	.37
4	.69	.04	.10	.08	.15	.14	−.16	.25
5	.66	.02	−.02	.19	.36	.17	−.05	.21
6	−.07	.57	.43	.05	−.04	.00	.24	−.11
7	.07	.58	.46	.08	−.01	.06	−.12	−.15
8	.10	.69	.28	.15	.15	.11	−.20	−.05
9	.02	.36	.32	−.27	−.06	−.12	−.25	.00
10	.07	.54	.22	.24	.00	.06	−.13	.16
11	.02	.14	.77	.24	.10	.06	−.08	−.16
12	.23	.08	.66	.15	.10	.00	−.07	−.18
13	.07	.29	.78	.12	.00	.04	−.16	−.08
14	.24	.26	.72	.13	.14	.11	−.28	.07
15	.11	.27	.51	−.04	−.13	.00	−.24	−.04
16	.07	−.14	−.03	.54	.21	.13	.23	.07
17	.10	.17	.07	.76	.41	.46	−.13	.26
18	−.07	.33	.26	.45	.17	.21	−.01	.16
19	.22	.34	.17	.51	.08	.12	.00	.29
20	.11	.03	−.01	.76	.38	.23	.17	.20
21	.09	.04	.00	.42	.64	.44	.04	.10
22	.31	.06	.07	.41	.75	.40	−.14	.21
23	.18	−.08	−.01	.31	.69	.25	.14	.14
24	.31	−.01	−.01	.40	.68	.35	−.17	.24
25	.17	.08	.13	.24	.59	.17	.20	.02
26	.04	.00	.01	.27	.20	.70	−.14	.12
27	.13	.02	.10	.31	.25	.66	−.25	.33
28	.19	.04	.03	.43	.39	.89	−.09	.18
29	.19	.05	.05	.37	.29	.82	−.08	.16
30	.06	−.04	−.04	.31	.17	.47	.24	.11
31	.02	−.24	−.22	.17	.05	−.05	.63	−.01
32	−.15	−.10	−.06	.13	.19	−.12	.62	−.18
33	.02	−.07	−.10	.09	.09	−.14	.44	−.11
34	.02	−.15	−.12	.28	.09	.07	.45	−.02
35	−.17	−.25	−.49	.01	−.01	−.02	.47	−.23
36	.19	.09	.01	.17	.14	.22	−.36	.70
37	.33	−.01	−.02	.27	.19	.28	−.14	.81
38	.23	.09	−.07	.06	.04	.15	−.19	.68
39	.23	.00	−.11	.08	.05	.19	−.20	.71
40	.24	.03	.22	.23	.17	.17	−.15	.57
41	−.06	−.04	.09	.09	−.01	−.03	.22	−.06
42	.31	.30	.39	.20	.20	.13	−.24	.39
43	.20	.15	.18	.07	.07	.08	−.08	−.09
44	−.19	−.10	−.13	.10	.02	−.16	.59	−.18

TABLE 11.10

Correlations among Factors (Φ)

	1	2	3	4	5	6	7	8
1	1.00	.08	.13	.14	.26	.15	−.17	.30
2	.08	1.00	.30	.01	.01	.04	−.28	.00
3	.13	.30	1.00	.14	.01	.06	−.20	−.14
4	.14	.01	.14	1.00	.39	.39	.13	.14
5	.26	.01	.01	.39	1.00	.32	−.02	.15
6	.15	.04	.06	.39	.32	1.00	−.20	.18
7	−.17	−.28	−.20	.13	−.02	−.20	1.00	−.23
8	.30	.00	−.14	.14	.15	.18	−.23	1.00

[11.4.3]
RESULTS

It will be noted in Table 11.8, the factor pattern or matrix of factor loadings with respect to the oblique factors, that the major loadings for factors A, S, and P are positive. These factors were reversed in direction in the orthogonal Criterion II solution shown in Table 11.4; that is, by chance they emerged from the orthogonal rotations 180° out of phase. They may be reversed at will by changing the signs on all the loadings for the given factors. The orientation for these factors in Table 11.8, however, is consistent with the direction of the factors as named. The reversal of the direction of factors A, S, and P from the Criterion II solution to the opposite and correct direction in the pattern matrix (Table 11.8) was accomplished here by reversal of all the signs in the columns of the transformation matrix (Table 11.6) which correspond to factors A, S, and P. Since the directional orientation for all the factors in Table 11.8 is consistent with the direction of the factors as named in the CPS taxonomy, the correlations among the factors, as shown in Table 11.10, have the proper signs for interpretation of the relationships among the CPS factors.

The major correlations to be noted in Table 11.10 are the two values of .39 between Activity versus Lack of Energy (A) and Emotional Stability versus Neuroticism (S) and Extraversion versus Introversion (E); these two latter factors correlate .32 with each other. Orderliness versus Lack of Compulsion (O) correlates .30 with Social Conformity versus Rebelliousness (C). Trust versus Defensiveness (T) correlates .30 with Empathy versus Egocentrism (P) and .26 with Emotional Stability versus Neuroticism (S). Masculinity versus Femininity (M) correlates −.23 with Empathy versus Egocentrism (P), −.20 with Extraversion versus Introversion (E), and −.20 with Social Conformity versus Rebelliousness (C).

Each of the eight rotated factors is presented in the following together with a brief description and the FHIDs with loadings of .3 or more in either the Criterion II orthogonal solution or in the oblique factor pattern solution.

I. *Trust versus Defensiveness* (T). Individuals who are high on this personality factor indicate that they believe more than the average person in the basic honesty, trustworthiness, and good intentions of other people. They believe that others wish them well and they have faith in human nature. Individuals who are low on T are cynical, defensive, suspicious, and have a low opinion of the value of the average man. Variables with loadings of .3 or more on this factor are as shown in Table 11.11.

TABLE 11.11

Variables with Loadings of .3 or More on Factor T

Variable	Orthogonal	Oblique
1. Lack of Cynicism	.68	.71
2. Lack of Defensiveness	.59	.60
3. Belief in Human Worth	.57	.53
4. Trust in Human Nature	.66	.68
5. Lack of Paranoia	.63	.63

II. *Orderliness versus Lack of Compulsion* (O). Individuals who are high on this factor tend to be very concerned with neatness and orderliness. They report being cautious, meticulous, and say they like to live in a routine way. Individuals who are low on this factor are inclined to be careless, sloppy, unsystematic in their style of life, reckless, and untidy. Variables with loadings of .3 or more on this factor are shown in Table 11.12.

TABLE 11.12

Variables with Loadings of .3 or More on Factor O

Variable	Orthogonal	Oblique
6. Neatness	.57	.47
7. Routine	.59	.50
8. Order	.69	.67
9. Cautiousness	.35	.25
10. Meticulousness	.54	.53
18. Need to Excel	.34	.31
19. Liking for Work	.34	.35

III. *Social Conformity versus Rebelliousness* (C). Individuals who are high on this factor depict themselves as accepting the society as it is, respecting the law, believing in law enforcement, seeking the approval of society, and resenting nonconformity in others. Individuals who are low on this factor are inclined to challenge the laws and institutions of the society, resent control, accept nonconformity in others and are nonconforming themselves. Variables with loadings of .3 or more on this factor are shown in Table 11.13.

TABLE 11.13

Variables with Loadings of .3 or More on Factor C

Variable	Orthogonal	Oblique
2. Lack of Defensiveness	−.21	−.30
7. Routine	.28	.30
11. Law Enforcement	.76	.80
12. Acceptance of the Social Order	.67	.66
13. Intolerance of Nonconformity	.72	.77
14. Respect for Law	.67	.70
15. Need for Approval	.44	.46
35. Tolerance of Vulgarity	−.44	−.46
40. Unselfishness	.23	.32
42. Response Bias Scale (*R*)	.31	.36

IV. *Activity versus Lack of Energy* (A). Individuals who are high on this factor report liking physical activity, hard work and exercise, having great energy and stamina, and striving to excel. Individuals who are low on A are inclined to be physically inactive, lack drive and energy, tire quickly, and have little motivation to excel. Variables with loadings of .3 or more on this factor are shown in Table 11.14.

TABLE 11.14

Variables with Loadings of .3 or More on Factor A

Variable	Orthogonal	Oblique
9. Cautiousness	.28	−.31
16. Exercise	−.51	.55
17. Energy	−.67	.71
18. Need to Excel	−.38	.38
19. Liking for Work	−.51	.55
20. Stamina	−.70	.74

The major orthogonal loadings are negative because the directional orientation of this factor in the orthogonal solution was opposite to what it should be to be consistent with the factor name.

V. *Emotional Stability versus Neuroticism* (S). Individuals who are high on this factor report being happy, calm, optimistic, stable in mood, and having confidence in themselves. Individuals who are low on the factor have inferiority feelings, are agitated, depressed, pessimistic, and have frequent swings of mood. Variables with loadings of .3 or more on this factor are shown in Table 11.15.

TABLE 11.15

Variables with Loadings of .3 or More on Factor S

Variable	Orthogonal	Oblique
17. Energy	−.30	.11
21. Lack of Inferiority Feelings	−.59	.54
22. Lack of Depression	−.69	.65
23. Lack of Agitation	−.66	.66
24. Lack of Pessimism	−.61	.56
25. Mood Stability	−.57	.60

As with the previous factor, the orthogonal loadings should be reversed in sign to make them consistent with the directional orientation of the factor as named.

VI. *Extraversion versus Introversion* (E). Individuals who are high on this factor depict themselves as outgoing, easy to meet, seeking the company of others, meeting strangers easily, and speaking before groups with little fear. Individuals low on the factor are reserved, seclusive, shy, cannot easily find things to talk about with others, and suffer from stage fright. Variables with loadings of .3 or more on this factor are shown in Table 11.16.

TABLE 11.16

Variables with Loadings of .3 or More on Factor E

Variable	Orthogonal	Oblique
17. Energy	.31	.10
21. Lack of Inferiority Feelings	.32	.25
26. Lack of Reserve	.68	.72
27. Lack of Seclusiveness	.60	.58
28. No Loss for Words	.84	.86
29. Lack of Shyness	.79	.82
30. No Stage Fright	.49	.52

VII. *Masculinity versus Femininity* (M). Individuals who are high on this factor report being rather tough-minded individuals who are not bothered by crawling creatures, the sight of blood, vulgarity, who do not cry easily, and who have little interest in love stories. Individuals who are low on this factor are inclined to cry easily, are bothered by blood and crawling things such as snakes and insects, are disturbed by vulgarity, and have a high interest in romantic love. Variables with loadings of .3 or more on this factor are shown in Table 11.17.

TABLE 11.17

Variables with Loadings of .3 or More on Factor M

Variable	Orthogonal	Oblique
25. Mood Stability	.27	.30
30. No Stage Fright	.28	.38
31. No Fear of Bugs	.61	.64
32. No Crying	.60	.63
33. No Romantic Love	.43	.43
34. Tolerance of Blood	.43	.44
35. Tolerance of Vulgarity	.34	.34
44. Sex (Male versus Female)	.54	.56

VIII. *Empathy versus Egocentrism* (P). Individuals who are high on this factor describe themselves as sympathetic, helpful, generous, unselfish, and interested in devoting their lives to the service of other people. Individuals who are low on this factor are not particularly sympathetic or helpful to others, tend to be concerned about themselves and their own goals, and are relatively uninterested in dedicating their lives to serving their fellow men. Variables with loadings of .3 or more on this factor are shown in Table 11.18.

TABLE 11.18

Variables with Loadings of .3 or More on Factor P

Variable	Orthogonal	Oblique
36. Sympathy	−.66	.65
37. Helpfulness	−.77	.79
38. Service	−.66	.67
39. Generosity	−.67	.68
40. Unselfishness	−.57	.60
42. Response Bias Scale (*R*)	−.40	.39

All the orthogonal loadings should be reversed in sign to make them consistent with the orientation of the factor as named.

[*11.4.4*]
DISCUSSION OF THE RESULTS

The orthogonal Criterion II solution (Table 11.4), the oblique factor pattern solution (Table 11.8), and the factor structure matrix (Table 11.9) all agree rather well in showing that in most cases the variables had major loadings on the factors they were expected to define and not elsewhere. The two variables that were most disappointing in this respect were FHIDs 9 and 35. Variable 9, Cautiousness, was expected to have a major loading on factor O, Orderliness versus Lack of Compulsion. It had loadings of .35 in the orthogonal solution and .25 in the oblique solution while exhibiting loadings of comparable magnitude in two other factors, C and A. The structure coefficient was highest for O, .36, but not much lower for C and A, .32 and −.27, respectively. This FHID, therefore, would be regarded as a prime candidate for revision or replacement in carrying out another study designed to improve the factor taxonomy and the test instrument.

Variable 35, Tolerance for Vulgarity, appears to be too complex factorially, measuring factor C as much as it measures factor M since its loadings and structure coefficient are of comparable magnitude in both factors. In a revision of the test instrument, this particular FHID would also be marked for revision or replacement. Variable 18, Need to Excel, has its highest loadings and structure coefficient on the factor it is supposed to define, factor A, but the loading is only .38 in both solutions, although the structure coefficient on factor A is .45. An attempt would be made to modify this FHID to obtain a higher loading on factor A in any new study.

The loadings for variables 33 and 34 are not very high on factor M, about .43, in both solutions, but they have no appreciable loadings on any other factors. This tends to make factor M somewhat broader and less homogeneous in composition compared to the other factors but this is not necessarily unacceptable. It would be desirable, however, to increase these loadings in a revised version of the taxonomy if at all possible.

The study reported above represents an example where the oblique solution gives very little more than the orthogonal solution in the way of useful information. In reporting the study, therefore, it would not be unreasonable to report only the orthogonal solution in detail with verbal comments on the nature of the oblique solution and its relationship to the orthogonal solution. All the oblique solution tables except perhaps the table of correlations among the factors could be omitted to save on costs where it appears to be important to do so.

[11.5] VALIDITY STUDIES

It has already been pointed out in Chapter 10 that good factor interpretation requires the investigator to go beyond a mere intuitive description based on a cursory analysis of the variables with major loadings on the factors. Hypothetico-deductive factor studies are an important part of a good program aimed at improved factor interpretation. In successive studies, new hypotheses are formulated about variables that should and others that should not appear on the factor if it is indeed what it is presumed to be. These hypotheses are tested empirically and conceptions about the factor construct are revised accordingly.

It is also important, however, to go beyond factor studies in an attempt to understand more fully the nature of a factor construct. An important research technique for this purpose is the validity study. Such studies can take many forms but they all share the following ingredients: (a) If the factor construct is as it is thought to be, data collected in a certain way can be expected to yield a predicted result. (b) The data are collected and the prediction is tested. (c) If the prediction is upheld, confirmatory evidence for the factor interpretation is obtained; if not, the factor construct may require revision. Thus, a prediction might be made that diagnosed paranoid schizophrenics should have lower than average scores on Trust versus Defensiveness (T) if it is measuring what it is presumed to measure. By administering the test to a sample of such individuals this hypothesis can be tested. Finding that paranoid schizophrenics do in fact have a significantly lower score than the "normal" individual on the average would be consistent with the hypothesis that the factor score is measuring what it is supposed to measure but, of course, would not prove it conclusively. As more and more hypotheses about the factor are confirmed in such studies, however, confidence in the factor interpretation is increased. This type of study is sometimes referred to as a "construct validation" study.

Another kind of validity evidence that can be advanced to support factor interpretations is that derived by comparing different measures of the same putative construct. If two different investigators have identified factors with the same or similar descriptions, determining that a high positive correlation exists between the measures of these two factors adds to the body of evidence supporting the factor interpretations. Such studies will be referred to here as "concurrent validity" studies.

[11.5.1] *CPS CONSTRUCT VALIDATION STUDIES*

The CPS instrument was published only a year ago at this writing; so there have been but a few construct validation studies carried out so far. One very simple study of this kind was the inclusion of "sex" as a variable in the factor study reported above. It was predicted that the variable 44, Sex, should have a substantial positive loading on factor M, Masculinity versus Femininity. Both loadings and the structure coefficient were above .5 on M for this variable. There are easier ways of determining sex than by giving the CPS, of course, but the verification of this deduction is a point in favor of the interpretation advanced for the M factor.

A doctoral dissertation study by Fabian, also reported in Fabian and Comrey (1971), has provided evidence for the construct validity of the Emotional Stability versus Neuroticism factor. As a part of the study, she administered an early version of the factor scale designed to measure S to 69 normal, nonneurotic general medical and surgical hospital patients, 31 neuropsychiatric outpatients, and 68 neuropsychiatric inpatients. All of the neuropsychiatric patients had been diagnosed as neurotics. The prediction was that the lowest mean Neuroticism score (or highest Emotional Stability score) would be shown by the general medical and surgical patients with the highest Neuroticism mean score occurring for the inpatient neurotics. Outpatient neurotics were expected to have an intermediate mean Neuroticism score. The mean Neuroticism scores for the normals, outpatients, and inpatients, respectively, were 85.1, 110.1, and 119.1, giving a highly significant F-ratio of 32.6 in a one-way analysis of variance test of the differences between means.

Comrey and Backer (1970) administered the CPS and a biographical inventory to 209 male and female volunteer students at UCLA. The questions on the biographical inventory were hypothesized to show a relationship in a predicted direction with one or more of the CPS. The most striking degree of confirmation occurred for the Social Conformity versus Rebelliousness (C) scale. Variables correlated significantly at the 1-percent level with Social Conformity versus Rebelliousness in this study were the following: amount of marijuana consumed, −.54; having some religious preference versus having none, .39; degree of participation in campus demonstrations, −.39; amount of addictive drugs used, −.36; extent of premarital sexual activity, −.27; amount of interest in joining the Peace Corps, −.26; number of mental problems listed, −.26; degree of unconventionality of dress, −.25; Jewish religion preferred versus other religion or none, .21; amount of tobacco consumed, −.20; amount of personal savings, .20; amount of trouble with police, −.19; married versus not married (females only), −.19; married versus not married (males only), −.18; number of times late to classes and/or appointments in a week, −.18; number of hours spent studying per week, .18. All of these relationships were predicted except the one concerning membership in the Jewish religion.

[11.5.2] *CONCURRENT VALIDITY STUDIES OF THE CPS*

Mentioned here briefly are two studies in which attempts were made to relate measures of the CPS factors to those from other systems. Comrey, Jamison, and King (1968) administered an early version of the CPS and the Guilford–Zimmerman Temperament Survey to the same subjects. Analyses of the Guilford–Zimmerman items were undertaken to find any FHIDs among the items that could be factor analyzed with the CPS FHIDs. Factor analysis of the matrix of intercorrelations of FHIDs from both tests led to the conclusion that Neuroticism (an earlier version of Emotional Stability versus Neuroticism) in the CPS test is very close to the Emotional Stability factor in the Guilford–Zimmerman test.

In another study (Comrey & Duffy, 1968), an earlier version of the CPS was administered along with the Eysenck Personality Inventory and the Cattell 16 PF to the same subjects. It was established that the Eysenck items factor analyzed into two factors, as expected, Neuroticism and Extraversion versus Introversion. It was further shown that these two factors are highly related to the CPS Neuroticism and Shyness factors, now renamed Emotional Stability versus Neuroticism and Extraversion versus Introversion, respectively.

The preceding studies by no means exhaust the possibilities for testing factor interpretations. Other interesting kinds of studies might involve trying to change factor scores experimentally by training. What kinds of experiences could be planned, for example, to increase levels of Trust versus Defensiveness? How can individuals with deviant M scores be trained to achieve regression toward the mean? How can people with very low Empathy versus Egocentrism scores be trained to place a higher value on satisfying the needs of others? Studies investigating these and related questions represent ways of checking the validity of CPS scales and in addition offer the possibility of obtaining socially useful findings.

Although the long series of factor studies and validation studies alluded to in this chapter represent a substantial research investment in the effort to interpret factors in the CPS taxonomy, it is clear that a great deal more remains to be done. The author hopes that the description of this program of test development by factor analytic methods will give the reader a lasting impression of the need to carry out factor analytic research in depth rather than to depend upon obtaining results of enduring value from a hastily conceived single study.

CHAPTER TWELVE

Computer Programs

The computations involved in factor analysis are so extensive that it has only been since the appearance of the electronic computer that this method has begun to gain the acceptance it deserves. This chapter provides a description of the programs developed by the author to execute various kinds of computations needed in factor analytic work. The programs described here are available from the publisher on magnetic tape. This avoids the inevitable errors in transcription made in punching program cards from a printed listing of the program. Such errors can be exceedingly difficult to locate. These programs were developed to carry out the operations preferred by the author in his own approach, however; so they do not represent a complete package to do all kinds of factor analytic computations. The reader will be able to find other sources of programs useful in factor analysis to supplement those described here. For example, the books on factor analysis by Horst (1965) and Guertin and Bailey (1970) both contain listings of programs. Cooley and Lohnes (1962) list programs useful in factor analysis as well as programs for other multivariate procedures. Nie, Bent, and Hull (1970) have published a package of programs for statistical computations in the social sciences, including several for factor analytic work. The BMD series of programs (Dixon, 1970a, b) have been particularly widely used and of course contain factor analysis programs. Computer programs have also appeared from time to time in such psychological journals as *Educational and Psychological Measurement* and *Behavioral Science*.

The computer programs described in this chapter were written in the FORTRAN IV language, although many of them were originally prepared in earlier languages and updated to FORTRAN IV language. Many of the programs require special job control language (JCL) cards to allow for the fact that tapes are used in the programs other than tape 5 for card input and tape 6 for printed output. If only the usual tapes 5 and 6 are used in the program, no notation of tapes used will be made in the program description. An example of the JCL cards for the factor analysis program is provided as a guide to the user. Other possibilities can be explored with a consultant at the computer installation where the program is to be used.

```
//GO.FT01F001 DD UNIT=SYSDA,DISP=NEW,SPACE=(TRK,(10,10)),
                 DCB=(BLKSIZE=7248,RECFM=VBS,LRECL=7244)
//GO.FT02F001 DD UNIT=SYSDA,DISP=NEW,SPACE=(TRK,(10,10)),
                 DCB=(BLKSIZE=7248,RECFM=VBS,LRECL=7244)
//GO.FT04F001 DD DUMMY
//GO.FT07F001 DD SYSOUT=B,SPACE=(TRK,1,RLSE),
                 DCB=(RECFM=FB,LRECL=80,BLKSIZE=3520)
//GO.SYSIN DD *
```

These cards follow the cards for the FORTRAN deck and precede the data cards. The cards above concerned with tape 7, the two cards immediately preceding the last one, will also serve for all those programs that call for the use of tape 7 for outputting punched cards. The number 1 that appears between TRK and RLSE in the first of these two cards represent the multiple of 88 cards, or a portion thereof, expected on output. If 88 cards or less are to be output, leave 1 here; if more cards will be output, increase this number to 2, 3, or whatever multiple of 88 is required to accommodate the amount of punched card output.

For each program to be described, the purpose and general nature of the program are presented first, followed by a description of the data cards to be input, including the control card, variable format cards, if any, and other special cards besides the basic data cards. Special tapes required are mentioned, if any, followed by the specification of the data card order and a description of the printed output of the program.

[12.1] PROGRAM 1. SCORING

This program was designed to compute raw scores by summing scores over smaller units, such as items on a test. Through the dimension cards (described below), the program is informed which items are to be included in the total score. The program will then add up those item scores and output a total score for each person on that variable. Scores may be computed for many different variables for a given person on the same run. If some of the items are negatively stated with respect to the variable name, to control for acquiescence response set, these item scores may be reversed before adding the scores into the total raw score. Items that are to be reversed are specified on the item reversal cards (described below). Composite scores based on a sum of subscores can be obtained with this program even though they are not items, however. After the scores for each person are computed on the different variables, they are printed out before the data for the next person are input for score computation. If desired, the computed scores may also be punched out on cards. These punched cards may be used as data input for other programs in this series.

Control Card. The control card is the first data card. The information to be punched on this card is summarized briefly in Table 12.1.

TABLE 12.1

Control Card Information for the Scoring Program

Columns	Information to be punched
1–3	N, the number of subjects or data objects, 999 maximum
4–6	NUDIM, the number of dimensions to be scored, 135 maximum
7–9	NSCORE, the number of items of data input per subject, 999 maximum
10–12	NCARDS, the number of cards needed to hold item numbers of the items of data to be scored on a given dimension or variable; this number may be 1, 2, 3, or 4 and is that number needed for the variable with the largest number of data items
13–15	NIREV, the number of data items to be reversed in direction, 001 minimum; if no reversal of data items is needed, use a dummy data item and reverse it
16–18	NMISP, the maximum data item score plus one; for example, if items are scored, NMISP would be 005 if the items were four-choice items with possible scores of 1, 2, 3, and 4
19–21	NCOL, the number of columns allowed for the scores in the punched card output; if the computed scores will be three digits, 003 must be punched here; if the computed scores will be two digits and only two columns per score are desired on punched output, punch 004 here; three columns may be allowed for two-digit numbers if desired by punching 003 here
22–24	NPUN, punch 001 to obtain both printed output and punched card output of the computed scores; punch 000 for only printed output

Variable Format Cards. The next three data cards following the control card are the variable format cards which specify how the data are arranged on the punched card data input. If item scores were the data, they might be punched one column per item, 72 items on a card, with columns 73–80 used for identification purposes. This would call for a format of (72I1). This is an integer format. This program accepts only integer formats. The number of data items indicated on the variable format cards must be compatible with the number of scores input per person as indicated in columns 7–9 of the control card. If all the format is contained on the first card within the first 72 columns, the second two cards will be blank, but they must be included.

Dimension Cards. For each total score variable or dimension to be scored by adding up component or item scores, one or more dimension cards must be included as data following the three variable format cards. The same number of dimension cards must be included for each variable, even though some variables may not require that many. The extras, if any, are left blank. The first entry on each set of dimension cards for each variable is the number of items to be scored on that dimension. This entry is placed in columns 1–3 of the first dimension card for the variable. Following this entry, the data-item numbers of the data items to be added up for this total score are punched in three column fields, ending with column 72 on card 1. If more items are to be scored than the 23 that can be fitted on the first card, continue on to a second card, and so on. The data-item numbers are based on the order in which the data items are input as specified on the variable format cards. The dimension cards are assembled in the order in which the variables are to be scored and printed out and placed after the control card.

Item Reversal Cards. These are the next data cards following the dimension cards. The data-item numbers of the data items to be reversed before adding up data items to obtain the total scores are punched on these cards, three columns per number and 24 numbers per card in columns 1–72. If more than 24 data items are to be reversed, their numbers are punched on a second, third, and subsequent cards until all such numbers have been punched, 24 per card. The program subtracts the data-item value for such items from NMISP, the entry in columns 16–18 of the control card, and places the result back in storage as a replacement for the original data-item value. At least one item reversal card with at least one data-item number is required by the program. If all data items are positive, add an extra dummy data item at the end of the data and reverse it to satisfy the demands of the program.

The Data-Item Cards. The data items to be operated on are punched on cards that follow the item reversal cards. All the data items for a single individual or subject are punched on consecutive cards according to the format specified on the variable format cards. These data items may be responses to test items, total scores over several items, or other kinds of numerical measures.

Tapes Used. In addition to the usual tapes for reading and writing, this program uses tape 7 for storage of computed total scores to be punched out as data for input to other programs. The JCL cards used with the program must allow for the use of this tape.

Data Card Order. The cards that constitute the data are arranged in the following order: control card, variable format cards, dimension cards, item reversal cards, and data-item cards.

Printed Output. The total scores for all the variables scored are printed out for the first person, then for the second person, and so on, until they have been printed out for the last person.

[12.2]

PROGRAM 2. REGRESSION PLOT

This program is designed to prepare a scatter plot for pairs of selected variables so that nonlinear regressions may be detected visually. A matrix of scores is input, such as that output from Program 1. Pairs of these variables may be selected for plotting. The scores for a given variable are scaled to be proportions of the total range from the lowest to the highest score. These proportions are plotted against each other. The distributional characteristics of the variables are preserved by this kind of transformation, revealing skewness as well as nonlinearity. Frequencies up to 9 are printed as numerals. Any cell in the scatter plot with a frequency higher than 9 will contain a letter from A to Z if the frequency is between 10 and 35. Any cell with a frequency of 36 or larger will contain a printout of the symbol *, an asterisk.

Control Card. The control card is the first data card. The information to be punched on this card is summarized briefly in Table 12.2.

Variable Format Cards. The five data cards following the control card are variable format cards describing the form in which the scores for the variables to be plotted are stored on the data cards. For example, (24F3.0) on the first variable format card would state that the scores are punched three columns per score in the first 72 columns of each card. The remaining four variable format cards would be blank in this case.

TABLE 12.2

Control Card Information for the Regression Plot Program

Columns	Information to be punched
1–3	N, the number of subjects or data objects, 800 maximum
4–6	NSCR, the number of scores per subject stored consecutively by subject, 50 maximum
7–9	JPLOT, punch 001 if only pairs of variables are to be plotted for which plot cards are read in; punch 000 if all pairs of variables are to be plotted

Score Cards. The scores for each subject are punched on one or more cards in the same format for every subject as shown in the five variable format cards. Cards for the first subject are followed by those for the second subject, and so on. These cards follow the variable format cards.

Plot Cards. If a value of 001 has been punched in columns 7–9 of the control card, a separate plot card must be prepared for each plot desired. For example, a plot card with 001 in columns 1–3 and 002 in columns 4–6 will cause variable 1 to be plotted with variable 2. The plot cards follow the score cards. They are omitted if 000 is punched in columns 7–9 of the control card.

End Card. If there are any plot cards used, an end card with 999 punched in columns 1–3 must follow the plot cards. This signals the program that there are no more plots to be made. This card is unnecessary if there are no plot cards, that is, if columns 7–9 of the control card contain 000.

Data Card Order. The cards that constitute the data are arranged in the following order: control card, five variable format cards, score cards, and optional plot cards and end card if only some of the plots are desired.

Printed Output. The scores input as data are printed out first with all the scores for the first subject, followed by all the scores for the second subject, and so on, until the scores for the last subject are printed out. Following the printout of the raw-score matrix, the scaled scores are printed out in transposed form. That is, the scaled scores for variable 1 are printed out for all subjects, then the scaled scores for variable 2, and so on. The scaled scores are proportions and are the values that are plotted. After the scaled score matrix is output, the plots themselves are printed out in the sequence 1–2, 1–3, ..., 1–n, 2–3, 2–4, and so on. If only some plots are called for, the plots are printed in order of the plot cards. The first-mentioned variable is plotted on the vertical axis and the second-mentioned variable is plotted on the horizontal axis.

[12.3]
PROGRAM 3. FREQUENCY DISTRIBUTION

This program will select from 1 to 15 variables from the data input and prepare a frequency distribution for each. The frequencies for each raw score are tallied and the centile equivalent calculated.

Control Card. The control card is the first data card. The information to be punched on this card is summarized briefly in Table 12.3.

TABLE 12.3

Control Card Information for the Frequency Distribution Program

Columns	Information to be punched
1–4	N, the number of subjects or data objects, for example, 0145, 999 maximum
5–7	NVAR, the number of variables or scores input per subject, 135 maximum
8–9	NFD, the number of frequency distributions to be prepared, 15 maximum
10–11	Punch 01 for one column per score on input, 72 scores per card in columns 1–72; punch 02 for two columns per score on input, 36 scores per card in columns 1–72; punch 03 for three columns per scores on input, 24 scores per card in columns 1–72

Variables-to-be-Distributed Card. Although 135 one-, two-, or three-digit integer scores may be input if desired, only 15 or fewer of these variables can be processed to produce frequency distributions in one pass through the machine. The variable numbers, from 1 up to 135, of the variables for which frequency distributions are desired are punched as three-digit numbers on the variables-to-be-distributed card; for example, 001, 005, 065 punched on this card would result in frequency distributions for the first, fifth, and sixty-fifth variables selected from the total number included in the data.

Data Score Cards. The data are punched by subject in columns 1–72 only and in one-, two-, or three-digit integer form. The cards for the first subject are followed by those for the second subject, and so on.

Data Card Order. The cards that constitute the data are arranged in the following order: control card, variables-to-be-distributed card, and data score cards.

Printed Output. The printed output consists of the frequency distributions, output consecutively in the order of the variable numbers on the variables-to-be-distributed card. The distributions are labeled consecutively beginning with the number 1, however, rather than with their numbers as shown on the variables-to-be-distributed card. Each score is listed with the frequency associated with it, from the lowest score to the highest score. The centile value corresponding to the score is also printed out.

[12.4]

PROGRAM 4. SCALED SCORE EQUIVALENTS

This program converts raw scores to corresponding scaled scores using a table that is input for conversion purposes. It can convert raw scores to centile scores or McCall *T*-scores, for example, or make any other kind of transformation called for by the table of equivalents. The table of equivalents is input to the program as data in the form of scaled score equivalent cards (described below). In essence, these cards contain a scaled score that is to be the equivalent for each raw score. The program reads the raw score, looks up the scaled score equivalent in the table, and outputs the appropriate scaled score. The most common usage of the program would be for getting centile or *T*-score equivalents for the raw scores with a large sample of subjects so that profile sheets could be made out to return to the individuals taking the tests. This conversion can be carried out in one operation for as many as 15 different raw scores per person.

Control Card. The control card is the first data card. The information to be punched on this card is summarized briefly in Table 12.4.

TABLE 12.4

Control Card Information for the Scaled Score Equivalents Program

Columns	Information to be punched
1–3	IMAX, the number of scores per person, 15 maximum
4–6	JCT, the subject number of the first subject for labeling
7–10	N, the number of subjects, a four-digit number less than or equal to 9999, for example, 0085

Scaled Score Equivalent Cards. The possible scaled scores are the 101 values from 0 to 100. For a given variable, the raw scores that are to correspond to these 101 scaled scores are punched on five scaled score equivalent cards in the format (24I3). That is, 24 three-digit numbers go on the first such card, the next 24 on the next card, and so on, until the final five go on the fifth card. The number punched in columns 1–3 of the first scaled score equivalent card is that raw score such that all raw scores less than or equal to it receive a scaled score of 0. The fifth number on the fifth card is a raw score such that all raw scores greater than or equal to it receive a scaled score of 100. The second number on the first scaled score equivalent card, in columns 4–6, contains the smallest raw score that is to receive a scaled score of 1. In columns 7–9, punch the smallest raw score that is to receive a scaled score of 2, and so on, until the fourth position on card five, in columns 10–12, contains the raw score that is the smallest raw score to receive a score of 99.

If a *T*-score equivalent transformation is desired and no *T*-score less than 20 nor greater than 80 is desired, a number lower than any score to be encountered would be entered as the raw score in the first 20 positions on the scaled score equivalent cards. A number higher than any raw score to be encountered must be entered in the last 20 positions. In between will be the raw-score equivalents that are the lowest raw scores to be given the corresponding *T*-scores between 20 and 80. There will be five scaled score equivalent cards for each variable to be converted to scaled scores.

Raw-Score Cards. One raw-score data card is prepared for each person with his first score in columns 1–3, his second raw score in columns 4–6, and so on. No more than 15 variables can be processed in one run.

Data Card Order. The cards that constitute the data are arranged in the following order: control card, five scaled score equivalent cards for each variable to be scaled, and raw-score cards, one per person.

Printed Output. A subject number is printed out first followed by a row of raw scores and then by a row of scaled scores for each subject. The subject numbers run consecutively starting with the number input on the control card in columns 4–6.

[12.5]

PROGRAM 5. FACTOR ANALYSIS

This program computes means and standard deviations, and correlations, extracts factors by the minimum residual method or the principal factor method (see Chapter 4), and rotates the extracted factors by the Kaiser Varimax method (see Chapter 7). If data are input as scores, the program will compute means and standard deviations and produce a correlation matrix for factor analysis. The program will also accept a correlation matrix as the basic data, however, in which case no means and standard deviations are computed. Estimated communalities may be input for the first factor calculations. If communalities are input, the method of factor extraction is by the minimum residual method with communalities, which yields a principal factor solution. If no communalities are input, the solution is by the minimum residual method proper which does not employ the diagonal cells in extracting the factors. These various options are controlled by appropriate entries on the control card.

The factor solution may be iterated, if desired, in which case the computed communalities from the first factor extraction sequence will be fed back as the estimated communalities for the second iteration. Even if the first iteration was carried out by the minimum residual method without communalities, the second and subsequent iterations will yield principal factor solutions since the diagonal cells are being utilized. If this option is desired, the solution is ordinarily iterated enough times to gain some stability in the communalities. From 10 to 15 iterations will usually be sufficient for most practical purposes. Iterating the solution obviously requires more computing time. It is the author's preference in most cases to obtain the minimum residual solution without communalities and without iteration. Some problems may require a principal factor solution, however; so this option is available in the program. Iteration of the communalities also may be needed in some cases. Where the minimum residual solution without communalities gives anomalous results, such as factor loadings over 1.0, which it can do at times, particularly with small matrices, it is necessary to estimate the communalities and recompute by the minimum residual method with communalities to give a principal factor solution.

Either of two iterative procedures may be chosen for obtaining convergence on a solution within the calculations for the extraction of a single factor. This iterative process is distinct from the iteration of the entire solution referred to above. The first of these iterative methods for extracting a factor was developed as the original form of the minimum residual solution (Comrey, 1962a) and will be referred to as the "averaging method" of iteration The. second iterative method is referred to as the "length-adjustment method." Both methods are described in Chapter 4. The length-adjustment method is generally preferred because it has a more definite and appropriate criterion for terminating factor extraction. When the iterative process by which the factor is computed terminates with paired vectors of opposite signs, the factor extraction process is automatically terminated. This would correspond to a factor with a negative eigenvalue. The averaging method of iteration may extract more small factors after this point, but such factors would not be regarded usually as being sufficiently large to warrant retention. The averaging method terminates by failure to reach convergence, thereby exceeding the maximum number of iterations allowed on the control card in columns 4–6.

The program user specifies how many factors he wishes to be extracted in columns 7–9 on the control card. The number of factors extracted will be equal to this number or less if the iterative process converges on vectors of opposite sign before this number of factors has been extracted. Failure to converge with the number of iterations specified in columns 4–6 of the control card will also result in fewer factors. The author usually specifies a maximum number of factors wanted large enough to be sure that all important factors will be extracted. It is better to err on the side of taking out too many. If the program fails to extract that many because of the convergence conditions encountered, no harm is done. Extracting too few factors can lead to difficulty.

The number of factors to be rotated by the Kaiser Varimax method is specified in columns 20–22 of the control card. It is often desirable to call for the rotation of fewer factors than have been extracted. One may expect about 10 meaningful factors, and hence call for the rotation of 10, but call for the extraction of 13 factors so that the last three can be inspected to determine if they are sufficiently large to warrant their inclusion in a rotated solution. If the number extracted is less than the number called for in columns 20–22, only the smaller number will be rotated. If no rotations are desired, specify 001 in columns 20–22 of the control card.

The criterion of convergence is specified in columns 10–15 on the control card. Successively iterated pairs of vectors must have all corresponding entries agree to less than or equal to this figure before the iterative process is considered to be converged. When the process reaches this point, the last vector in the iterative sequence is taken as the extracted factor. The smaller the value specified here, the more iterations will be required to attain convergence. Values that are too large permit termination of the iteration before adequate convergence has taken place. Values that are too small may make it impossible to obtain convergence in the allowed number of iterations, if at all. Some value between .0050 and .0005 will suffice for most practical problems.

It may be desirable to retain some of the computed matrices on punched cards to be input as data in future calculations. This option may be exercised by placing a 1 in column 24, 25, or 26 to receive the correlation matrix, the minimum residual matrix, or the Varimax matrix, respectively, as punched card output in the format (12F6.3). This means that the matrix entries are punched 12 to a card, with six columns per entry and three figures to the right of the decimal point. Cards for the first column of the matrix are punched first, followed by those for the second column, and so on, until the cards for the last column are punched. If these punched cards are not needed, zeros are placed in columns 24, 25, and 26.

In the iterative process to obtain a given factor, many intermediate vectors are computed before convergence takes place. Normally only the final converged vector is output. If the intermediate vectors computed in the iterative process are desired for some reason, they will be printed out by placing a 1 in column 28 of the control card. If these are not wanted, as they usually are not, place a zero in this column.

If a correlation matrix rather than scores is input, this is indicated by placing a 1 in column 23 of the control card. If scores are to be input, a zero is punched in column 23.

If the entire factor solution is to be iterated several times, the amount of output becomes rather voluminous, requiring many printed pages, if all the results are printed for each of the iterated intermediate solutions. Where many iterations are called for by specifying more than 01 in columns 32 and 33 of the control card, it is possible to limit the amount of printed output to just the final computed factor matrices for each cycle by specifying a 1 in column 31 of the control card. If the entire printed output of all information for each cycle is desired, a zero is punched in column 31. All information would be called for if only one cycle of factor extraction is being requested, that is, if 01 is placed in columns 32 and 33.

Control Card. The control card is the first data card. The information to be punched on this card is summarized briefly in Table 12.5.

Variable Format Cards. The program in its present form calls for the input of five variable format cards immediately following the control card. These cards inform the program how the data are stored, regardless of whether scores are input or whether a correlation matrix is input. Several cards are included because complicated formats require skipping columns on the data cards, making the format rather lengthy to specify. Only the first 72 columns are used on any card of the variable format cards. A simple example would be the punching of the following on card 1: (18F4.3). This would be a suitable format for inputting a correlation matrix. Each correlation would be allowed four columns on the card to accommodate a sign and a three-digit number. The decimal point need not be punched. The first 18 correlations in column 1 would be punched on the first card, the next 18 on card 2, and so on, until all correlations are punched. The second column would start with a new card, and so on.

A typical format for inputting scores instead of a correlation matrix would be the following: (36F2.0). This would call for two columns for each score, 36 scores per card. An individual's scores would be punched in consecutive fields of two digits, 36 on card 1, the next 36 or part thereof on card 2, until all scores are punched. The scores for the second person start on a new card. If any of the scores runs to three digits, of course, this format would not be acceptable. Integer formats may not be used for data input to this program; for example, (36I2) is unacceptable.

TABLE 12.5

Control Card Information for the Factor Analysis Program

Columns	Information to be punched
1–3	N, the number of variables, 115 maximum
4–6	NITER, number of iterations to be allowed without convergence before terminating run for failure to converge, 200 maximum
7–9	MNFAC, maximum number of factors to be extracted
10–15	ERROR, criterion of convergence; a value between 0.0050 and 0.0005 is appropriate for most cases
16–19	NCASES, the number of subjects, cases, or data objects
20–22	MFROT, the maximum number of factors to be rotated
23	ISCMC, punch 1 if a correlation matrix is to be input as data; punch 0 if scores are to be input and the correlation matrix computed
24	IRMPO, punch 1 if the correlation matrix is to be output on punched cards; punch 0 if no card output is desired
25	IMRPO, punch 1 if the minimum residual matrix is to be output on punched cards; punch 0 if no card output is desired
26	IVMPO, punch 1 if the Varimax matrix is to be output on punched cards; punch 0 if no card output is desired for this matrix
27	ISH2, punch 1 if communality estimates are to be input to the program as data using format (18F4.3); punch 0 if no estimated communalities are to be furnished to the program
28	ITOP, punch 1 if all intermediate trial vectors before convergence are to be printed out; punch 0 if only the converged vectors are to be printed out; 0 is usually punched here
29	ITYPE, punch 1 for the length-adjustment type of iteration; punch 0 for the averaging type of iteration; 1 is usually punched
30	JTP, a punch of 1 formerly permitted tape input of data and 0 for other forms of data input; the program is not used with this option at present so 0 is punched here
31	IOS, punch 1 for limiting the amount of printed output to only the summary final matrices; punch 0 to obtain all the usual output; 1 is punched ordinarily only when several cycles are run to obtain converged communalities
32–33	IZAG, the number of cycles of the factor solution to be run; if no iteration of communalities is desired, 01 will be punched here to obtain the minimum residual solution without communalities; for several cycles to obtain converged communalities, 10 to 15 is ordinarily sufficient as an entry in this control card field

Cards for Scores or Correlation Coefficients. The first six data cards are taken up with the control card and the variable format cards. The data proper begin with the seventh card. These cards will be either a correlation matrix, in which case a 1 must be punched in column 23 of the control card, or these cards will contain scores on all the variables for each subject, in which case a 0 will be punched in column 23.

Cards for Estimated Communalities. If estimated communalities are to be input, leading to a principal factor-type solution, or a minimum residual solution with communalities, a 1 is punched in column 26 of the control card and the estimated communalities are punched on cards to follow the cards for the other data. These communalities must be punched in the format (18F4.3), that is, 18 per card in order from first to last, allowing four columns per entry. The first column in each field is left blank unless unities are to be inserted, in which case the punched entry would be 1000. No decimal point need be punched.

Tapes Used. In addition to the usual tapes for reading and writing, this program uses tapes 1 and 2 as temporary storage scratch tapes. Tape 4 is referred to in the program for tape input but is treated as a dummy tape in the use of the program without tape input. Tape 7 is used for storage of matrices for punched card output. The JCL cards used with the program must allow for the use of these tapes.

Data Card Order. The cards that constitute the data are arranged in the following order: control card, five variable format cards, cards for scores or for the correlation matrix, and cards for the communalities if estimated communalities are to be input.

Printed Output. Printed output begins with a line giving three entries from the control card, NITER, MNFAC, and ERROR. Following this, the means and standard deviations are printed out if scores are input. The correlation matrix is printed out next in 19-column sections. Prior to the calculation of each factor, the matrix is examined for the largest residual entry and its absolute value is printed out. Also, the sums of squares and the standard deviation of the residuals above the main diagonal are output before the computations for the factor are begun. Following the output of the number of the factor being computed, the largest discrepancies between entries in pairs of vectors are output, one for each iteration cycle within a single factor computation sequence. If convergence is obtained, the last value will be less than the ERROR entry on the control card and the converged factor data will be output. First, the sum of squares of the factor loadings for the factor will be output. Following this, the factor loadings will be output. The communalities are output next. These include the sums of squares of previous factor entries plus the squares of the present factor entries. After this, the cycle begins anew for another factor.

If convergence is not reached in the number of iterations allowed, the last two trial vectors will be output and factor extraction will stop. If the length-adjustment iterative process is used and convergence takes place on vectors of opposite sign, these last trial vectors will be printed out showing equal but opposite signed values. In the event that all intermediate trial vectors are called for, by a 1 being placed in column 28 of the control card, these vectors will be printed out before the highest absolute residual is printed.

After the last factor has been computed and printed out, the residual matrix is printed. The minimum residual matrix is printed next, although the values have already been output one factor at a time. The last column of the minimum residual matrix contains the communalities although it is not labeled as such.

The remainder of the printout concerns the Varimax rotations. The differences between rotated communalities and unrotated communalities are output as a check. They should all be zero. Following this, the values of the Varimax variance sum after each cycle of rotations are output. These values should increase and stabilize at a maximum value. The Varimax matrix is the last one to be printed out. Again, the last column contains the communalities. The program returns to start calling for more input to do another factor analysis so an error message indicating insufficient data may be expected at the end of the printout.

If suppression of the intermediate output is called for by placing a 1 in column 31 of the control card, much less output will be obtained. After the correlation matrix, each minimum residual factor matrix computed will be output with communalities in the last column. Following this, the usual information for the Varimax rotations and the final Varimax rotated matrix will be printed out with communalities in the last column. The remainder of the printout described above will be suppressed.

The Varimax subroutine used as part of this program was written by Lynn Hayward as a term project while he was a student in the author's class in factor analysis. Essentially the same program is used for Varimax rotations in the BMD series of programs (Dixon, 1970a, b) since Hayward was working on those programs at that time.

[12.6] PROGRAM 6. TANDEM CRITERIA

This program was designed to carry out orthogonal analytic rotations by the Tandem Criteria method (see Chapter 7). The program operates on an unrotated factor matrix such as that punched out during the execution of program 5, or the cards may be punched specially for the input to this program. The user may elect to compute any one of four kinds of solutions: (a) unnormalized Criterion I solution; (b) unnormalized Criterion II solution;

(c) normalized Criterion I solution; or (d) normalized Criterion II solution. The normalized solutions correct all communalities to 1.0, do the rotations, and then rescale the results to the original communalities. This procedure has the effect of giving equal weight to every variable in the rotations. Such a procedure is justifiable and probably preferable if all or most of the variables have substantial communalities. In the case where many variables have very low communalities, however, it does not seem appropriate to inflate their communalities to 1.0 and permit them equal weight in the rotation process. Under these circumstances, the author would prefer the unnormalized solution.

Criterion I should be applied first in carrying out the rotational solution, making certain to include too many rather than too few factors since Criterion I has the capacity to extract any useful variance from extra factors without the tendency to build them up at the expense of the major factors. Criterion I tends to put as much variance on one factor as possible, consistent with the principle that variables with high loadings on the same factor should be substantially correlated. Thus, if a general factor appears in the data, Criterion I can allow it to appear, but it will not permit a general factor to be produced by placing a factor midway between two uncorrelated clusters. If two correlated factors appear in the data, Criterion I will tend to put more variance on one of the two, leaving the remainder of the variance for a smaller second factor.

Criterion II will attempt to distribute the variance on as many factors as possible, as long as the variables with substantial loadings on each factor are sufficiently correlated with each other. For this reason, Criterion II cannot be applied where extra factors of small variance are included in the matrix to be rotated. With the inclusion of such extra small factors, Criterion II may even split a major factor into two parts, superimposing one part of the major factor onto one of the small extra factors. If there are only k major factors, however, and exactly k factors are rotated by Criterion II, the variance will be distributed rather evenly among the factors without building up one factor at the expense of another. Criterion II rotations, therefore, are ordinarily applied after Criterion I rotations have been completed and a determination has been made of the number of major factors that the solution contains. Then, only this number of factors is rotated by Criterion II. Thus, Criterion I is applied first and the results inspected to permit elimination of factors of negligible importance. Then, the surviving factors are rotated by Criterion II. It may be necessary to try several solutions with Criterion II to see which is the best since it may not be possible to determine exactly how many factors should be retained from an inspection of the Criterion I solution. The Criterion I solution will usually permit a determination of the number of factors that should be retained within rather narrow limits.

Control Card. The control card is the first data card. The information to be punched on this card is summarized briefly in Table 12.6.

TABLE 12.6

Control Card Information for the Tandem Criteria Program

Columns	Information to be punched
1–3	N, the number of variables, 114 maximum
4–6	M, the number of factors to be input, 29 maximum
7–9	IC, the number of cycles of rotation desired; each factor is rotated with each other factor during each cycle; eight to ten cycles is normally sufficient, for example, 008 to 010
10–12	ITAPE, punch 000 if data are to be input on cards; punch 001 if data are to be input on tape; use of this option for data input requires program modification
13–15	KIK, punch 000 if rotation by Criterion I is desired; punch 001 if rotation by Criterion II is desired
16–18	INC, punch 000 for a "raw" or unnormalized solution; punch 001 for a normalized solution
19–21	IPUN, punch 000 if no punch out of the rotated factor matrix is desired; punch 001 if punch out of the rotated factor matrix by format (12F6.3) is desired; this punched output may be taken from the Criterion I solution to be used as data input for the second-stage Criterion II solution; any unwanted factors may be physically removed by removing the appropriate punched cards from the deck punched out

Variable Format Cards. When the unrotated matrix is input on punched cards, the format in which the cards are punched must be specified on the three variable format cards that are included as the next three data cards. If the entire format fits on the first card, in columns 1–72, the second and third cards must still be included even though they are blank. A typical format punched on the first card would be (12F6.3) since this is the format in which these programs punch out matrices to be used in further computations. If the data cards were punched up from a printed matrix, a different format might be used. The first column of the unrotated matrix appears on the first few cards, followed by the second column, and so on, until the last column appears on the final cards in the deck.

Cards for Unrotated Factor Matrix. The punched cards containing the matrix of factors to be rotated follow the three variable format cards. Each factor is punched on the same number of cards, a new factor starting a new card. The order of the factors may be rearranged by rearranging the order of the cards. This is often done when rotating by Criterion II because the Criterion

I factors that are input as data for the Criterion II rotations may require rearranging to be in order of descending size with respect to the total sums of squares of the factor loadings. Sometimes a particular factor may be dropped and the Criterion II rotations done again. This requires removal of the cards for that particular Criterion I factor from the deck before performing the Criterion II rotations.

Tapes Used. In addition to the usual tapes for reading and writing, this program uses tape 7 for storage of matrices for punched card output. The JCL cards used with the program must allow for the use of this tape.

Data Card Order. The cards that constitute the data are arranged in the following order: control card, three variable format cards, and cards for the matrix to be rotated.

Printed Output. The first printed output is the reproduced correlation matrix, obtained by summing the inner products of the factor loadings over all the factors for each pair of variables. Following this, the unrotated factor matrix input as data is printed out, with the communalities in the last column. The sums of squares of the entries in each column are output next and the sum of the communality values. If the normalized criteria are used, the scaled correlation matrix and the scaled unrotated factor matrix are printed out. These matrices are based on adjusting the communalities to unity for purposes of the computations. The differences between unrotated and rotated communalities follow and should be zero. The Tandem Criteria function values for each complete rotational cycle, rotating each factor with each other in one cycle, are output next. These values increase and stabilize for Criterion I rotation; they decrease and stabilize for Criterion II rotations. Following these values, the rotated factor matrix is output with either Criterion I or Criterion II factors, depending on which method was called for. The sums of squares of the entries in each factor are output following the matrix. The communalities for the variables are printed in the last column of the rotated matrix, and the sum of these communality values is printed out as the last item of information output.

[12.7] PROGRAM 7. VARIMAX ROTATION

This program is designed to accept an unrotated factor matrix as data and perform a normal Varimax rotation. This program utilizes the same Varimax subroutine, written by Lynn Hayward, that was used in Program 5.

Control Card. The control card is the first data card. The information to be punched on this card is summarized briefly in Table 12.7.

TABLE 12.7

Control Card Information for the Varimax Rotation Program

Columns	Information to be punched
1–3	N, the number of variables, maximum of 115
4–6	L, the number of factors

Variable Format Card. The second data card specifies the format in which the unrotated factor matrix is punched on cards. A typical format would be (12F6.3) since the unrotated matrix cards punched from program 5 would be in this form. Integer format is not acceptable.

Cards for Unrotated Factor Matrix. These cards are prepared in the same way as for input to Program 6, that is, by columns of the matrix. These cards follow the control card and the variable format card.

Data Card Order. The cards that constitute the data are arranged in the following order: control card, one variable format card, and cards for the unrotated matrix.

Printed Output. The unrotated matrix input as data is printed out first. Following this, the differences in communalities between the unrotated and rotated matrices are output. These should be zero. Next, the variance function values for the cycles of rotation are output, as in the Program 5 output. Finally, the rotated Varimax matrix is printed out.

[12.8] PROGRAM 8. HAND ROTATION

This program permits the rotation of factors or reference vectors relative to each other in any way desired, utilizing any preferred criterion of rotation. For convenience, the term "factor" will often be substituted for the term "reference vector" in the discussion that follows. Any factor may be rotated obliquely relative to any other factor, or orthogonal rotations may be carried out. Any orthogonal unrotated matrix may be input as data to this program.

This is usually a matrix of extracted, unrotated factors, although it could be an orthogonal rotated matrix, such as a Varimax matrix or a matrix output from one of the Tandem Criteria rotation treatments. A transformation matrix may be input as data with the unrotated matrix if some rotation of the unrotated matrix has already taken place and the object is to carry out additional rotations. This transformation matrix contains the effects of all rotations that have been performed up to this time. Any further rotations performed will result in a modification of this transformation matrix to take into account the new rotations. If no rotations have been performed on the input matrix, even though it may itself be a matrix resulting from previous rotations, it is not necessary to input a transformation matrix. An identity matrix will be supplied by the program as the beginning transformation matrix. Subsequent rotations will modify this identity matrix.

This program permits the plotting of factors against one another. These graphs are inspected to determine what rotations are to be performed, if any. Use of the program may be solely for plotting factors without rotation, or factors may be rotated and graphing of the resulting factors with each other called for so that the next rotations may be planned. If plots are desired for a given factor with the other factors, its factor number is shown on the plot card, which is one of the data cards described subsequently. The factor in question will be plotted with all the other higher numbered factors in order up to and including the factor number specified on the control card as the last factor to be plotted. Plots for more than one factor may be called for. To obtain all plots, use NP = NFAC − 1 in columns 16–18 of the control card, LAST = NFAC in columns 19–21 of the control card, and list all but the last factor on the plot card.

Rotation cards are input as data cards to the program to describe any rotations to be performed. Each axis that is to be rotated requires one rotation card. If no rotations are to be performed, no rotation cards are input. An end card, another data card, is placed after the last rotation card, signaling the program that no more rotations are to be performed. The sequence of operations is as follows: (a) The unrotated matrix, that is, the matrix to be operated on, is read in together with a transformation matrix if previous rotations have been performed already; (b) the unrotated matrix is postmultiplied by the transformation matrix to get the rotated factors—this is an identity matrix supplied by the program if no transformation matrix is input; (c) the transformation matrix is modified to take into account any rotations called for on the rotation cards, and the original unrotated matrix is multiplied by this modified transformation matrix to give the new rotated matrix; (d) the transformation matrix is premultiplied by its transpose to give the matrix of correlations among the reference vectors (simple axes); (e) the transformation matrix will be punched out if 001 is punched in columns 10–12 of the control card, providing the cards for a matrix that can be fed into the program as data along with the unrotated matrix for the next series of rotations; (f) plots of the rotated factors with each other will be made if 001 has been punched in columns 13–15 of the control card. It should be noted that the plots can be called for without actually carrying out any rotations.

Control Card. The control card is the first data card. The information to be punched on this card is summarized briefly in Table 12.8.

TABLE 12.8

Control Card Information for the Hand-Rotation Program

Columns	Information to be punched
1–3	N, the number of variables, 115 maximum
4–6	NFAC, the number of factors, 50 maximum
7–9	Punch 001 if a transformation matrix is to be input by the format (12F6.3), punched by columns; punch 000 if no transformation matrix is input and the program is to supply the identity matrix for this purpose
10–12	Punch 001 if the computed transformation matrix encorporating the current rotations is to be punched out by columns in the format (12F6.3); punch 000 if the matrix is not to be punched out
13–15	Punch 001 if some plots are to be made of the factors against each other; punch 000 if no plots are to be made
16–18	NP, the number of factors to be plotted with the other factors, where NP may not exceed N − 1
19–21	LAST, the number of the last factor to be plotted with the other factors with lower numbers; LAST must be greater than NP

Blank–Dot–Asterisk Card. In column 1, there is no punch; in column 2, punch a dot, or period; in column 3, punch an asterisk. The dot is used to make a border for the plot and the asterisk marks a data point. Other symbols may be substituted for these if desired.

Unrotated Factor Matrix. This matrix is input by columns in the format (12F6.3). Each new column starts a new card. This is the matrix with which the rotations start. It is always input for each series of rotations. It is only the transformation matrix that changes as the series of rotations progresses. The transformation matrix is an identity matrix at the start. The "unrotated" matrix could, of course, be a matrix that has been rotated before in some other rotational sequence. It is merely the matrix that is the starting point for this series of rotations.

Transformation Matrix. This matrix is not input for the first series of rotations as a rule since the series of rotations ordinarily starts with an unrotated matrix. In this case, the program supplies the identity matrix as the starting transformation matrix. The first series of rotations occurs as a result of modification of this identity transformation matrix (see Chapters 5 and 6) as a function of the rotations desired. This modified transformation matrix is input for the next series of rotations to be modified further for the purpose of performing additional rotations. The transformation matrix may be punched out in each series of rotations to be used as input for the next series of rotations. The format is (12F6.3) and the matrix is punched by columns.

Rotation Cards. For each rotation to be performed, a card is prepared as shown in Table 12.9. If an orthogonal rotation is desired, this requires two

TABLE 12.9

Rotation Card Information for the Hand-Rotation Program

Columns	Information to be punched
1–3	IXVAR, the number of the factor that is to be rotated
4–6	IYVAR, the number of the factor against which the factor is to be rotated
7–12	TAN, the tangent of the angle of rotation; for example, −0.250 would call for the rotation of the factor listed in columns 1–3 away from the factor listed in columns 4–6 by an angle for which the tangent is .250; if the entry had been +0.400, the first-named factor would have been rotated *toward* the second-named factor by an angle with a tangent of .400; note that the second-named factor is not rotated

rotation cards. The first rotates one factor against the other by a certain angle and the second rotates the other factor against the first by the same angle but with an opposite sign to maintain orthogonality of the rotated factors. If only one factor is rotated at a time, the solution will become oblique.

End Card. Each rotation card results in one rotation of one factor. A factor may be rotated only once in each set of rotations as a rule. To carry out more than one such rotation risks overshooting the mark. Each factor may be rotated once in each series with reasonable safety. After the last rotation card, an end card is placed with 999 punched in columns 1–3. This signals the program that there are no more rotations to be carried out.

Plot Card. The numbers of the factors for which plots with other factors are to be obtained are punched on this card in numerical order using the format

(2413). Each of these factors will be plotted against factors with higher numbers up to and including the factor designated as the last factor to be plotted in columns 19–21 on the control card.

Tapes Used. In addition to the usual tapes for reading and writing, this program uses tape 7 for storage of the transformation matrix for punched card output.

Data Card Order. The cards that constitute the data are arranged in the following order: control card, blank–dot–asterisk card, cards for the unrotated factor matrix, cards for the transformation matrix if any, rotation cards if any, end card, and plot card if any plots are to be done.

Printed Output. The unrotated factor matrix input as the starting point for the rotation process is printed out first followed by the starting transformation matrix. If a transformation matrix has been input, it will be output. If none has been input, the identity matrix will be printed out as the starting transformation matrix. The contents of the rotation cards, if any, and the end card will be printed out next. The rotated factor matrix is output next followed by the modified transformation matrix used to obtain the rotated factor matrix. The **C** matrix, the correlations among the reference vectors, is output next. Finally, the plots of the factors with each other are printed last, if any plots have been called for.

[12.9]

PROGRAM 9. ANALYTIC OBLIQUE ROTATION

This program is used to make improvements in the simple structure solution through additional analytic rotations and then to compute the pattern matrix, the structure matrix, and the matrix of correlations among factors (see Chapter 6). The program performs either of these functions separately or both of them together. Thus, if a hand oblique solution has been obtained using the previous program, the unrotated matrix and the transformation matrix may be input as the data for this program. Then, the solution may be adjusted further analytically, or merely the computation of the pattern, structure, and $\mathbf{\Phi}$ matrices called for. It is also possible to merely do the analytic rotations, output the transformation matrix, and use the previous program to plot the rotated factors for inspection before considering the rotation process terminated. This solution may even be adjusted further by hand rotations, using Program 8, and the results returned to Program 9 for calculation of the final matrices.

The analytic rotation procedure, developed by the author, is basically a trial-and-error method for improving simple structure. For each factor, a trial rotation toward and away from each other factor in turn is made. A count of the number of points in the hyperplane, for example, $+.10$ to $-.10$, is made for the present factor position and for the two trial positions. If one of the trial positions has a higher hyperplane count than the present position, the new factor position is accepted provided it does not bring the solution to a degree of obliquity exceeding the maximum specified on the control card. The program cycles through all the possible rotations and then repeats the process as often as called for on the control card. The program starts off with a search angle that is specified on the control card. This search angle is ordinarily made to be rather modest in size since the intent is to adjust an existing solution rather than to obtain an entirely new one. The size of this angle is progressively reduced on successive repetitions of the rotational cycle so that smaller and smaller search angles are used in the late stages to make only fine adjustments. It is possible to use this program starting with an original extracted, unrotated factor matrix, using a large initial search angle. The author's experience has been that poor results are obtained in this way. It is better to start with a Tandem Criteria solution or a Varimax solution, or some other first approximation solution, and then only make comparatively small changes in the character of this initial solution by means of this analytic program. When applied with relatively small initial search angles, this method will usually increase the hyperplane count for an initial solution considerably without grossly modifying the nature of the solution input. It operates to "clean up" the solution relative to simple structure criteria.

Control Card. The control card is the first data card. The information to be punched on this card is summarized briefly in Table 12.10.

Unrotated Factor Matrix Cards. The unrotated factor matrix is punched on cards by columns in the format (12F6.3). These cards ordinarily would be output by a previous program. Each column starts on a new card.

Transformation Matrix Cards. The transformation matrix, if any, is punched on cards by columns in format (12F6.3). These cards ordinarily would be output by a previous program. Each column starts on a new card.

Tapes Used. In addition to the usual tapes for reading and writing, tape 7 is used to store matrices that will be punched out on cards.

Data Card Order. The cards that constitute the data are arranged in the following order: control card, unrotated matrix cards, and transformation matrix cards.

TABLE 12.10

Control Card Information for the Analytic Oblique Rotation Program

Columns	Information to be punched
1–3	N, the number of data variables, 115 maximum
4–6	NFAC, the number of factors to be rotated, 50 maximum
7–9	NITER, the number of rotation cycles desired; during each cycle, each factor is rotated with each other factor; eight to ten cycles is normally sufficient
10–12	HYPW, the hyperplane width desired; .10 punched here gives a width of .20 that is, from −.10 to +.10, the typical value to be used; a smaller value may be used with large samples, if desired
13–15	TAN, the tangent of the initial search angle; the trial positions of the reference vectors will be toward and away from the other factor by an angle with this tangent; for adjusting a rotated solution, one would rarely use a value higher than .20 here and would often prefer a smaller initial search angle
16–18	PROP, the proportion of the tangent of the search angle that is to be used for the next cycle; for example, if .80 is used, and if the initial search angle tangent is .20, the tangent of the search angle for the second cycle of rotations will be .16, for the third about .13, and so on
19–21	RMAX, the largest correlation between reference vectors to be tolerated; if a trial rotation yields a higher hyperplane count but would cause a correlation between reference vectors higher than this value, the rotation will not be carried out
22	Punch 1 if the transformation matrix is to be punched out in format (12F6.3); punch 0 if no punch out of transformation matrix is desired
23	Punch 1 if a transformation matrix is to be input following the matrix of unrotated factors; punch 0 if no transformation matrix is to be input
24	Punch 1 if analytic rotation of reference vectors is to be omitted; punch 0 if analytic rotation of reference vectors is to be carried out
25	Punch 1 to bypass the computation of the pattern matrix, structure matrix, and the **Φ** matrix; punch 0 if these matrices are to be computed and output
26	Punch 1 if the **Φ** matrix of correlations between factors is to be punched out; punch 0 if this matrix is not desired; these cards may be input to the factor analysis program for a second-order factor analysis if desired

Printed Output. The unrotated matrix, or previously rotated matrix, input as data is printed out first followed by the oblique reference vector projections. The transformation matrix is output next. This matrix multiplied by the unrotated matrix produces the oblique reference vector projection matrix. The next matrix output is the matrix of correlations among the reference vectors, **C**. The hyperplane counts for the factors for each cycle of rotations are output next. The oblique reference vector projections following analytic rotations, the new transformation matrix, and the new **C** matrix follow if analytic rotation has been called for. Finally, the pattern matrix, the **Φ** matrix, and the structure matrix are printed out in that order if called for.

[12.10]
PROGRAM 10. MULTIPLE REGRESSION

This program was designed to compute the multiple correlation and the β weights for predicting a given criterion variable from several predictor variables. The multiple correlations and β weights may be obtained for several criterion variables on one run, using the same predictor variables in each analysis. The data consist of a correlation matrix in which the predictor (independent) variables constitute the lower numbered variables. The criterion (dependent) variables follow in order after the predictor variables. The portion of the matrix that involves correlations among the criterion variables may be left blank, although cards must be input for this section of the matrix, blank or not. The program extracts the matrix of correlations among the predictors from this larger matrix and inverts it. This inverse matrix is multiplied by the columns of correlations of the predictor variables with the criterion variables to obtain columns of β weights, one column for each criterion variable. The squared multiple correlation is obtained as the inner product of the β weights and the correlations of the predictor variables with the criterion variable. (See Chapter 10 for a description of this method of obtaining the multiple correlation and the β weights.) Care must be taken to see that the matrix of correlations among the predictors is nonsingular, that it *has* an inverse. A correlation matrix will be singular, and hence will have no inverse, when linearly dependent variables are included in the same matrix. For example, if odd-item scores, even-item scores, and total scores are included as three variables in the same matrix, linear dependence exists because the total score is the sum of the other two.

Control Card. The control card is the first data card. The information to be punched on this card is summarized briefly in Table 12.11.

Variable Format Cards. The next three data cards following the control card are the variable format cards, which specify how the data are arranged on

TABLE 12.11

Control Card Information for the Multiple Regression Program

Columns	Information to be punched
1–3	N, the number of predictor variables
4–6	NC, the number of criterion variables; N+NC must be less than or equal to 50

the punched card data input. Only columns 1–72 may be used. The data consist of a full correlation matrix including all the predictor variables first followed by the criterion variables. Entries are punched both above and below the diagonal except that the section of the matrix for correlations among the predictors may be left blank. The diagonals may be left blank also as unities will be supplied for these cells by the program. The cards are punched for the first column of the matrix, followed by those for the second column, and so on. A typical format might be (12F6.3), which would allow six columns for each correlation coefficient. This would give a punched value, for example, of -0.358 for a single entry. If it is desired to remove a certain predictor that is punched on the cards for a second run, the variable format may be changed to eliminate that variable from the appropriate row of the correlation matrix. The cards for that column, however, must be physically removed from the deck.

Data Cards for the Correlation Matrix. The cards for the correlation matrix follow the three variable format cards and are prepared in accordance with the instructions given above. Each column of the correlation matrix is started on a new card. It is helpful to label the cards for each column of the matrix by the variable they represent. Then, if a predictor variable or a criterion variable is to be omitted on a given run, the cards for that variable or those variables can be removed easily from the deck and the variable format cards adjusted to eliminate the appropriate row or rows of the matrix.

Data Card Order. The cards that constitute the data are arranged in the following order: control card, variable format cards, and cards for the correlation matrix.

Printed Output. The matrix of correlations among the predictors and criterion variables is printed out first just as input. Following this, the inverse of the matrix of correlations among the predictor variables is printed out. Finally, the multiple correlation and the β weights are printed out for each of the criterion variables.

[12.11]

PROGRAM 11. CORRELATION MATRIX LIST

This program accepts as input a correlation matrix punched by columns (or by rows) in the format (12F6.3) and prints the matrix in a form convenient for photocopying. The zeros and decimal points that appear in the usual printout from the other programs are removed so that only a three-digit number without a decimal point is printed. Thus, instead of printing −.007, only −7 would be printed with the 7 in the last column. The number .457 would be printed as 457. The punched card output option from the various programs involved in factor analytic computations provide decks of cards in the format (12F6.3). Thus, in the factor analysis program, the correlation matrix may be punched out in this format if desired for listing with Program 11.

Control Card. The first data card is the control card. The number of variables is punched in columns 1–3. The cards for the correlation matrix deck follow the control card. Printed output is as described above.

[12.12]
PROGRAM 12. FACTOR MATRIX LIST

This program lists a factor matrix for photocopying just as Program 11 lists a correlation matrix. The control card has the number of variables punched in columns 1–3 and the number of factors punched in columns 4–6. The cards for the factor matrix, in format (12F6.3), follow the control card.

References

Browne, M. W. On oblique Procrustes rotation. *Psychometrika*, 1967, **32**, 125–132.

Burke, C. J. Additive scales and statistics. *Psychological Review*, 1953, **60**, 73–75.

Carroll, J. B. An analytic solution for approximating simple structure in factor analysis. *Psychometrika*, 1953, **18**, 23–38.

Carroll, J. B. Biquartimin criterion for rotation to oblique simple structure in factor analysis. *Science*, 1957, **126**, 1114–1115.

Cattell, R. B. Parallel proportional profiles and other principles for determining the choice of factors by rotation. *Psychometrika*, 1944, **9**, 267–283. (a)

Cattell, R. B. Psychological measurement: Normative, ipsative, interactive. *Psychological Review*, 1944, **51**, 292–303. (b)

Cattell, R. B. *Factor analysis*. New York: Harper, 1952.

Cattell, R. B. (Ed.) *Handbook of multivariate experimental psychology*. Chicago: Rand McNally, 1966.

Cattell, R. B., & Cattell, A. K. S. Factor rotation for proportional profiles: Analytic solution and an example. *British Journal of Statistical Psychology*, 1955, **8**, 83–92.

Cattell, R. B., Cattell, A. K. S., & Rhymer, R. M. *P*-technique demonstrated in determining psychophysiological source traits in a normal individual. *Psychometrika*, 1947, **12**, 267–288.

Cattell, R. B., & Dickman, K. A dynamic model of physical influences demonstrating the necessity of oblique simple structure. *Psychological Bulletin*, 1962, **59**, 389–400.

Cattell, R. B., & Foster, M. J. The rotoplot program for multiple single-plane, visually guided rotation. *Behavioral Science*, 1963, **8**, 156–165.

Cattell, R. B., & Muerle, J. L. The "maxplane" program for factor rotation to oblique simple structure. *Educational and Psychological Measurement*, 1960, **20**, 569–590.

Chesire, L., Saffir, M., & Thurstone, L. L. *Computing diagrams for the tetrachoric correlation coefficient*. Chicago: Univ. of Chicago Press, 1933.

CLIFF, N. Orthogonal rotation to congruence. *Psychometrika*, 1966, **31**, 33–42.

COMREY, A. L. A factor analysis of items on the MMPI Hypochondriasis scale. *Educational and Psychological Measurement*, 1957, **17**, 568–577. (a)

COMREY, A. L. A factor analysis of items on the MMPI Depression scale. *Educational and Psychological Measurement*, 1957, **17**, 578–585. (b)

COMREY, A. L. A factor analysis of items on the MMPI Hysteria scale. *Educational and Psychological Measurement*, 1957, **17**, 586–592. (c)

COMREY, A. L. A factor analysis of items on the MMPI Psychopathic Deviate scale. *Educational and Psychological Measurement*, 1958, **18**, 91–98. (a)

COMREY, A. L. A factor analysis of items on the MMPI Paranoia scale. *Educational and Psychological Measurement*, 1958, **18**, 99–107. (b)

COMREY, A. L. A factor analysis of items on the MMPI Psychasthenia scale. *Educational and Psychological Measurement*, 1958, **18**, 293–300. (c)

COMREY, A. L. A factor analysis of items on the MMPI Hypomania scale. *Educational and Psychological Measurement*, 1958, **18**, 313–323. (d)

COMREY, A. L. A factor analysis of items on the F scale of the MMPI. *Educational and Psychological Measurement*, 1958, **18**, 621–632. (e)

COMREY, A. L. A factor analysis of items on the K scale of the MMPI. *Educational and Psychological Measurement*, 1958, **18**, 633–639. (f)

COMREY, A. L. Comparison of two analytic rotation procedures. *Psychological Reports*, 1959, **5**, 201–209.

COMREY, A. L. Factored homogeneous item dimensions in personality research. *Educational and Psychological Measurement*, 1961, **21**, 417–431.

COMREY, A. L. The minimum residual method of factor analysis. *Psychological Reports*, 1962, **11**, 15–18. (a)

COMREY, A. L. A study of thirty-five personality dimensions. *Educational and Psychological Measurement*, 1962, **22**, 543–552. (b)

COMREY, A. L. Personality factors compulsion, dependence, hostility, and neuroticism. *Educational and Psychological Measurement*, 1964, **24**, 75–84.

COMREY, A. L. Scales for measuring compulsion, hostility, neuroticism, and shyness. *Psychological Reports*, 1965, **16**, 697–700.

COMREY, A. L. Tandem criteria for analytic rotation in factor analysis. *Psychometrika*, 1967, **32**, 143–154.

COMREY, A. L. *The Comrey personality scales*. San Diego: Educational and Industrial Testing Service, 1970. (a)

COMREY, A. L. *Manual for the Comrey personality scales*. San Diego: Educational and Industrial Testing Service, 1970. (b)

COMREY, A. L., & AHUMADA, A. An improved procedure and program for minimum residual factor analysis. *Psychological Reports*, 1964, **15**, 91–96.

COMREY, A. L., & AHUMADA, A. Note and Fortran IV program for minimum residual factor analysis. *Psychological Reports*, 1965, **17**, 446.

COMREY, A. L., & BACKER, T. E. Construct validation of the Comrey personality scales. *Multivariate Behavioral Research*, 1970, **5**, 469–477.

COMREY, A. L., & DUFFY, K. E. Cattell and Eysenck factor scores related to Comrey personality factors. *Multivariate Behavioral Research*, 1968, **3**, 379–392.

COMREY, A. L., & JAMISON, K. Verification of six personality factors. *Educational and Psychological Measurement*, 1966, **26**, 945–953.

COMREY, A. L., JAMISON, K., & KING, N. Integration of two personality factor systems. *Multivariate Behavioral Research*, 1968, **3**, 147–160.

COMREY, A. L., & LEVONIAN, E. A comparison of three point coefficients in factor analyses of MMPI items. *Educational and Psychological Measurement*, 1958, **18**, 739–755.

COMREY, A. L., & MARGGRAFF, W. A factor analysis of items on the MMPI Schizophrenia scale. *Educational and Psychological Measurement*, 1958, **18**, 301–311.

COMREY, A. L., & SCHLESINGER, B. Verification and extension of a system of personality dimensions. *Journal of Applied Psychology*, 1962, **46**, 257–262.

COMREY, A. L., & SOUFI, A. Further investigation of some factors found in MMPI items. *Educational and Psychological Measurement*, 1960, **20**, 779–786.

COMREY, A. L., & SOUFI, A. Attempted verification of certain personality factors. *Educational and Psychological Measurement*, 1961, **21**, 113–127.

COOLEY, W. W., & LOHNES, P. R. *Multivariate procedures for the behavioral sciences*. New York: Wiley, 1962.

DIXON, W. J. (Ed.) *BMD biomedical computer programs*. Berkeley: Univ. of California Press, 1970. (a)

DIXON, W. J. (Ed.) *BMD biomedical computer programs. X-series supplement*. Berkeley: Univ. of California Press, 1970. (b)

DUFFY, K. E., JAMISON, K., & COMREY, A. L. Assessment of a proposed expansion of the Comrey personality factor system. *Multivariate Behavioral Research*, 1969, **4**, 295–308.

EBER, H. W. Toward oblique simple structure: A new version of Cattell's Maxplane rotation program for the 7094. *Multivariate Behavioral Research*, 1966, **1**, 112–125.

EYSENCK, H. J. *The structure of human personality*. London: Methuen, 1960.

EYSENCK, H. J., & EYSENCK, S. B. G. *Personality structure and measurement*. San Diego: Knapp, 1969.

FABIAN, J. J., & COMREY, A. L. Construct validation of factored neuroticism scales. *Multivariate Behavioral Research*, 1971, **6**, 287–299.

FERGUSON, G. A. The concept of parsimony in factor analysis. *Psychometrika*, 1954, **19**, 281–290.

FLEISHMAN, E. A. & HEMPEL, W. E. Changes in factor structure of a complex psychomotor test as a function of practice. *Psychometrika*, 1954, **19**, 239–252.

FRUCHTER, B. *Introduction to factor analysis*. New York: Van Nostrand-Reinhold, 1954.

GILBERSTADT, H., & DUKER, J. *A handbook for clinical and actuarial MMPI interpretation*. Philadelphia: Saunders, 1965.

GORSUCH, R. A comparison of Biquartimin, Maxplane, Promax, and Varimax. *Educational and Psychological Measurement*, 1970, **30**, 861–872.

GREEN, B. F. The orthogonal approximation of an oblique structure in factor analysis. *Psychometrika*, 1952, **17**, 429–440.

GUERTIN, W. H., & BAILEY, J. P. *Introduction to modern factor analysis*. Ann Arbor: Edwards, 1970.

GUILFORD, J. P. Creativity. *American Psychologist*, 1950, **5**, 444–454.

GUILFORD, J. P. When not to factor analyze. *Psychological Bulletin*, 1952, **49**, 26–37.

GUILFORD, J. P. *Fundamental statistics in psychology and education*. (4th ed.) New York. McGraw-Hill, 1965.

GUILFORD, J. P. *The nature of human intelligence*. New York: McGraw-Hill, 1967.

HARMAN, H. H. *Modern factor analysis*. Chicago: Univ. of Chicago Press, 1967.

HARMAN, H. H., & JONES, W. H. Factor analysis by minimizing residuals (Minres). *Psychometrika*, 1966, **31**, 351–368.

HENDRICKSON, A. E., & WHITE, P. O. Promax: A quick method for rotation to oblique simple structure. *British Journal of Statistical Psychology*, 1964, **17**, 65–70.

HORST, P. *Factor analysis of data matrices*. New York: Holt, 1965.

HOWARTH, E., & BROWNE, J. A. An item-factor-analysis of the 16 PF. *Personality*, 1971, **2**, 117–139.

HURLEY, J. R., & CATTELL, R. B. The Procrustes program: Producing direct rotation to test a hypothesized factor structure. *Behavioral Science*, 1962, **7**, 258–262.

JAMISON, K., & COMREY, A. L. Further study of Dependence as a personality factor. *Psychological Reports*, 1968, **22**, 239–242.

JENNRICH, R. I., & SAMPSON, P. F. Rotation for simple loadings. *Psychometrika*, 1966, **31**, 313–323.

KAISER, H. F. The Varimax criterion for analytic rotation in factor analysis. *Psychometrika*, 1958, **23**, 187–200.

KAISER, H. F. Computer program for Varimax rotation in factor analysis. *Educational and Psychological Measurement*, 1959, **19**, 413–420.

KELLEY, T. L. Comment on Wilson and Worcester's "note on factor analysis." *Psychometrika*, 1940, **5**, 117–120.

LAWLEY, D. N., & MAXWELL, A. E. *Factor analysis as a statistical method.* London: Butterworth, 1963.

LEVONIAN, E., COMREY, A. L., LEVY, W., & PROCTER, D. A statistical evaluation of the Edwards Personal Preference Schedule. *Journal of Applied Psychology*, 1959, **43**, 355–359.

MCDONALD, R. P. Numerical methods for polynomial models in nonlinear factor analysis. *Psychometrika*, 1967, **32**, 77–112.

MARKS, A., MICHAEL, W. B., & KAISER, H. F. Comparison of manual and analytic techniques of rotation in a factor analysis of aptitude test variables. *Psychological Reports*, 1960, **7**, 519–522.

MOSIER, C. I. Determining a simple structure when loadings for certain tests are known. *Psychometrika*, 1939, **4**, 149–162.

NEUHAUS, J. O., & WRIGLEY, C. The quartimax method: An analytic approach to orthogonal simple structure. *British Journal of Statistical Psychology*, 1954, **7**, 81–91.

NIE, N. H., BENT, D. H., & HULL, C. H. *Statistical package for the social sciences.* New York: McGraw-Hill, 1970.

SAUNDERS, D. R. A computer program to find the best-fitting orthogonal factors for a given hypothesis. *Psychometrika*, 1960, **25**, 199–205.

SCHÖNEMANN, P. H. A generalized solution of the orthogonal Procrustes problem. *Psychometrika*, 1966, **31**, 1–10.

SCHÖNEMANN, P. H., & CARROLL, R. M. Fitting one matrix to another under choice of a central dilation and a rigid motion. *Psychometrika*, 1970, **35**, 245–255.

THOMSON, G. H. *The factorial analysis of human ability.* (4th ed.) Boston: Houghton, 1951.

THURSTONE, L. L. *Multiple factor analysis.* Chicago: Univ. of Chicago Press, 1947.

THURSTONE, L. L. An analytic method for simple structure. *Psychometrika*, 1954, **19** 173–182.

TRYON, R. C., & BAILEY, D. E. *Cluster analysis.* New York: McGraw-Hill, 1970.

TUCKER, L. R., KOOPMAN, R. F., & LINN, R. L. Evaluation of factor analytic research procedures by means of simulated correlation matrices. *Psychometrika*, 1969, **34**, 421–459.

ZIMMERMAN, W. S. A simple graphical method for orthogonal rotation of axes. *Psychometrika*, 1946, **11**, 51–55.

Index